Politics and Technology in the Soviet Union

Politics and Technology in the Soviet Union

Bruce Parrott

The MIT Press
Cambridge, Massachusetts
London, England

This book was set in Palatino
by The MIT Press Computergraphics Department
and printed and bound by The Murray Printing Co.
in the United States of America.

Library of Congress Cataloging in Publication Data

Parrott, Bruce, 1945–
   Politics and technology in the Soviet Union.
   Bibliography: p.
   Includes index.
   1. Technological innovations—Soviet Union.
2. Technology and state—Soviet Union. I. Title.
HC340.T4P29 1983   338'.06   82–22953
ISBN 0–262–16092–7

The Russian Institute of Columbia University sponsors the *Studies of the Russian Institute* in the belief that their publication contributes to scholarly research and public understanding. In this way the Institute, while not necessarily endorsing their conclusions, is pleased to make available the results of some of the research conducted under its auspices. A list of the *Studies of the Russian Institute* appears at the back of the book.

To James H. Parrott and Mary K. Parrott

with love and thanks

Contents

Acknowledgments

Many persons have helped in the writing of this book. I am especially indebted to Seweryn Bialer. It was he who supervised my work on the subject while I was in graduate school, and in subsequent years he has provided further aid. His stimulating ideas and rigorous reading of several drafts have done much to shape the final work. Herbert Dinerstein, my colleague at the Johns Hopkins School of Advanced International Studies, offered insightful suggestions about both substance and style. His advice about the psychology of scholarship was a boon in handling the ups and downs of a long research effort. Loren Graham generously read three drafts of the manuscript and offered many constructive criticisms. Thane Gustafson's perceptive comments likewise helped me bring a complex subject into clearer focus. I am also grateful to Zbigniew Brzezinski, Abraham Brumberg, and Eric Willenz for their help.

Several persons have read and commented on parts of the manuscript. David Holloway, Ronald Amann, Philip Hanson, Julian Cooper, and Kendall Bailes all contributed ideas from their extensive knowledge of Soviet science and technology. Franklyn Griffiths, Grey Hodnett, and Werner Hahn gave me the benefit of their close understanding of Soviet politics. From Jim Reardon-Anderson, Gordon Schloming, and Robert Osgood came key questions about the analytical approach and presentation of findings. Michael Kaser, Terese Sulikowski, and Erik Hoffmann made useful comments on early chapter drafts. None of these friendly critics would agree with everything the book contains, but all have made it better than it otherwise could have been.

As every researcher knows, book-writing has a logistical side. Two scholars, Gordon Rocca and Rensselaer W. Lee, kindly simplified my task by sharing their files with me. Peter Promen and Linda Carlson of the S.A.I.S. library and many staff members of the Library of Congress gave essential aid in tracking down elusive research materials. I am also indebted to a series of capable research assistants: Elizabeth Anthony, Marie Zehngebot, Tom Gross, Michael Matera, Mark Koenig, Christopher Naylor, Robert Paris, and Laura Hastings. Over a span of several years they performed many laborious tasks connected with the project. I wish, too, to thank Sharon Mee, who typed draft after draft with gratifying precision, and Janice Delbert, who skillfully typed the final round of revisions.

Several institutions provided shelter and support while I worked on the book. I am grateful to St. Antony's College, Oxford, and particularly to Ronald Hingley and Dennis O'Flaherty, for two years of associate membership in the early 1970s. More recently, Johns Hopkins furnished an atmosphere conducive to scholarly pursuits and gave me a year's leave for uninterrupted research and writing. The Rockefeller Foundation supplied the handsome fellowship that made this indispensable year possible. In addition, the State Department's Office of External Research supported part of the work on the Brezhnev period.

Finally, I owe a special debt of gratitude to the persons whose main contributions to this work are emotional rather than intellectual. My parents, to whom the book is dedicated, offered unstinting encouragement and assistance for a graduate career that was both tumultuous and long. Sara Foose Parrott, my wife, contributed in innumerable ways: by tolerating the seemingly endless metamorphoses of my schedule for completing the book, by her enduring belief in the value of the undertaking, and by her thoughtful counsel during difficult stretches of writing. Above all, she helped by being a strong and intelligent companion. Ronald Tiersky, Alan Kogan, and Gerhard Adler likewise helped sustain me to the end of the endeavor. Without their aid at crucial junctures, the book might not have been completed.

Politics and Technology in the Soviet Union

1   The Central Issues

In the years since the Bolsheviks set out to industrialize the Soviet Union, American views of the Soviet regime and its technological capacities have fluctuated sharply. In the 1920s many observers anticipated that the need for economic growth and technological advance would cause the Bolsheviks either to transform their dictatorship and reestablish a large foreign presence in the economy or to be overthrown.[1] In the 1930s, as the capitalist West plunged into the Great Depression and the USSR entered the era of central planning, a number of Americans revised this view and concluded that the Soviet Union's rapid industrial expansion reflected an unprecedented technological dynamism, although others persisted in believing that the system was incompatible with the long-term requirements of modern technology.[2] The skeptics received a shock in 1949, when the Soviet regime built and exploded an atomic device much more rapidly than many Westerners had thought possible. Even more dramatically, the launching of Sputnik in 1957 unleashed a wave of American apprehension that the USSR might soon surpass the United States technologically, if indeed it had not already done so.

In the late 1960s and 1970s the pendulum swung back. Many American observers began to study the technological weaknesses of the Soviet system and to view these weaknesses as a central motive for the Soviet shift toward détente with the West. Soviet diplomatic overtures, in this appraisal, were a stratagem designed to overcome the inherent defects of the command economy; Western technology was to serve as a surrogate for painful domestic economic reforms. In its most unqualified version, this idea convinced a number of policymakers that the United States could openly demand and obtain basic Soviet political concessions in exchange for American technology.

Notwithstanding their volatility and questionable accuracy, these wide swings of American opinion do reflect an underlying judgment that is surely correct—namely, that the interaction between Soviet politics and technology is a matter of fundamental intellectual and political significance. The subject bears directly on our understanding of the way the pressures of technology are shaping, or failing to shape, the internal structure of the USSR, as well as the structure of modern industrial societies in general. The topic is equally important for our understanding of the connections between domestic affairs and foreign relations. Thanks to a wealth of primary sources relevant to both domestic and foreign policymaking, analysis of the interaction between Soviet politics and technology can disclose a great deal about such linkages.[3] The interaction is also of vital concern to Western public officials and business executives who must make certain assumptions about it in order to formulate their own policies toward the USSR.

Despite its compelling interest, however, the relationship between Soviet politics and technology has thus far received relatively little scholarly attention. A small group of Western researchers has revealed a great deal about the operation of Soviet scientific and technological institutions per se, but has devoted less attention to the political dynamics of these institutions and the general political implications of their performance.[4] Some pioneering works have shed valuable light on the political dimensions of science and technology, particularly in the USSR's early years,[5] and writers with a broader-gauge interest in contemporary Soviet politics have recently sought to incorporate the subject into more general treatments of the Soviet system.[6] Nonetheless, many facets of the subject remain to be explored. In view of its inherent intellectual importance and political topicality, it deserves a great deal of further study.

This book analyzes three aspects of the interaction between Soviet politics and technological performance. The first is how technological progress has fitted into official Soviet thinking about the USSR and its relations with the rest of the world. The second is the evolution of the regime's technological strategy: the mix of foreign and domestic technology by which the regime has tried to achieve its goals, and the avenues by which it has sought to obtain technology from other countries. The third is the political and administrative dynamics of the domestic institutions through which the regime has striven to create and introduce new industrial technology.[7] Each chapter examines these three themes in relation to a particular historical period, starting from 1928, when the First Five-Year Plan was launched, and ending in 1975, when most of the Brezhnev administration's policies had taken shape.

Any inquiry into the broad relationship between Soviet politics and technology, of course, raises many complex intellectual issues, and the themes outlined above cannot encompass them all. Nevertheless, the themes highlight some of the most important dimensions of the relationship. By tracing these focal problems across several decades, we can better understand how the pursuit of technological progress has figured in Soviet domestic politics and foreign relations. We can also judge more accurately whether recent technological problems and policies are genuinely new, thereby obtaining a firmer grasp of their implications for the future.

Technological Progress in Offical Thought and Policy

Technological advance has always occupied a central place in the public pronouncements of the Soviet political elite. As Marxists, the members of the elite have assumed that technological progress is equivalent to social progress, at least in a socialist system, and they have ranked it among their highest political priorities. Moreover, the Leninist view of the Western world has sharpened this commitment. According to the Leninist theory of imperialism, advanced capitalist societies, because they have reached a stage of economic development in which they are controlled by a handful of monopolists intent on private gain through foreign expansion, are inherently warlike.[8] Members of the Soviet elite have thus prized technological advance as a means of countering the enduring threat to the USSR that they perceive to be rooted in the internal political and economic structure of Western regimes. Technological innovation has also derived political salience from the official notion that socialism is capable of attaining a higher level of economic development than is possible under capitalism. Lenin articulated this widespread attitude when he argued that imperialist economies were exhibiting a tendency toward declining rates of technological progress. The USSR, he asserted, would triumph over its imperialist rivals and achieve full communism when its labor productivity surpassed Western levels.[9] In one form or another, all these ideas have been fixtures of the official outlook since the 1920s.

In Soviet political life, however, nuances in the expression of official ideas can have large policy consequences. This book closely examines how Soviet politicians and observers have handled such ideas. Precisely how have commentators depicted the political and technological challenge from Western capitalism, and how strongly have they asserted that the USSR's technological dynamism is superior? Has the treatment of these ideas varied significantly over time and among individuals? If so, why? Here we come to the reason for describing published Soviet statements on these topics as "official thought" rather than "ideology." Soviet public utterances do, of course, contain a great deal that is ideological, in the sense of self-serving ideas asserted as truth even though they are demonstrably false or are not susceptible to empirical tests.[10] But applying this label to all Soviet theories about politics and economics can be costly. Some Western observers have conceived of Soviet ideology as a body of precepts that, while it conditions elite perceptions and determines policy choices, remains unaffected by ongoing historical developments. Others instead have regarded it as a sham which does not affect Soviet officials' perceptions or policies. Although there are certainly occasions when one or the other interpretation captures the truth about Soviet pronouncements, it is an error to assume that either captures most of the truth most of the time. Frequent recourse to the concept of ideology encourages Western scholars to neglect the reciprocal interplay between official views, unfolding historical events, and policy choices. It is therefore wiser to regard official Soviet ideas as part of an ongoing corpus of thought, leaving open the possibility that this body of thought is subject to modification by events and to different policy interpretations by its exponents.

As we trace the history of Soviet theories about the challenge from capitalism and the comparative technological dynamism of the USSR, we will explore the connection of these ideas to policymaking. As a rule, every country's images of its own capabilities and its surrounding environment influence the choices it perceives and makes.[11] Our analysis will examine the part that such images, and alterations in them, have played in the formulation of Soviet technological policies. For example, if public depictions of the belligerence of the capitalist world have exhibited long-term changes or internal discrepancies at particular times, what role have these variations had in decisions about the development of military versus nonmilitary technologies? If there have been variations in appraising the USSR's rate of technological advance vis-à-vis rival Western states, how have these differences figured in decisions concerning domestic reforms of science and industry? Similarly, what part have assessments of capitalist hostility and Soviet technological performance played in decisions about importing Western technology and negotiating strategic arms limitation agreements? These are some of the principal questions we will try to answer about the interaction between politics and technology at the upper levels of the regime.

In doing so, we must bear in mind that the statements of Soviet officials, like those of political officials elsewhere, are intended to serve various political functions that are sometimes contradictory. Official pronouncements may be aimed at clarifying the regime's current situation and long-term prospects, in which case factualism and frank discussion of problems are called for. On the other hand, official statements may be intended to legitimize the regime in the eyes of the political elite, the general citizenry, or foreign sympathizers; when this is so, problems must be played down. Yet another function may be to influence the views and actions of the country's foreign opponents by denying that it has any weaknesses whatsoever. Finally, public pronouncements may be intended to exalt the accomplishments of the official speaking or to denigrate those of his rivals. The presence of such different motives makes interpreting the official attitudes toward Soviet technological progress a complex task. Individual commentators may disagree not only about the real technological capacities and deficiencies of the USSR but about the purpose that statements on the subject should serve. We will have numerous opportunities to observe how commentators have manipulated the corpus of official thought to suit their own ends.

This book explores the hypothesis that Soviet attitudes toward technological advance can be divided very roughly into two broad tendencies: "traditionalist" and "nontraditionalist." These two tendencies have always had a wide intellectual foundation in common. Both have firmly asserted that Marxism-Leninism provides the most accurate methods of political and economic analysis. Both have accepted the idea that socialism is the correct form of organization for Soviet society. Each has held that there are basic conflicts between the socialist and capitalist worlds. For this reason, as well as because of an intrinsic Marxist enthusiasm for material progress, each has agreed that rapid technological development must be a central goal of the Soviet system.

At the same time, however, the two tendencies have differed significantly. The traditionalist tendency has emphasized the aggressiveness of the USSR's capitalist competitors and treated the development of Soviet military technology as a matter of overriding priority. It has asserted vigorously that the Soviet Union is technologically more dynamic than the West and can therefore surpass the West economically in the foreseeable future. In keeping with these judgments, the traditionalist view has regarded economic autarky from the West as desirable. The nontraditionalist tendency, on the other hand, has muted the theme of imperialist aggressiveness and sometimes shown a concern that heavy stress on Soviet military technology unduly hampers technological progress in nonmilitary spheres. Adopting a more generous view of the growth of Western productive capacities, the nontraditionalist view has manifested less certainty that the USSR can surpass the West technologically, and it has attached considerable value to technological ties with the West as a means of meeting Soviet economic shortcomings.

In proposing to weigh this hypothesis against the historical evidence, I wish to emphasize that it posits only differing tendencies, not hard and fast divisions, within the elite. Within very short periods, some commentators have made contradictory statements on the issues mentioned above, probably because they have had different audiences and purposes in mind. Some persons have also clearly shifted from one outlook to the other over time.[12] Moreover, at any given moment many officials have occupied a middle ground combining elements of the traditionalist with elements of the nontraditionalist outlook.

Nonetheless, a carefully drawn distinction between the traditionalist and nontraditionalist tendencies raises fruitful questions for historical analysis. If there is firm evidence of such differing outlooks, has either been completely dominant in any period? Could we, for example, say that the traditionalist view reigned supreme in the Stalinist years, and that the nontraditionalist view triumphed in the early 1970s? It is also important to know whether attitudes have moved consistently in one direction or ebbed and flowed. Further, how have historical trends and events—be they the Great Depression, the Grand Alliance and Lend Lease, or the declining Soviet growth rates of the 1970s—affected the balance between the two schools of thought?

Finally, it is worth inquiring about the internal process by which changes from traditionalist to nontraditionalist attitudes (or vice versa) have occurred. Have the changes resulted from an elite consensus about the way external events and domestic trends should be evaluated? Or have they been accompanied by serious controversy? By closely studying the historical development of these viewpoints on technological advance, we should be able to illuminate the manner in which specific policies have been formulated.

Technological Strategy

Each chapter begins with an analysis of the trends in official thought and their policy implications. Next each chapter examines the policies themselves, starting with technological strategy and then turning to the organization of domestic science and industry. In this book the term "technological strategy" refers to choices between domestic and foreign technology and to decisions about the avenues by which foreign know-how should be acquired.[13] The examination of such choices should help illuminate how the USSR's internal political dynamics, its economic needs, and its external environment have interacted to shape a very important kind of governmental policy.

Like the leaders of other states, Soviet officials have had to make hard decisions about the best technological strategy for their country.[14] To begin with, how much technology should be acquired abroad, and how much should come from the expansion of indigenous research and development (R & D)? The attraction of relying heavily on foreign technology is that it may allow the recipient country to avoid many false starts and wasteful expenditures on duplicative R & D. This tack may also permit an underdeveloped country to begin introducing sophisticated technologies long before it has completed the time-consuming process of educating and institutionalizing a large scientific and engineering community. In this way the country may benefit from the "advantages of backwardness."[15] On the other hand, heavy dependence on foreign technology may retard the creation of the more modest number of scientists and engineers that the country requires to make intelligent choices and adaptations of foreign technology.[16] Or it may prevent the growth of links between existing domestic research organizations and industry, as domestic researchers succumb to a "xenophilia" that diverts their attention from national technological needs to the research priorities of their foreign scientific audience, and as domestic enterprises become attached to external sources of proven technology.[17] Moreover, imported technology may be ill suited to the balance of labor and capital available in the recipient country, and the suppliers may be unwilling to adapt it to local requirements.[18]

Apart from choosing between domestic and external sources of technology, the Soviet regime has had to pick the means of acquiring the foreign technology it sought. Should foreign direct investment be relied on, for example, as a channel for assimilating foreign equipment and know-how? Such investment eases capital shortages, offers expert management, and stimulates the investor to provide up-to-date technology. On the other hand, it gives foreign companies and governments administrative influence on the economy and furnishes no timetable for removing that influence. From this angle, simple imports of capital goods may seem preferable, since they come with fewer strings attached; but they are also less likely to embody the most modern technology available.

Two other alternatives to foreign investment are technical-assistance contracts and international exchanges of personnel. Such arrangements, while reducing foreign leverage on the economy, still permit the transfer of technology through persons rather than only through machines and documents. This is a major benefit, since face-to-face contacts are usually a critical part of successful technology transfers.[19] Yet technical-assistance contracts may not give suppliers as strong an incentive to furnish steady new infusions of technology as does direct investment; noncommercial exchanges of personnel are even less satisfactory on this score. Moreover, if the recipient country sends native scientists and engineers to be trained in the supplier country, it heightens the risk that many will emigrate, depriving their homeland of the talents needed to increase its capacity for innovation.[20] Equally important, contacts with foreigners and foreign cultures may prompt native specialists to adopt ideas which call into question their government's legitimacy.

Foreign technology may also be assimilated by copying designs from single models of imported machines (a process known as "reverse engineering") and by purchasing foreign scientific and technical literature. Such channels are appealing because they are inexpensive, require minimum foreign cooperation, and threaten less collateral political contamination than do more active forms of technology transfer. One disadvantage is that to work at all, they require a fairly sophisticated domestic R & D community to follow and assimilate foreign technical ideas. Even more serious, these channels are likely to institutionalize lags in the transfer of technology. In dynamic sectors, by the time native specialists have extracted the key principles of a technology from foreign equipment or publications and put copies into production, foreign producers will be manufacturing a more advanced version of the technology.

Several factors shape a country's choice of technological strategy. One, of course, is the country's level of economic development in comparison to other states. The lower the level, the more likely the country is to depend heavily on foreign technology acquired through a wide range of channels. But political considerations also have a large and sometimes contrary impact. Take, for example, the international political environment. The more hostility a regime perceives from other states, the more reluctant it will be to give them diplomatic leverage by seeking and depending on their technology. Internal political considerations also play a role. The internal order may be bolstered by technological borrowing in some cases but undermined in others.[21] Technological borrowing is especially suspect in countries where the political elite emphatically regards the domestic way of life as superior to that of other countries. Regimes are also wary of widespread technological contacts when leaders doubt their society's internal cohesion and ability to withstand the disintegrating influences of foreign cultures.[22]

As we explore the way such considerations shaped technological strategy in the Stalin and post-Stalin years, we should bear in mind that Soviet technological choices after 1928 were in part responses to the experiences of the Tsarist and pre-1928 Bolshevik regimes. The Tsarist government followed a technological strategy that, on the whole, perpetuated a gap between domestic scientific research and practical applications. Although it supported a few illustrious scientific establishments for the sake of national prestige, the government made little effort to connect their work to economic undertakings. A comparatively high level of achievement in the basic sciences existed alongside distinctly limited accomplishments in creating and particularly in applying indigenous technologies. Despite the interest of some Russian scientists in using their ideas for economic ends, few fruitful applications of indigenously developed technology took place during the 1800s, and most researchers continued to view their chief task as studying and solving scientific problems deemed important by their foreign scientific peers.[23] There were some accomplishments in introducing native technologies, but there were many more failures.[24]

The principal causes of this situation were the government's efforts to segregate society at large from the unsettling influence of scientific ideas, together with the absence of any strong commercial impulse within Russia's small industrial class and economically stagnant nobility to support engineering studies.[25] The regime compensated for these weaknesses by turning to foreign sources for most of its new technology. Toward the end of the nineteenth century it began to invite extensive foreign investment, and many of the enterprises that sprang up after 1890 relied on their foreign owners for new products and methods. As a rule these foreign companies had no research laboratories in Russia;

the technologies and many of the specialists who applied them came from abroad.[26] This technological strategy appealed to the Tsarist elite because it allowed the regime to draw on foreign capital as well as foreign knowledge, thereby avoiding a wrenching reorganization of domestic institutions and a massive reallocation of domestic resources to more productive uses. But it achieved these ends at the cost of perpetuating a gap between domestic science and industry and increasing foreign influence in the economy—a price that the extensive economic disruption caused by the outbreak of World War I showed to be high.

When it took power in 1917 the Bolshevik regime introduced fundamental changes into Russia's technological strategy. Nationalizing industry and repudiating Tsarist foreign debts, the new leaders condemned the role foreign investment had played in prerevolutionary Russia. Moreover, even in the impoverished conditions of the Civil War and early 1920s, they strove vigorously to build up Soviet research institutes as a source of technology for industry.[27] Building such links was difficult, however, and the country also faced a painful shortage of capital for investment in proven technologies. For these reasons Lenin and his colleagues still entertained the hope that they could attract assistance from the industrial West on terms which would not grant ownership rights to the foreign participants or infringe on the powers of the new regime. However, Western hostility and the inbred Bolshevik suspicion of capitalism frustrated this hope. The Bolsheviks found the vast majority of the foreign offers to set up concessions in the USSR unsatisfactory, and the contribution of concessionary enterprises to the output of large-scale industry remained tiny, although the limited number of new technical ideas introduced in the 1920s were acquired through concessionary and technical-assistance agreements with foreign firms.[28] The new regime had discarded the Tsarist technological strategy but had not yet arrived at a workable alternative.

Responding to these problems in the late 1920s, Stalin and his lieutenants swung the country toward a new technological strategy based on the rapid growth of domestic R & D, a stress on short-term technological borrowing through nonconcessionary channels, strong pressure for a quick transition to indigenous technologies, and a draconian willingness to squeeze the resources for industrial investment out of popular consumption. In the next chapter we will examine these policies and their implementation. How much technology from foreign sources did the regime actually obtain? Which channels of technology transfer did it emphasize, and how did this emphasis change over time? We will also consider how the emerging features of the Stalinist political system affected the transfer process. Did the system's increasingly xenophobic character restrict technological exchanges with foreign suppliers to a noticeable degree? How was the mounting campaign for political isolation from the West reconciled with the pressures for quick assimilation of Western technology? Finally, we will ask whether the regime overcame the long-standing gap between research and production. How successfully did it negotiate a switchover from foreign sources of technology to domestic R & D? Did the structure of the political system permit the creation of institutions that could systematically generate and apply indigenous technologies?

Later chapters focus on how Stalin's successors evaluated and modified his technological strategy. To what extent did they alter the balance between domestic R & D and foreign know-how? What issues were raised by the attempt to open up the country more fully to outside sources of technology? Was the apparent trade-off between domestic political security and more rapid technological advance a point of controversy? Not least important, how well did the post-Stalin policies actually work? Was the system, still centrally directed and comparatively insular in its foreign relations, able to assimilate greater amounts of foreign technology effectively?

Domestic Institutions and Technological Innovation

We have seen that the analysis of technological strategy shades into questions about the functioning of domestic R & D. After scrutinizing Soviet technological strategy, each chapter examines the domestic institutions intended to promote technological innovation. These include R & D establishments and industrial enterprises as well as the party and government hierarchies that oversee them. It is the responsibility of these organizations to adapt and diffuse technologies acquired abroad and create indigenous technologies where necessary.

In principle there are several prerequisites for rapid innovation that a research and production system ought to satisfy.[29] One is to mobilize large amounts of material and human resources for R & D and for investment in proven technologies. This problem may be especially difficult for countries in the early phases of industrialization, since they are usually short of both capital and skilled specialists. A second requirement is to build institutional mechanisms to coordinate the activities of the many participants in the innovation process. Because many R & D projects follow an unpredictable path, both intellectually and organizationally, a far-flung collection of researchers, industrial designers, and managers must be able to adjust quickly to one another's decisions.[30] A third requirement is to stimulate the exploration of alternative solutions to major research and technological problems. Even when highly motivated to solve a problem, a group of specialists may not select the line of inquiry that ultimately proves most fruitful, and once committed to a given approach, it may resist shifting its attention to alternatives. Closely related is the need for large exchanges of information among individual specialists and among R & D organizations. Exchanges among professionals from adjacent specialties foster a particularly high level of scientific and technological creativity; and information acquired outside an R & D organization frequently plays a major role in stimulating innovations inside it.[31] Finally, a country's institutional arrangements must provide strong positive incentives for individual researchers and groups of specialists to achieve high levels of technological performance. While some specialists find inherent pleasure in research and industrial management, innovation often entails arduous labor as well. External reinforcement is therefore essential, particularly if the expansion of a country's R & D effort draws into such careers a growing proportion of persons who feel no inner "calling" to achieve top-flight technological results.[32]

Beginning in the late 1920s, the Stalinist regime sought to meet these requirements through the central planning of research and production on a mobilizational basis. Frequently setting its sights above what was socially or economically feasible, the regime drove industrialization forward at an extraordinary pace. At its height, this effort engendered what one critic has called "bacchanalian planning," in which the resource constraints on growth were disregarded.[33] Under Stalin, a small number of leaders and officials monopolized administrative authority; often they exercised it arbitrarily, making no effort to be consistent with established bureaucratic rules or their own past decisions. While the regime attached considerable importance to technical expertise, it also judged its specialists by harsh political criteria. Top officials stridently attacked any specialists who trespassed on questions of general economic policy, and they frequently made specialists scapegoats for their own economic miscalculations.[34] As for the exchange of information, the regime, relying heavily on political enthusiasm to fuel the industrialization campaign, sometimes transformed the public media from channels of accurate information into propaganda vehicles and restricted the publication of economic data—particularly data unflattering to its public image—on grounds of national security.[35] In comparison with Western practice, this mobilizational system of central planning imposed far greater restrictions on the autonomy of individual scientific and industrial organizations and drastically limited the scope of market relations.[36]

This book analyzes how the Stalinist system satisfied or obstructed the conditions for technological innovation outlined above. Certainly the party-state apparatus demonstrated a massive capacity to extract and channel resources from agriculture to the industrial sectors favored by the political elite. Did it also generate sufficient resources to underwrite the R & D needed for the creation of indigenous technologies? Western scholars have argued, with a good deal of empirical support, that plentiful slack resources are necessary as a cushion against the unforeseen difficulties which usually accompany innovation, and that without reserves of this kind technological advance is extremely difficult.[37] Given its deep commitment to mobilizing all available resources for purposes of investment, did the Soviet bureaucratic apparatus regularly provide such slack?

In addition, the book examines how the central authorities coordinated the large number of specialized R & D units set up during the Stalinist industrialization campaign. How readily did top officials, who operated within a political culture which asserted that they possessed supreme economic and political wisdom, accept advice from scientists and engineers advocating specific technological innovations? How eagerly did these specialists propose such improvements? Perhaps they were encouraged by the regime's strong commitment to technological progress; but perhaps its emphasis on protoreligious enthusiasm and its tendency to single out scapegoats diminished their inclination to push innovations. If political mobilization was not an adequate incentive for the advancement of technology, what other incentives did the regime establish, and how well did they work? Despite the prevailing claim that top officials had the best grasp of economic problems, it is obvious that in many cases the specialized knowledge required for technological progress was actually in the hands of their expert subordinates. In the absence of market competition as a means of discerning what was technologically possible, were the central authorities able to elicit high levels of cooperation and performance from the various specialized hierarchies on which they relied for innovation?

Clearly the coordination of R & D depended not only on the center's administrative capacity but on the circulation of information within the system. To what degree did the Soviet commitment to the public ownership of technological ideas promote wide dissemination of information among specialists? How well did information flow between bureaucratic hierarchies? And how was its movement affected by the atmosphere of international and internal class struggle that Stalin assiduously fostered to buttress his dictatorship?

The book also considers how the Stalinist preoccupation with the threat of war shaped the organization of technological innovation. Was the administration of R & D in the weapons industries different from R & D in nonmilitary sectors? How did the levels of innovation in the military and civilian sectors compare with each other? And was there any serious effort to promote technological spin-offs from military to civilian industries?

Later chapters examine the efforts of Stalin's successors to improve the overall performance of Soviet research and production. Their underlying aim has been to move from "extensive" growth, based primarily on the mobilization of increasing quantities of labor and capital, to "intensive" growth, in which improvements in the quality of technology and other inputs play a heightened role. Although over the last half-century many industrial nations have made this transition,[38] Soviet growth remained extensive during Stalin's reign. The country's remarkable economic expansion was due above all to the mobilization of underutilized land, labor and capital, and to massive transfers of such resources between sectors, especially from agriculture to industry.[39] By the 1960s, however, many of these underutilized resources had been channeled into productive economic activity. Moreover, Stalin's successors, wishing to make the political system more stable and secure, abandoned most of his draconian methods for squeezing out new supplies of capital and labor. The post-Stalin elite has consequently confronted the necessity to tap new sources of economic growth. One means of attaining more intensive growth is to establish institutions that can achieve greater output from the technologies already in use. But an even more important means is to set up institutions that can accelerate the introduction of new technologies, thereby providing steady improvements in productivity over the longer term.

Stalin's successors have thus felt a need to recast the inherited system of mobilizational administration into a system combining features of two other forms of administration, the "mechanistic" and the "organic," which are more conducive to intensive growth.[40] The characteristics of mechanistic administration include a comprehensive, internally consistent system of rules, together with a precise division of tasks based on those rules. In mechanistic organizations initiative remains largely in the hands of administrative leaders, but the leaders' opportunity for arbitrary action is limited by the widespread expectation that they will interpret and apply the corpus of rules consistently.[41] A central concern of the leaders is to avoid wasteful expenditures and ensure that all money and manpower are productively employed at all times. In keeping with this goal, information flows mostly along vertical lines, dominated by instructions from the top and standardized reports from the bottom.[42] Western students of administration have suggested that organizations of this type are well suited to maximize the economic yield from relatively stable technologies, since gradual modification of the rules allows progressive refinement of the division of labor, routinizes past learning about the technology in use, and permits major economies of scale.[43] The available evidence generally supports this conclusion.[44]

To promote timely innovation in rapidly changing fields of technology, on the other hand, the organic pattern of administration is more effective. Organic institutions are more flexible than mechanistic ones. Based on a limited number of rules that are consistent but far from comprehensive, they have a more loosely defined internal division of labor and depend more on the spontaneous cooperation of highly trained specialists.[45] Leaders of such organizations must supervise many experts, some doing work that the leaders can understand only imperfectly, if at all. This bifurcation of administrative authority and technical expertise is most pronounced in R & D agencies, but it exists in many manufacturing organizations as well.[46] In organic institutions, leaders respond to this limitation on their effective administrative reach by adopting a more consultative style. They relax the effort to combat waste and compel the full employment of resources through rules, since they cannot codify these things in complex specialties. To capitalize on the expertise within the organization, leaders give fewer orders and solicit more policy advice from subordinate specialists; they also encourage extensive lateral communication among specialists working in different parts of the organization. The organic pattern is especially conducive to innovation because it reduces the lag in adapting formal rules to new technologies, narrows the administrative grounds for resistance from persons threatened by innovation, and stimulates interaction among the specialists who possess the knowledge required to generate new technology. In contrast, mechanistic organizations' centralization of authority, procedural rigidity, and restrictions on specialists' initiative all retard the pace of innovation.[47]

This book explores how Stalin's heirs have tried to modify the administration of research and production along such lines and to what degree they have succeeded. At first glance, it seems that they have tried to heighten the productivity of existing technology at industrial plants by substituting a more mechanistic pattern of administration for the old mobilizational methods of overseeing enterprises; and that they have sought more organic methods of administering the R & D organizations most closely involved in creating new technologies. Have they managed to reconcile the quest for more efficient use of existing technologies with the search for more organic administration where rapid innovation is required? Post-Stalin institutional changes may conceivably have increased the margin of slack resources for innovation, but the mechanistic aversion to "waste" may have tied up these resources in other uses. The decline of the Stalinist mobilizational style may have increased the initiative of R & D specialists, but the mechanistic features of the industrial system may have raised new barriers to innovation-minded specialists. We must likewise consider whether cooperation and information exchange among specialists have improved as the Stalinist siege mentality has gradually given way to a calmer organizational atmosphere.

It is also vital to learn how the larger political and economic context, which often determines the evolution of formal organizations, has affected efforts to improve Soviet scientific and industrial administration.[48] Recent studies indicate that in America, for example, market competition has played a key role in compelling business leaders to recast the administrative structure of corporations. The first step in this process was the reorganization of many corporations along the mechanistic lines of "scientific management"—sometimes against the preferences of their patriarchal leaders—in the early twentieth century.[49] Later, as competition based on innovation became critical to their survival, corporations dealing with rapidly changing technologies shifted toward the organic pattern of administration, while companies in sectors with relatively stable technologies continued to rely on the mechanistic form.[50] Market competition has thus helped force the transition to new forms of corporate administration, particularly organic forms. Moreover, market competition has apparently been an essential condition allowing business leaders to introduce organic forms of administration which simultaneously make their subordinates more autonomous, that is, freer of day-to-day supervision, and more responsible for high levels of economic performance, because their overall achievements can be measured against those of rival companies.[51] Were each company a monopoly, such decentralization would be extremely difficult, since specialists would lack a strong stimulus to perform well and leaders would lack reliable summary criteria which could signal that the specialists' performance was bad.

These findings suggest important questions about recent Soviet efforts to upgrade the organization of technological innovation. Without widespread market competition among research and production organizations, has the leadership been able to impose effective discipline on the specialized hierarchies responsible for innovation? Has it been able to set criteria that reward successful innovations and accurately distribute blame among the many agencies involved in failed attempts? Or have the very number and complexity of proposed innovations exceeded the administrative capacity of the central planners, making their ability to direct the innovation process more nominal than real?[52]

Along with the substantive effectiveness of recent reforms, we will scrutinize the political dynamics of the reform process. Where has the impetus for reform come from? Have party officials resisted all changes meant to speed innovation, as some Western notions of the conservative "apparatchik" would imply, and have industrial officials favored such changes? And what of the scientists and technical personnel? Some observers have depicted them as vigorous proponents of fundamental institutional and political change.[53] This idea is plausible if we assume that professional autonomy is an overriding goal of Soviet scientists and engineers. But we know that many Western specialists operate in corporate settings that seriously restrict their professional freedom, and they seem to accept these limitations.[54] It is therefore possible that many Soviet specialists have also come to terms with the restrictions on their autonomy. Indeed, they may conceivably have resisted major R & D reforms precisely because such reforms would expose them to a painful new discipline and new dangers of professional failure. In short, to what extent have specialists put personal security above technological achievement? And what does this indicate about the strength of the bureaucratic support available to reform-minded politicians and about the tactics which such leaders must follow? These, then, are the questions that the book explores.

2   The Stalinist Foundations, 1928–1941

Although some elements of the Soviet regime have changed markedly since the late 1920s and 1930s, that period still constitutes the epoch when, with enormous pain and suffering, many enduring attitudes and institutions were formed. Hence it is with these years that our analysis must begin. Taking as a point of departure the questions posed in the introduction, this chapter examines three aspects of the early Stalinist system: the place of technological progress in official Soviet thought, the nature of the regime's technological strategy, and the bureaucratic apparatus through which the ruling elite sought to promote rapid innovation.

Socialism, Capitalism, and Technological Progress

In the Stalinist tradition of official thought that crystallized during the late 1920s and 1930s, three central ideas defined the relationship between politics and technology. First, there was the preeminent goal of preserving and strengthening the Soviet Union as a bastion of socialism. Gradually the new elite's universalistic Marxist outlook was suffused by a deep Russian nationalism, hardening its determination to transform the USSR into a great industrial and military power. Second, there was the belief, more pronounced under Stalin than under Lenin, that the regime was engaged in an all-out struggle for survival domestically and internationally. This attitude was epitomized by the Stalinist thesis that the USSR was encircled by aggressive capitalist enemies, making a future war inevitable and requiring that the Soviet state's coercive power be constantly increased until a world revolution eliminated the capitalist threat. Third, it was believed that the USSR's socialist system was radically superior to capitalism in generating economic and technological growth. Stalinists were plainly aware that the USSR was inferior to the West in terms of current technological capacities, and this awareness contributed to their anxieties. But they also harbored an intensely optimistic conviction that the Soviet system possesed an unparalleled technological dynamism which would enable it to surpass the West in a relatively short historical interval.

The Emergence of the Stalinist Tradition

From the beginning of the Bolshevik regime Lenin and his colleagues wished to raise the technological level of the economy, but most of their energies were initially absorbed by the more immediate need to reestablish commerce and industry in the wake of a devastating civil war. Moreover, many of their hopes were fixed on the notion of an impending wave of international revolution that would swamp their capitalist enemies, end the diplomatic isolation of Soviet Russia, and create other socialist regimes able to provide generous technological aid for the country's development. Noting that the pace of Soviet industrialization hinged not only on domestic events but also on the course of the world revolution, the Twelfth Party Congress in 1923 remarked hopefully: "The overthrow of the bourgeoisie in any of the advanced capitalist countries would very rapidly be reflected in the whole tempo of our economic development, multiplying the material-technical resources of socialist construction."[1]

By the mid-1920s this prospect had faded. In 1925 and 1927, Party Congresses regretfully acknowledged that there had been a "partial stabilization of capitalism" in Europe, along with "partial progress of technology" and advances in the administration of the capitalist economy.[2] For the time being the USSR was clearly on its own, and the output of its rehabilitated economy was approaching prewar levels beyond which further growth would be more difficult.

In response to these conditions the political leadership adopted the slogan of building socialism in one country and started to lay greater stress on industrial expansion. The Fourteenth Party Congress at the end of 1925 called for an upgrading of industry's technological level, but cautioned that this effort should not exceed the available financial means.[3] The Fifteenth Party Conference a year later took a large step away from this moderate line by challenging Soviet industry to overtake the industry of the Western powers "in a relatively minimal historical period."[4] During the next two years the time-frame of industrialization was compressed still more drastically. By 1929 the regime was speaking without qualification of the need to "overtake and surpass" the economic levels of the advanced capitalist states through a "maximum mobilization" of the country.[5] Constantly repeated, this slogan became the central theme of the Stalinist drive to industrialize at breakneck speed.

Along with the new stress on rapid industrialization came a heightened appreciation of the importance of science and technology. In early 1927 the Central Committee, calling for new industrial plants based on the latest scientific achievements, rebuked those who underestimated the "decisive significance of the improvement of technology" for further economic development. Later in the year, the Fifteenth Party Congress urged a larger role for "scientific technology" and instructed officials not to hesitate to make the expenditures necessary to improve production methods. A 1928 Central Committee resolution similarly pointed out that the goals articulated by the Congress demanded "the closest tie of science, technology, and production" and a "decisive convergence" of research with the country's economic tasks.[6]

The most compelling justification for forced industrialization came from the Stalinist depiction of the USSR's international environment. In 1927 Stalin proclaimed that *"the period of 'peaceful co-existence'"* with the West *"is receding into the past,* giving place to a period of imperialist assaults and preparation for intervention against the USSR." The next year the Soviet-dominated Communist International asserted that "the principal and fundamental tendency" of the Western powers was "to encircle the USSR and conduct counterrevolutionary war against her in order to strangle her." An imperialist attack on the USSR, it stated, was "inevitable."[7] In 1930 the Sixteenth Party Congress attributed this danger to the mounting internal contradictions of capitalism. The partial stabilization of capitalism in the mid-1920s, it stated, was giving way to a sharpening of all the contradictions of the imperialist system, thereby increasing the likelihood of war.[8] In painting this ominous picture of the outside world, the Stalinists were striving to intensify domestic anxiety as a weapon against their party rivals. By linking the foreign menace to the specter of widespread subversion by class enemies at home, they were able more easily to intimidate the domestic critics of their economic policies.[9] But the champions of forced industrialization were also responding to what they saw as a real danger, at least over the long term. In 1931, for example, Stalin delivered an impassioned speech in which he rejected any slowing of the pace of industrialization. Old Russia, he proclaimed, had been "beaten continuously" by enemies ranging from the Mongols to the Poles and the French. "We lag behind the advanced countries by 50 to 100 years. We must make up this distance in 10 years. Either we do this or they crush us."[10] The party's programmatic statements reflected the same sense of crisis.[11]

These passages show that the Stalinist regime viewed its race with the West as an all-out struggle in which industrial power, and especially military power, would be decisive. In the short run, the regime might modulate its forecasts of inevitable war for diplomatic purposes, as it did during the mid-1930s in an effort to rally foreign opposition to the growing assertiveness of Nazi Germany. This was part of the stratagem of seeking a "breathing spell," that is, a postponement of military conflict with the West in order to accumulate added military and economic strength.[12] But the regime continued to believe that it was encircled by hostile powers. This situation, as Stalin explained it, required the state to keep amassing military might, as well as to perform the two other critical functions of developing the economy and protecting against imperialist subversion.[13] Elite members remained convinced that sooner or later, capitalism's multiplying internal tensions would unavoidably trigger a war which would pit one or more Western states against the USSR. During the 1930s this expectation exerted a pervasive influence on Soviet technological policies.

If the image of mounting imperialist aggressiveness was one essential justification for the industrialization drive, another was the increasingly negative Soviet depiction of capitalism's faltering economic prospects. Stalin's policy of overtaking the West in one gigantic campaign rested not only on the argument that imperialist intentions made it necessary, but also on the assertion that the economic superiority of the Soviet system was so marked that the campaign could succeed. Thus the Sixteenth Congress coupled its stress on the rising aggressiveness of capitalism with an attack on theories that exaggerated capitalism's economic dynamism. Recent trends, said the Congress, refuted the reformist notion that an "organized capitalism," that is, a capitalism free of some of the contradictions described by Marx, was emerging in the West.[14] One target of this censure was Bukharin, who in the late 1920s had suggested that capitalist states could avoid debilitating internal class conflicts and muster their scientific and economic resources more effectively than in the past.[15] The Bukharinite theory implicitly challenged the possibility of surpassing the industrial West in one compressed industrial surge, and this was partly why it drew heavy fire.[16] The political requirements of Stalin and his allies, however, were not the only reason for the official reassessment of capitalism's economic future.

The reassessment was also provoked by an authentic capitalist crisis, which shattered established views of capitalism all over the world. The severity of the Great Depression strongly reinforced the Stalinist argument that surpassing the West was both more urgent and more feasible than had previously been thought.

The first Soviet explanation of the Western economic collapse emphasized capitalism's inability to mobilize its economic resources for sustained growth. At the Sixteenth Party Congress, Stalin explained that the crisis arose from the contradiction between capitalism's tremendous productive potential and the limited capacity of its impoverished masses to pay for goods and thereby realize that potential.[17] To buttress his argument he juxtaposed growth figures for selected industries in several capitalist states with figures for the USSR. The contrast was vivid, and the comparisons he drew at the Seventeenth Party Congress four years later were still more so: since 1929 capitalist industrial output had fallen sharply, whereas Soviet output had grown about twofold.[18] The conclusion that capitalism was inferior to socialism in the mobilization of resources seemed inescapable. It encouraged V. M. Molotov, Chairman of the Council of People's Commissars and a close ally of Stalin, to argue that comparisons with the West were a criterion of Soviet economic performance which had recently assumed "the greatest political significance."[19]

As the scope of the Depression became clear, Soviet observers concluded that capitalism's difficulties in mobilizing resources were now so pronounced that they undermined its technological creativity as well. Citing Western cuts in governmental research budgets and corporate efforts to suppress innovations, one authority on foreign R & D argued that "capitalism is beginning in an absolutely regular and systematic way to brake the development of technology and obstruct technological progress."[20] Similar ideas appeared in the party's programmatic statements, which remarked that there was a growing Western tendency to abolish research institutions and regard technological progress as superfluous.[21] Even skeptics of the industrialization campaign now granted some truth to this view. In 1931, Bukharin wrote that in the past capitalist crises had given added impetus to the search for new technology. But in present capitalist systems, he said, "a different tendency can be observed: a slackening of technical progress. . . . This practice and theory of *technical regression* began to flourish towards the end of the period of so-called 'industrial prosperity.' "[22]

The decline of Western technological dynamism encouraged the Stalinist leadership, always eager to claim unique virtues for the Soviet system, to assert the USSR's superiority in this sphere. In 1932 the Seventeenth Party Conference referred somewhat ambiguously to "a series of the most major scientific technological achievements" realized during the First Five-Year Plan; it implied, but did not directly state, that these advances had been generated from domestic R & D. More vigorously, the Seventeenth Party Congress two years later stated that "significant successes have been achieved in the . . . development of [Soviet] scientific-technological thought, which has independently solved a series of the most major technological problems."[23] These assertions were coupled with an especially forceful view of the technological levels Soviet industry would soon reach. Noting that in several sectors the USSR had already created enterprises "which leave behind the level of European technology," the 1932 Party Conference claimed that during the Second Five-Year Plan the country would "move into first place in Europe in a technological respect."[24] Much of the basis for this confidence lay in the Soviet system's exceptional capacity to mobilize resources. As one industrial planner put it, by working from a pre-established plan "we can . . . apply technology at reconstructed and new enterprises in unlimited amounts and also more rapidly than is possible with the existing method of mobilizing capital in capitalist countries."[25]

Socialist Superiority: Anxiety and Reaffirmation

Given its utility in legitimizing the Stalinist elite, however, the notion of the USSR's superior technological dynamism was not always asserted as unequivocally as we might expect, particularly in the mid-1930s. One apparent reason is that after 1932 most of the capitalist powers began to rebound from their earlier economic collapse before slipping into a major new recession in 1937.[26] Another reason is that flat assertions of socialist superiority gave rise to complacency and reduced the pressure for rapid technological change. In 1934 one commentator remarked that the belief that no capitalist technological progress had occurred since 1929 was "widely held" in the USSR, and he cautioned that this was untrue. Another authority similarly labeled this belief "a most harmful self-deception, one of the most dangerous manifestations of the conceit and bragging" warned against by Lenin and Stalin.[27] The chairman of the State Planning Commission (Gosplan) called for a determined struggle against overconfident "hat-throwing" prompted by recent successes in surpassing the West technologically. While the Depression had caused the breakdown of the Western economies as a whole, he said, it had also "driven individual very strong enterprises to struggle for self-preservation . . . by developing their production technology. Therefore the technology of individual enterprises . . . in capitalist countries has continued to stride forward, and the danger of falling behind it threatens every branch of our industry."[28] Public expressions of concern on this score seemed to peak at mid-decade.[29]

Not surprisingly, Soviet anxiety focused particularly on military technology. This preoccupation came through clearly in a statement in 1936 by the head of the Central Committee's Department of Scientific-Technological Inventions and Discoveries. Not only were capitalist firms competing technologically with each other, he warned; so were capitalist states. Germany and America, for example, were conducting "a whole series" of secret military research projects that constituted "a technological weapon against other countries, and ultimately against our Land of Soviets." The "particular comrades" who viewed capitalism's crisis "in an oversimplified manner" should remember this danger.[30] Such cautionary words represented an effort to prevent a general belief in the technological superiority of socialism from crippling attempts to catch up in crucial sectors, and they may well have prompted the changes that some officials attempted to make in the administration of R & D in 1935–6.[31]

Despite these signs of anxiety, Soviet spokesmen in the late 1930s forcefully reaffirmed the superior technological dynamism of the socialist system. Encouraged, perhaps, by the onset of a new industrial downturn in the West, they depicted the capitalist economic crisis late in the decade as "immeasurably sharper than in 1929," and the contrast between socialism and capitalism as "immeasurably deeper and clearer." The limited Western technological progress occurring under these conditions could not offset superior Soviet rates of development.[32] As an editorial in the party's main journal put it, "the greatest advantages of communism over capitalism are shown by the fact of [its] historically unprecedented introduction of new technology into the national economy of the USSR." The Soviet Union had accumulated an "enormous [amount of industrial] technology; the prospects for its development are unlimited; in our country there are no capitalist trammels and chains of private property fettering the development of productive forces."[33]

Near the end of the 1930s Stalin gave his personal imprimatur to this view. Earlier in the decade he had already sought to strengthen his political position by identifying himself closely with the record-breaking feats of Soviet airmen, which were often timed to coincide with heightened intraparty conflict.[34] At the Eighteenth Party Congress in 1939 he used the theme of technological progress to buttress the legitimacy of the Soviet system as a whole. Stating that the reconstruction of industry and agriculture on the basis of the most contemporary technology had been completed, he summed up the system's technological level and innovative capacity: "One may say without exaggeration that, from the viewpoint of production processes, from the viewpoint of the saturation of industry and agriculture with new technology, our country is the most advanced in comparison with any other country, where old equipment weighs down production and obstructs the process of introducing new technology."[35] The USSR, he claimed, now occupied "first place," not simply in Europe, but in the world. In terms of technological levels as well as rates of economic growth, "we have already overtaken and passed the major capitalist countries."[36] This assertion was not intended to license any slackening of the industrialization drive. Stalin coupled it with an acknowledgment that the USSR still trailed some other states in per capita industrial output, and he called for this gap to be rapidly closed. But his words did suggest that the present technological dynamism of the Soviet system and its massive capacity to mobilize resources would be sufficient for the USSR to overtake the West in every respect, given some additional time. Proceeding from this assumption, the regime announced a decision in early 1941 to formulate a fifteen-year plan that would enable the USSR to surpass its capitalist rivals in per capita output.[37] Despite the warnings against technological complacency and "hat-throwing" that continued to appear occasionally,[38] the Stalinist outlook clearly assumed the superior technological dynamism of the Soviet system.

The official adoption of this perspective during the 1930s served to justify an all-out push to boost the level of Soviet technology, and it thereby helped legitimize the extraordinary power that the ruling elite exercised over the rest of society. But it also placed a serious challenge before the regime. Could the USSR actually overtake its capitalist rivals technologically in the foreseeable future, and by what means?

Domestic and Foreign Technology in Soviet Strategy

The Stalinist attempt to overtake the West industrially had a major impact on the country's technological strategy. After the Revolution, Lenin and his cohorts had renounced Tsarism's heavy dependence on foreign investment, and they had sought to improve domestic R & D. But they had also imported considerable amounts of Western capital goods and relied on foreign technology in other forms as well.[39] Between 1922 and 1926 the government had concluded about 150 concession agreements providing either for the operation of foreign companies in the USSR or for joint-stock undertakings with combined Soviet and foreign management.[40] By 1927 the regime had also signed twenty-one technical-assistance contracts with foreign firms.[41] Insofar as the transfer of unembodied technology is measurable in monetary terms—and the equivalence is by no means clear—it was far smaller during the 1920s than before the Revolution.[42] Still, whatever the absolute scale of technology transfers from the West, they remained the main source of new industrial ideas for the USSR.[43] The Bolsheviks had sharply reduced the influence of foreign investors and managers on the economy. But they had not accelerated the rate of technological change, nor had they significantly diminished the country's reliance on technology created abroad.

Trends During the First Five-Year Plan

By the late 1920s the Stalinists' determination to slash Soviet dependence on foreign technology was clear. In late 1925 the Fourteenth Party Congress, accenting the "whole series of new dangers" posed by economic ties with the West, had urged that the USSR begin producing its own machines in order to avoid becoming "an appendage of the capitalist world economy."[44] This commitment became more and more intense as Stalin and his backers defeated their rivals, and it led to a massive expansion of domestic research and technical manpower. According to Western estimates, spending on scientific research rose almost sixfold between fiscal 1927–8 and 1932,[45] and enrollments in institutions of higher technical education nearly quadrupled in the same period.[46] In comparison with most developing countries, the early Stalinist regime channeled an unusually large share of its resources into R & D related programs.

The very urgency of the industrialization campaign, however, meant that the country could not wait for this investment in research and education to generate independent technological momentum. For rapid progress in the short term, more rather than less technological borrowing was needed. In 1927 the Fifteenth Party Congress called frankly for the "widest use of Western European and American scientific and scientific-industrial experience," a policy that Stalin strongly reaffirmed at the Sixteenth Party Congress three years later.[47] The goal of such borrowing was to be rapid technological independence from the capitalist world, as the Fifteenth Congress made clear when it contrasted its position to that of Trotsky, who had come to advocate a lengthy period of dependence on foreign economic ties to facilitate industrialization.[48] In the Stalinist scheme of things, the period of stepped-up technology transfers was to be intense and short. The incorporation of other nations' accumulated experience would allow the USSR to skip the intermediate stages separating it from technological modernity, and then to forge ahead on the basis of its own R & D achievements.

If this was the policy, commentators nevertheless differed markedly in interpreting it. Some critically disposed representatives of scientific and engineering circles emphasized the need for an immediate shift to technology created through domestic research. Arguing that Soviet science must be on a higher level than Western science, one spokesman called technological borrowing a "great inadequacy" that would hurt the USSR in the competition with capitalism.[49] A group of prominent chemists complained more bluntly that the regime's economic plans "in large part mechanically borrow existing forms of foreign technology which are routine and in many cases outdated."[50] These criticisms suggest a tendency to which we will later return: whether moved by objective judgment or professional pride, Soviet scientists and engineers were sometimes unwilling to accept that foreign technology was better, and they sometimes opposed wholesale technological borrowing even when the political authorities favored it.

More orthodox interpreters of official policy, by contrast, felt that the essential first step toward self-sufficiency was a massive draft on foreign technology with a minimum of indigenous innovation. Anastas Mikoian, Commissar of Trade and a close associate of Stalin, remarked that by failing to use all available Western technology, "we will lose ten times more than it is necessary to pay for technical assistance." Soviet specialists intent on being original, he complained, wanted "to discover America, which has already long been discovered." This was "not only a useless but a very expensive exercise. Our own scientific thought must develop. But there is no reason for us . . . to invent that which has already been discovered and tested abroad."[51] Another writer argued that "in the design of factory construction . . . it is necessary to lean cautiously on the strength of Russian designers [and] *not to allow them by self-education to go through the school and history of errors and achievements* already passed through by other countries." American technology could serve as the model of Soviet construction "for a long time." "The *privilege of a young* developing *country*," he concluded, "should be used by us on all sides, and all *the achievements of our predecessors* must be taken *ready-made*."[52] During the First Five-Year Plan this view prevailed, and the party instructed economic officials and engineers to concentrate on assimilating foreign technology rather than on devising novel designs.[53]

In keeping with this strategy, Soviet reliance on foreign technology during the First Five-Year Plan was extremely heavy. As the industrialization campaign got under way, imports of producer goods climbed steeply; their annual average volume for 1928–31 exceeded the comparable figure for 1924–7 by about 50 percent.[54] Trade provided nearly four-fifths of all the machine tools installed in 1932, and almost 15 percent of total gross investment during the First Five-Year Plan.[55] Between 1927 and 1932 the number of concessions to foreign companies dropped substantially, from 68 to 24, but the number of foreign technical-assistance contracts more than compensated, mushrooming from 30 in 1928 to 124 in 1931.[56] Largely because of these contracts, the number of foreign specialists at work in industry and construction rose to about 1,000 in early 1930 and more than 9,000 in late 1932.[57]

The contribution of foreign companies and personnel was indispensable to the industrialization drive. Their importance can be grasped by considering their part in constructing the modern new plants that were the heart of the First Five-Year Plan. During fiscal 1928–9 foreign firms executed designs for one-quarter of all the capital construction newly undertaken in that year.[58] Since these statistics pertain to a period in which the number of technical-assistance agreements reached slightly more than half the 1931 level,[59] there is reason to believe that the role of foreign designers in industrial construction grew still more during late 1929 and 1930.

The metallurgical sector illustrates the meaning of these figures. *Giprotsvetmet*, the design organization primarily responsible for planning enterprises in nonferrous metallurgy, signed a contract with a U.S.

corporation for assistance in designing ore-dressing plants.[60] French influence was strong in the planning and construction of aluminum plants, American influence predominant in zinc and lead production.[61] An American company played a central role in designing and building two of the largest Soviet copper enterprises, and American engineers occupied many of the top administrative posts at operating smelters.[62] *Gipromez*, the design organization in charge of planning iron and steel plants, was heavily staffed with foreign experts by 1929 and continued to be so until at least 1933.[63] An American company designed the huge Magnitogorsk iron and steel complex, modelling the installation on the U.S. Steel plant in Gary, Indiana, and another U.S. firm supervised the construction of more than fifty iron and steel enterprises.[64]

The pattern of technological borrowing manifested in metallurgy was not the exception but the rule during the First Five-Year Plan. The chemical industry benefited from numerous agreements with Western firms.[65] The automobile and tractor industries depended even more on Western technical aid.[66] Western corporations offered extensive assistance to the Soviet electrical equipment and fuel sectors,[67] and they played a lesser but not inconsequential role in developing several machine-building plants.[68] Thus, while Soviet scientists and engineers created a few novel technological achievements during this period in areas like synthetic rubber, industry as a whole remained extremely dependent on foreign sources for new technology.[69]

Political Crosscurrents

Such technology transfers required large exchanges of specialists between the USSR and the West, and it was in this realm that tensions between the strategy of borrowing and the Stalinist political system first appeared. Although the party leadership was seeking to acquire more foreign technology, it simultaneously began to assert that the capitalist West was directing a major campaign of sabotage and espionage against the USSR. Because the stratum of intellectuals and specialists inherited from the Tsarist era was alleged to be a central instrument of this campaign, the process of technology transfer became the focus of contradictory economic and political pressures.

At the start of the First Five-Year Plan the party authorities promoted fuller international exchanges of specialists. A Central Committee resolution in 1928 underlined the need for more foreign trips by scientists and engineers and called for more visits to the USSR by foreign specialists; this position was reiterated by both Molotov and Stalin in 1930.[70] The idea of wider foreign contacts elicited a favorable response, especially among scientists. The Third All-Union Congress of Scientific Workers called for more foreign scientific travel. A high science official revealed that the question of joining international scientific unions was now being considered, and the Academy of Sciences resolved to establish a bureau that would facilitate personal contacts with foreign scientists.[71] Engineers also greeted the policy of increased foreign contacts.[72]

At the same time, however, the campaign of xenophobia unleashed by Stalin and his associates to consolidate their power was making foreign contacts a source of potential political incrimination. As early as 1925 the party had gone on record as requiring that foreign specialists working at Soviet plants "meet the elementary requirements of the Soviet public."[73] But this muted note of caution was mild compared to the accusations that were leveled against foreign capitalists and their representatives during the First Five-Year Plan. In 1928, as part of an attack on his rivals from the Right Opposition, Stalin staged a show trial of several Soviet engineers and economic officials. This incident became known as the Shakhty affair. The Shakhty defendants, it was charged, had conspired with foreign intelligence services and former owners of Russian enterprises to disrupt Soviet defenses and prepare for foreign military intervention. Their actions, declared the Central Committee, marked "new forms and new methods of struggle of the bourgeois counterrevolution against the proletarian state."[74] Less than two years after the defendants were convicted, members of the Academy of Sciences were accused of counterrevolutionary activity, although the purge of the Academy produced no show trial. In 1930 the Industrial Party trial resulted in harsh sentences for the accused, several of whom were prominent industrial researchers and engineers.[75] In this recriminatory atmosphere denunciations of bourgeois "wreckers" in science and industry became widespread, and several thousand specialists were arrested and imprisoned.[76]

The intention behind the arrests was to break resistance to Stalin's power and to deflect dissatisfaction over the economic disruptions resulting from forced industrialization. But the intensity and xenophobia of this campaign encouraged charges that impinged on the process of technology transfer. At the Sixteenth Party Conference in 1929, one speaker, saying that a case of wrecking more extensive than the Shakhty affair had been uncovered, claimed that it involved "direct assistance to foreign firms in their negotiations with the USSR about concessions and also in deals for the supply of equipment." Similar accusations recurred with some frequency.[77] Given the charges of collusion with foreign capitalists that were made in the show trials, it was only natural that suspicion should spread to many dealings with foreign firms, making those dealings more difficult.

The mood of suspicion that clouded such transactions contributed to an increase of Soviet defections. As the domestic political atmosphere worsened in the late 1920s, a growing number of citizens on foreign assignment chose not to return home. At the Sixteenth Party Congress G. K. Ordzhonikidze, then a prominent supporter of Stalin, revealed that the number of foreign trade officials who had refused to return from their posts abroad had climbed from an annual average of thirty-two in 1927–8 to sixty-five in 1929 and forty-three in the first half of 1930 alone.[78] The number of defectors in these years thus constituted more than 8 percent of the 2,500 Soviet trade representatives stationed abroad in 1929.[79] Although no fuller statistics are available, these defections probably exemplified a trend among all specialists traveling outside the country after the assault on bourgeois "wreckers" got under way.[80] If this is so, it means the regime faced a serious loss of scientific and technical talent for which its own policies were largely to blame.

At the end of 1929 Lazar Kaganovich, a candidate member of the Politburo and an ally of Stalin, broached the question of the relation between Western technology transfers and wrecking. "Of course," he told the Central Committee, "we must decisively emphasize that there can be wreckers and fascists among foreign engineers." Several recent cases proved this, necessitating "very great vigilance" toward Western specialists. The USSR, Kaganovich said vigorously, must obtain assistance from the best foreign engineers but should not engage "good-for-nothings or riff-raff" as technical consultants.[81] The defensive tone of this remark suggests that critics of the industrialization campaign may have pointed out the inconsistency between extensive technological borrowing and Stalinist allegations of widespread Western sabotage.[82] But even if they did, it did not deter Kaganovich from asserting that the proper answer to wrecking was to continue exchanges of technical personnel while screening the Western and Soviet participants more closely.[83] Ordzhonikidze concurred, saying that "only the very best, the most faithful, mature officials" should be sent abroad.[84]

Despite the Stalinists' appetite for foreign technology, their growing emphasis on the danger of subversion gradually undermined communication between Soviet and Western specialists. As indicated above, the number of foreign specialists in the USSR increased steadily until 1932. Until about 1931 contacts with foreign industrial specialists outside the USSR also grew[85] but then declined. Whereas 900 industrial specialists traveled abroad in fiscal 1928–9, in 1931 only 485 did so.[86] About the same time, long-term foreign postings became more difficult to obtain. In 1930 Ordzhonikidze revealed that the number of trade officials stationed abroad had recently been cut by more than 40 percent.[87]

Foreign travel also became harder for scientists. In fiscal 1927–8, 143 government-subsidized and 244 unsubsidized foreign trips were authorized for scientists; there were about 2,000 applicants for these slots.[88] In fiscal 1928–9, although the government substantially increased the foreign currency set aside for such trips, the number of openings for self-financed trips was sharply reduced, and the number of scientists traveling abroad fell to about 140.[89] According to an underground (samizdat) account written by a Soviet scientist in the post-Stalin period, "after 1929 the participation of Soviet scientists in international scientific meetings and symposia practically ceased altogether."[90] It therefore seems likely that foreign travel by scientists diminished from the early 1930s onward.[91] The establishment of political restrictions on travel was attested to by the indignant response from the central trade union bureau for scientific workers, which resolved "to consider absolutely abnormal the refusal to issue a foreign passport to a scientist who has been offered a foreign trip by the Commissariat of Enlightenment." Nevertheless the restrictions remained in force.[92] They illustrate the suspicion with which Stalinists viewed foreign contacts—even those that might have strengthened Soviet technological capacities.

Stalinist suspicions did not yet extend to the purchase and dissemination of foreign scientific and technological literature. Around the start of the First Five-Year Plan, the government sharply increased the acquisition of such materials. Between 1925 and 1931, for example, the number of books and journals purchased annually by the library system of the USSR Academy of Sciences increased from 27,722 to 42,148.[93] In view of the regime's emphasis on borrowing foreign technology during this period, it is reasonable to assume that purchases by the central economic organs grew as well. During this period expanded periodical bibliographies for science and technology were published to guide specialists to pertinent foreign as well as domestic research, and short descriptions of foreign technological developments were prominently featured in industrial journals.

By 1932, then, some general tendencies were becoming apparent. While building up its R & D establishment for the future, the Stalinist regime drew heavily on foreign technology for the short term. But the attempt to benefit from the "advantages of backwardness" was beset by conflicting economic and political pressures. Even as they stressed the importance of foreign technology, the political authorities were gradually circumscribing the modes of technology transfer by assailing "survivals of capitalism" in the USSR and attacking the alleged links of these remnants with foreign imperialists. The liquidation of many concessions and the shift to technical-assistance agreements were meant to deprive foreign firms of administrative leverage within the economy, and the controls over foreign travel were intended to limit personal contact with foreigners. Meanwhile, the acquisition of foreign technological literature was greatly expanded. Together, these steps foreshadowed a transition from negotiable transfers, in which the contributor has the power to cut off the flow of technology, to nonnegotiable transfers, in which the process of transfer is outside his control; and from person-based transfers, in which the collateral transmission of political ideas is likely, to impersonal modes of transfer, where it is not.[94] It remained to be seen whether such measures were compatible with the volume of foreign technological borrowing the political authorities wished to maintain.

Trends during and after the Second Five-Year Plan

The period coinciding roughly with the Second Five-Year Plan (1933–7) witnessed attempts to create a greater amount of indigenous technology. In 1934 the Seventeenth Party Congress stated that by the end of the Plan the USSR would be transformed "into a technologically and economically independent country and into the most technologically advanced state in Europe." Toward this end, Soviet "scientific-technological and inventive thought must become a powerful weapon in the introduction of new technology."[95] The regime backed up these words with material support. Between 1932 and 1935 expenditures on research grew by an estimated 60 percent, and capital investments in science during the Second Five-Year Plan were slated to triple the level during the First.[96] By mid-decade the USSR was spending almost twice as large a proportion of its national income on scientific research as was the United States.[97] This impressive allocation of resources to R & D exemplified the regime's determination to mobilize domestic resources for self-sufficient development.

The new emphasis on creating indigenous technology evoked a strong favorable response from many specialists. Although technological nationalism had existed earlier among scientists and engineers, the attitude was now more widely and vigorously expressed. Summarizing the proceedings of a conference of research institutes for the metal-fabricating industries, one author proclaimed that "in scientific research our possibilities are completely unlimited—and this at a time when the capitalist world is closing down its scientifc research . . . as a result of its internal contradictions."[98] In a foreword to the findings of a conference of industrial designers, another asserted that the time for drawing on foreign technology "is already past." In 1927–31, when the USSR had followed such a policy, the "majority of machines taken by us from the foreign market turned out to embody factors completely unnecessary and harmful to us." In a large number of cases, foreign practice had proved "unacceptable."[99] As the journal of the metallurgical industry put it not long afterward, "assimilating new types of production, our Soviet machine-building industry is not simply copying foreign designs, but is developing designs of machines which are peculiar to our socialist economy."[100] Aside from the motives of national and professional pride, many specialists probably welcomed the greater emphasis on indigenous technology for another reason. As we shall see, the first steps away from heavy reliance on foreign technology coincided with a relaxation of the regime's political hostility toward native R & D personnel.

This enthusiasm for indigenous technology was not universally shared, however. Despite the party's formal commitment to achieve technological independence, some top officials remained convinced that technological borrowing still offered major savings in capital investment. In 1935, for example, the chairman of Gosplan reversed the priority that most commentators were by then according indigenous technology. Although native technical thought had a part to play in the country's industrial expansion, he said, "the use of foreign technology has exceptionally important significance." The USSR should draw especially heavily on foreign technology in the sectors where it lagged, such as chemistry, oil refining, metal-rolling and light industry. This approach was "all the more important because we can multiply each improvement by millions and billions of rubles of capital investments in our grandiose construction."[101] These words echoed the party's

earlier desire not to have its R & D personnel "rediscover America" at
the cost of slower growth. The tension between achieving technological
independence and using the best technology available was reflected
in trends after 1933.

In keeping with the increased stress on independence, negotiable
technology transfers diminished. During the Second Five-Year Plan,
imported capital goods constituted only about 2 percent of gross investment, and fewer than 10 percent of all the machine tools newly
installed in 1936–7 were produced abroad.[102] Between 1932 and 1935
thirteen of the remaining concession agreements were annulled by
either the USSR or the foreign signatory, leaving only eleven still in
effect in 1936.[103]

Meanwhile the number of technical-assistance contracts, which had
risen sharply before 1931, began to fall as well. Following a special
government decision on this question in May 1931, the number of
agreements with foreign companies dropped from 124 to 74 in early
1932, dipping further to 46 in early 1933.[104] Circumstantial evidence
suggests that the reduced emphasis on foreign technology may have
helped improve the political standing of native R & D specialists. A
month after the government took its decision, Stalin called a halt to
the assault on "wreckers" within the technical intelligentsia. He kept
his future options open by arguing that the danger of wrecking would
persist as long as capitalist encirclement existed, but he stated that it
was time to shift the regime's prevailing line toward specialists from
"repression" to "attraction and concern." It seems plausible that one
motive for the new line was to create more favorable conditions for
domestic R & D.[105]

Although substantial, the cutback of negotiable transfers nevertheless
had its limits. Technical-assistance agreements in particular still played
a significant role. During the Second Five-Year Plan the government
changed some existing agreements with foreign companies into con-
sulting contracts with individuals,[106] and it signed some new agreements
with companies. Since Soviet historians have published no figures on
agreements with companies after 1933, we are left with a less exact
picture of their scope than is desirable. But we can gain a general idea
by considering all the agreements discovered by Antony Sutton that
can be dated with some precision. Judging by these incomplete data,
approximately half as many technical-assistance contracts with foreign
companies were in effect during the Second Five-Year Plan as during
the First.[107] This is consistent with annual Soviet expenditures on foreign

technical assistance, which dropped in 1935 and 1936 to slightly more than half the level of 1929–30.[108] Also, a Western source reported in 1936 that about 1,000 foreigners were working in the arms industries,[109] and there must have been additional specialists working in nondefense sectors as well. It therefore seems that the number of assistance contracts did not fall much below the level reached in early 1933. Like the earlier agreements, those signed later in the decade aided a range of industrial branches.[110]

The post-1931 reduction of negotiable technology transfers, especially imported goods, resulted from the interplay of economic circumstances and Soviet preferences. Its precise timing was due to the USSR's deteriorating balance of payments and its inability to export enough to sustain high imports.[111] The situation was worsened by the Western campaign against Soviet "dumping" of exports on foreign markets and by the disadvantageous credit and sales terms offered by some Western manufacturers to Soviet buyers.[112] But better opportunities for Soviet exports in the early 1930s probably would not have prevented the strong swing toward autarky that became evident during the second half of the decade.[113] Although some commentators obliquely attempted to differentiate economic independence from autarky and to justify continued reliance on trade,[114] the political tide was running heavily against this view. Already in 1930 persons who advocated long-term reliance on imports were being excoriated as "wreckers" who sought the "subordination of the country to foreign capital."[115] These autarkic pressures were intensified by the Stalinist fear of imperialist intervention against the USSR. Thus, between 1933 and 1938 total Soviet trade turnover stagnated at less than one-third of the 1930 level, even though by 1935 the commercial terms offered by Western lenders showed a marked improvement and the country was generally running a sizable balance-of-trade surplus.[116] Officials might hold out the prospect of more trade as a diplomatic blandishment to foreign governments,[117] but the prevailing inclination after 1930 was toward autarky.

The decisions of 1931–2 did not bring full technological autarky, but they did signify a shift toward nonnegotiable forms of technological borrowing. While dependence on assistance contracts remained strong in process technologies like oil refining and in especially complex products like precision instruments, the rapidly expanding supply of native technical manpower allowed the importance of such channels to be reduced. In many industrial fields the opportunities for technology transfer through copying and emulation, rather than through imports and assistance agreements, had grown substantially by the mid-1930s. Hence "reverse engineering" of products from foreign samples and reliance on printed descriptions of foreign inventions became progressively more important.[118]

The significance of these methods is shown by heavy industry. Earlier we noted the expansion of the domestic production of machine tools that took place after 1933, sharply reducing Soviet dependence on imports of this kind. The operational order that launched this expansion was issued in 1933 by Ordzhonikidze, the Commissar of Heavy Industry. In it he criticized the limited assortment of machine tools being made domestically—the number of types was about forty—and ordered that two hundred new types be put into production by the end of the Second Five-Year Plan. The decree instructed the Commissariat's design bureaus and production administrations "in selecting the designs of the new types of machine tools to take as a basis, after their comparative study, the best American or European models."[119]

Three years later, a major conference on R & D in heavy industry demonstrated that the emphasis on copying still characterized the activities of the Commissariat. A. A. Armand, then responsible for coordinating research within the Commissariat, commented that both theoretical and applied research still lagged behind West European and American levels. While the industrial research institutes had done "a series of good projects," he noted, "a significant part of them were done only after it became known that analogous projects had been done abroad."[120] The case was put more forcefully by the heads of some of the institutes themselves, who called for better facilities and more administrative freedom. The director of the main institute of automotive technology contended that further steps must be taken in order for the automotive industry "not to lag behind the level of world technology and not simply to copy foreign models."[121] More broadly, the head of the Commissariat's Institute of Physicochemical Problems asserted that changes were necessary "in order for industry not to work with a lag, buying and copying everything abroad." The representative of another research organization said that up to that time the institutes had "occupied themselves exclusively with copying," and argued it was necessary "to reorient ourselves toward creative work."[122] There was a clear note of frustration in these comments. The speakers obviously felt that a transition to more original R & D was advisable, but that there were serious obstacles to achieving such a change. Some scientific and engineering leaders, at least, were impatient to be done with reliance on foreign technology and the emulation of foreign research.

Policy and Politics

Although by the mid-1930s most politicians and specialists agreed in principle on the need to create a greater amount of indigenous technology, it was very difficult in specific cases to make an effective transition from foreign know-how to native R & D. Ideally, such transitions should have occurred when the opportunity offered by neogotiable and nonnegotiable technology transfers had been exhausted. In practice, several knotty problems, ranging from bureaucratic obstacles to professional pride and excessive patriotism, complicated the process. Thanks partly to administrative impediments, specialists sometimes tried to devise original designs before they had assimilated the relevant foreign ideas. One authority on R & D commented that economic organizations were using the opportunities offered by technical-assistance agreements with foreign firms "very badly"; he added that foreign travel by Soviet specialists was often arranged on the spur of the moment, thus denying them adequate time to prepare for the trip.[123] The head of the technical council overseeing heavy industry likewise complained about the poor use of foreign technical-assistance contracts.[124] These difficulties were compounded by the disdainful attitude of some designers and managers, whom one engineer labeled "amateur handicraftsmen." A person of this type "considers it a scandal for him simply to have to learn the achievements of world technology. He assumes that always and in all cases his direct obligation is to introduce 'something new,' something 'of his own,' even if that would worsen the quality of the plan or design."[125] As a result of such attitudes, specialists who had been sent abroad for as long as six months were occasionally thrust back into their old jobs without having an opportunity to sum up the findings of their trip and suggest applications.[126] But even when designers and producers drew on foreign technological ideas, as the majority still did in the mid-1930s, there was no guarantee that they would know and use the most advanced concepts. Frequent failures to use the latest foreign ideas, said the writer just quoted, were "no less a brake on technological progress" than was the wholesale neglect of foreign R & D, and "our designers often know only outdated models of foreign technology."[127] Production personnel often told their designers, "don't think up something, give me the kind of machine I saw in Magdeburg or

Hannover in 1928," that is, seven years before, and they neglected the importance of updating the foreign technology that they ordered copied.[128]

The problem of moving effectively from foreign technology to native R & D was undoubtedly intensified by the lack of large-scale contacts in the mid-1930s between R & D personnel and their foreign counterparts. Greater international contact would have made Soviet scientists and engineers more aware of the latest foreign technology that might be copied, thereby providing better benefits from technology transfers. More contact would also have helped guarantee that Soviet specialists had a full grasp of foreign technology when they set out to create novel designs, and that the technologies they devised were therefore actually better than the foreign ones they might have copied. Without extensive foreign contacts it was difficult to ensure that the industrialization campaign was receiving a maximum yield either from technology transfers or from the growing emphasis on native R & D.

This kind of international communication was difficult to foster, however, because of the Stalinists' deep suspicion of foreign contacts. The presumption of a clear connection between external contacts and political disloyalty gradually strengthened during the 1930s. For a year or two, from 1931 to 1933, the presumption was relaxed; as the official attitude toward Soviet "bourgeois specialists" became more conciliatory, so did that toward foreign specialists.[129] But in 1933 the suspicion of foreign contacts began to intensify once more. In March 1933 the regime arrested on espionage charges several British engineers working for the Metro-Vickers Company as part of its technical-assistance group in the USSR; twelve Soviet specialists were arrested as parties to the conspiracy. All the accused were found guilty, and the Soviet defendants received stiff prison sentences.[130]

Arguing that show trials of this kind were economically self-defeating, a few top leaders, particularly Ordzhonikidze, resisted the desire of harder-line Stalinists to expand the scope of the prosecutions. It seems plausible that this resistance prevented the Metro-Vickers incident from assuming the scale of the Industrial Party affair three years before.[131] Nevertheless, the unpredictability of such accusations made personal association with foreign specialists dangerous, and the nature of the charges tended to discredit the idea of importing foreign equipment and know-how.[132] Moreover, in the following year another show trial identified foreign specialists as the instigators of a fire that destroyed part of the huge new "Uralmash" plant. In this instance no foreign specialists were brought to trial, and the press went to some lengths to avoid naming the companies allegedly involved.[133] But some writers emphasized the similarity of the case to the Industrial Party and Metro-Vickers affairs; the publicity for the trial dwelt on the theme of foreign specialists as a source of sabotage; and five of the eleven Soviet defendants were sentenced to death.[134]

The Stalinist regime also hampered technology transfers by restricting foreign travel. During the mid-1930s some specialists, particularly engineers, continued to travel outside the country. But these trips were usually short inspection tours. They did not give R & D personnel sufficient time to study foreign technology carefully while abroad, and the lack of sustained contact with foreign specialists made it very difficult for them to stay abreast of foreign trends once they returned home.[135] Moreover, the total number of specialists traveling abroad almost certainly declined, and for many international travel became impossible. Perhaps the most dramatic illustration was the government's refusal in 1934 to allow a gifted young physicist, Petr Kapitsa, to return from a Soviet vacation to the long-term research project that he had been conducting in Cambridge, England.[136] Similar obstacles faced other researchers. For example, the six Soviet scientists invited to an international congress on glass and ceramics in Milan did not take part in the proceedings but "only sent their reports."[137] In addition, political obstacles apparently blocked some efforts within the Academy of Sciences to widen participation in international scientific organizations. In 1930 and 1935 the USSR joined two unions belonging to the International Council of Scientific Unions (I.C.S.U.). But despite support within the Academy for joining other member unions,[138] the USSR remained outside six other major members of the I.C.S.U. and did not join the I.C.S.U. itself until 1955.[139]

Prominent specialists resisted the imposition of such restrictions. At the meeting on R & D in heavy industry held during the summer of 1936, at least four participants cited the lack of foreign travel as a brake on technological progress.[140] About the same time, N. P. Gorbunov, Permanent Secretary of the Academy of Sciences, asserted that the Academy was not making sufficient use of world science to help the industrialization drive. In order to do so, he claimed, it was "necessary broadly to introduce the practice of long trips by our young scientists to the laboratories of the world's major scholars, and also to invite outstanding foreign scholars to our laboratories." Foreign trips, he added, should also be used as an incentive to reward researchers for high-quality research achievements.[141] Similar complaints were heard about the opportunities for foreign travel available to engineers.[142]

This position was noticeably at odds with the growing isolationism of the Stalinist leadership. Early in July 1936 *Pravda* published an article that accused a prominent mathematician and member of the Academy, N. Luzin, of "servility" to foreign science and of working "for the benefit . . . of the present masters of fascist science."[143] For a moment it appeared that the attack would expand into a general campaign against "servile" enthusiasm for foreign science, which was alleged to be "a massive phenomenon among Soviet scientists."[144] But after one such foray the criticisms remained centered on Luzin, whose expulsion from the Academy was demanded.[145] The Academy presidium responded by appointing a special commission to investigate the charges against him.[146] Judging by rather substantial changes in the commission's composition during the month between the first attacks on Luzin and the commission's final report, its deliberations appear to have generated political controversy.[147]

When the commission reported, it became evident that the leadership of the Academy was trying to preserve something of the position which its Permanent Secretary had articulated a few months before. The commission granted that Luzin had had "an essentially disloyal" attitude toward Soviet power, as reflected by his decision to spend "whole years" doing research abroad, his publication of articles in foreign rather than Soviet journals, and his "condescending" attitude toward Soviet science. The Academy Presidium tactfully agreed and admitted that "scientific society" had the right to demand Luzin's expulsion from the Academy. But it sought to deflect political wrath from Luzin by simply warning him that similar conduct in the future would provoke expulsion. Moreover, the Presidium reiterated, albeit in attenuated form, its interest in ties with foreign science. Soviet science had made a great contribution to socialist construction, it said, and the USSR's achievements had been connected in large measure with "our collaboration in world scientific progress."[148] This assertion clearly ran against the grain of the charges directed at Luzin by *Pravda*.

The Academy's defensive position in the case provoked a storm of criticism from above. Emphasizing that it was *Pravda*, not the Academy, which had uncovered the wrongdoing, an editorial in the Academy's own *Vestnik* flayed its leaders for their "absolute inactivity" in the face of what was now asserted to be "a terrorist group" within the Academy. In view of *Pravda*'s reference to the persistence of "Luzinism" and the need for "the greatest vigilance" even after the Presidium had acted, there can be little doubt that the *Vestnik* editorial was written and inserted into the journal on the orders of higher political authorities.[149] Sounding the same theme of inactivity by the Academy, the head of the Central Committee's Department of Scientific-Technological Inventions and Discoveries joined in condemning "the servility of our scholars before foreigners."[150]

Under such sustained political pressure, the Academy gave ground, though on a slightly different front. It acted against alleged disloyalty among its members, but it chose a case that may have appeared more clear-cut to some academicians. The case involved two chemists, V. N. Ipat'ev and A. E. Chichibabin, who had overstayed their allotted time abroad and whose return the political authorities had long desired.[151] The Academy urged the two to return, and when they refused, it expelled them in December 1936.[152] There were some indications of veiled resistance to this move. Shortly before the Academy's December meeting its acting president, V. L. Komarov, had publicly promised that a "firm decision" on this case would be adopted by the Academy "unanimously."[153] In the event, although sixty-three academicians voted for expulsion, six abstained; and several voted against Komarov and the other members of a slate of officers up for election by the Academy membership.[154]

Despite this concession to Stalinist political views, the pressure against the Academy continued to build. At the beginning of 1937 a major new campaign opened to cleanse the Academy of the "enemies of the people" recently discovered in its "leading posts."[155] No doubt the discovery had been prompted in part by the Academy's rearguard defense of foreign scientific relations. The new campaign directed special animosity toward the Institute of History of Science and Technology, which had allegedly inculcated servility before foreign science and completely neglected the "role of Russian science in world progress."[156]

But more was at stake than whether scientists should be allowed to visit and communicate with their foreign colleagues. A central motive behind the attacks on Luzin and the Academy was, or soon became, the position of Bukharin, one of the principal leaders of the party opposition to Stalin in the late 1920s. In August 1936 Stalin tried to institute an investigation of Bukharin on charges of anti-Soviet conspiracy, but the attempt failed.[157] This left Stalin with the task of chipping away more of Bukharin's support, and one way of doing so was to attack his standing within the Academy of Sciences. This tactic must have appealed to Stalin and his allies for several reasons. At the height of their struggle against Bukharin in the late 1920s, the Stalinists had shown a fear that Bukharin's political ideas might receive active support among Soviet scientists and engineers; faint traces of this fear can be found in Stalinist statements made in the mid-1930s.[158] Although there is no evidence that he acted on it in the 1930s, Bukharin too seems to have entertained this notion, and his election to the Academy Presidium in late 1935 guaranteed him the kind of personal access to influential scientists and engineers that he had earlier enjoyed as head of the science department of the Commissariat of Heavy Industry.[159] The Academy was thus a natural target for the Stalinist drive to bring Bukharin into the dock on charges of treason.

Whether or not the Luzin affair was instigated with this purpose in mind, it laid the basis for an attack on Bukharin, as the campaign against "servility before foreign science" turned explicitly against "wrecking" by him and his scholarly associates in the Academy.[160] One piece of circumstantial evidence suggests that Bukharin may have sought to curb the campaign against Luzin in its early stages. *Izvestiia*, of which Bukharin was still editor in early August 1936, reprinted the Academy's moderate resolution on Luzin without comment; *Pravda*, by contrast, did not reprint the resolution, and warned that "Luzinism" still existed in scientific circles.[161] But *Izvestiia*'s oblique approval of the Academy's position hardly staunched the Stalinist attack. Early in 1937 numerous officials of the Academy came under fire for "the most substantial inadequacies and errors in the work of the Presidium."[162] Then in May 1937, two months after the forces opposing a broad expansion of the political purges had been defeated by Stalin at a crucial plenum of the Central Committee,[163] Bukharin was expelled from the Academy amid warnings that not all his accomplices there had yet been unmasked.[164] A month later N. P. Gorbunov, the Permanent Secretary, lost his post and was removed from the Presidium.[165] By one of the ironic twists so common in the Stalin years, Luzin himself was not expelled from the Academy even though Stalinists now charged that there were "whole nests" of counterrevolutionaries in many Academy institutes and continued to use the accusation of "Luzinism" to intimidate Soviet scientists.[166]

A major result of the campaign against servility before foreign science was a further attenuation of the already limited personal contacts of Soviet scientists and engineers with non-Soviet specialists. This tendency was clearly foreshadowed in the accusations heaped on the Academy by the political authorities as part of the campaign. It should never be forgotten, maintained one editorial, that the political conspiracy recently unmasked by the secret police had "broadly used the exceptional position of the Academy." To communicate with each other, the conspirators had made "innumerable trips . . . under the banner of scientific trips of the Academy." This fact, suggested the editorial, indicated that the Academy leadership had failed to ensure "the necessary supervision and direction" over scientific travel.[167]

While there are no published figures on foreign scientific and technical contacts in the late 1930s, there is no doubt that they were extremely limited. According to a careful recent Soviet samizdat study, "most Soviet scientists were denied communication with foreign colleagues, whether by attendance at international meetings or by correspondence."[168] A Western account indicates that scientists who had received permission to attend several international congresses abroad in the late 1930s were prevented from actually attending.[169] Personal contacts with foreign specialists inside the USSR were also apparently cut back. A comprehensive official chronology of Soviet cultural life lists only one major international scientific meeting in the USSR from 1936 through 1941, as compared with five from 1928 through 1935.[170] In short, the personal avenues for technology transfer were drastically restricted.

Perhaps more surprisingly, the political leadership had begun to suspect even scientific and technological publications. During the Stalin years a special system was set up for screening foreign technological literature and removing suspect political material—in 1953 the secret police sections at enterprises were reportedly in charge of this activity[171]—but it is not clear whether the system was in effect before World War II. At any rate, despite the efforts of Academy officials to keep expanding the volume of incoming literature,[172] the amount of printed foreign scientific and technological information actually received declined sharply in the purge years just before the war. In 1940 the library system of the Academy received about 25,000 foreign books and journals.[173] This amounted to only 37 percent of the 68,419 books and journals received in 1936,[174] marking a dramatic decline.

These diverse political and economic pressures had complex effects on technological strategy in the late 1930s. The party leadership was clearly committed to making a general transition from Western technology to indigenous R & D. Yet because of the difficulty of doing so, political leaders and industrial commissars still pressed for the duplication of foreign designs in many cases. In 1939, the Council of People's

Commissars examined Soviet machine-tool production and concluded that the output of sophisticated machine tools was "strongly lagging behind the development of the economy." After laying out a timetable for manufacturing several categories of new tools, the Council stipulated that the four commissariats involved in creating the machines should be "guided by . . . broad use of the experience of the best foreign machine-tool plants in design and production, taking as patterns the latest models of machine tools proven in production."[175] In 1940 the Council ordered comparable steps for the rapid production of sophisticated press forge equipment. The Commissariat of Heavy Industry was instructed to design and manufacture the new equipment "with consideration of the experience of advanced foreign firms, copying the best foreign models."[176]

In the late 1930s, however, the emulation of foreign designs was easier to order than to achieve. Political xenophobia, the technological nationalism of many specialists, and the difficulty of following foreign technological trends all complicated the effort. In some instances, specialists concentrated on creating their own designs in spite of the instructions of their supervisors. Citing archival sources, a recent Soviet study reports that in the machine-tool sector, where reverse engineering was especially feasible, only 40 percent of the new machines introduced in 1940 were based entirely or partially on foreign models; the remaining 60 percent were based on purely Soviet designs.[177] A careful Western study of the same industry likewise shows that 40 percent of the main models introduced between 1938 and 1940 were based on foreign designs, as compared with 95 percent in 1928–32 and 75 percent in 1933–7.[178] These data, of course, pose serious problems of interpretation. Identifying technological origins is a difficult task at best, and the nationalist mood of the late 1930s may have encouraged Soviet designers to conceal the foreign origins of some of their work. Economic officials and observers at the time, however, believed the tendency was real. At the end of 1938 one of them lamented that designers of textile machines often tried to devise machines that had already been created abroad.[179] In 1941 the journal of the machine-tool industry, complaining that foreign technology was being used "unsatisfactorily," branded the shortage of foreign technological information at major plants "absolutely impermissible," and called for "extremely serious study" of recent developments abroad.[180] It thus seems that real movement away from foreign technological borrowing did occur, often against the orders of the central economic agencies, and that acute nationalism and inadequate foreign contacts contributed strongly to this trend.

The difficulty of shifting from extensive foreign technological borrowing to the creation of sophisticated indigenous technologies was illustrated by an analysis of Soviet industrial automation that appeared in the main party journal. Discussing R & D policy in this broad area, the writer called for "still more careful study" of capitalist technology. Assimilating the maximum amount of Western technology remained a critical task, he asserted, because the USSR still trailed the West in many industrial sectors. Moreover, Western technology was not an immobile target; it was continuing to develop.[181] At the same time, however, the USSR must adopt a new orientation toward technological change by increasingly shifting "the center of gravity to the independent thought of our scientists, designers, [and] inventors." Failure to do so would make it "impossible to overtake the technology of the advanced capitalist countries. . . . In this struggle, stagnation and every effort to confine oneself to copying, which inevitably lags behind, are . . . a step backward."[182] At present, said the writer, economic officials paid attention only to those foreign methods that had been proved in regular production and gave little heed to new foreign methods of automation that were still in the developmental stage.[183]

Obviously it was not easy to find transitional arrangements that could draw on both foreign technology and domestic R & D. Apparently Soviet officials and specialists tended either to foreswear creative initiative for the sake of copying what had already been proven abroad or else to ignore foreign technology in a determination to be creative at any cost. Neither extreme was appropriate. Closer attention to foreign technology had to be combined with an ability to improve on foreign achievements through the flexible use of domestic R & D. Only in this way would the shift from foreign to domestic sources of technology involve progress rather than regression.

To summarize, Soviet technological strategy, buffeted by the conflicting forces unleashed during the 1930s, followed no absolutely consistent course. Yet certain general tendencies are discernible. The party leadership, intent on surpassing the West, at first drew massively on Western science and technology, but it also dramatically expanded domestic R & D. Spurred by the overriding importance of military power and the problem of obtaining foreign military know-how, the regime quickly moved beyond the strategy of borrowing in weapons development. By the late 1930s Soviet designers were producing weapons that approached, if they did not already equal, those of the USSR's international rivals.[184] Despite the party's commitment to complete technological independence, however, there were few equivalent achievements in civilian technology, and top economic officials frequently fell back on foreign copying for new methods and products in nonmilitary sectors.[185] Outside the military sphere it proved unexpectedly hard to make a smooth transition from Western technology transfers to the creation of advanced indigenous technologies.

Bureaucracy and Indigenous Innovation

Relations among the country's evolving institutions for science and industrial production reflected the search for a workable technological strategy. To achieve independence from foreign technology, the regime had undertaken an enormous expansion of R & D resources. But it also had to use the resources effectively, and here organization was the key. The problems that troubled indigenous innovation during the 1930s stemmed not from a general neglect of domestic science and technology, but from the institutional pattern according to which the national R & D effort was organized.

In considerable measure, that pattern perpetuated the divorce between scientific investigation and practical activity inherited by the regime. Under Stalin the government pushed scientists toward much greater concentration on topics with clear social utility. Nonetheless, the two burgeoning spheres of research and production remained largely isolated from one another except in the field of military weaponry, where the regime managed to build strong links between them. One sign of this isolation was the growing proportion of Soviet scientific research conducted within the USSR Academy of Sciences, together with a diminution of the proportion conducted within the industrial commissariats. By the end of the decade it was apparent that even though the Academy had become a major center for applied research, the organizational ties between its scientific programs and the commissariats' industrial practice were weak. The Academy was heavily criticized for this shortcoming, but much of the responsibility rested elsewhere, for the same weak linkage between research and production was visible inside the commissariats themselves. There, R & D bodies regularly met pressures that discouraged significant innovation, and factories tended to be cut off from the limited amount of creative R & D these organizations were able to perform.

Beginning with the Academy and then turning to the commissariats, we will examine these obstacles to indigenous innovation. To what degree were they due to the industrialization campaign's initial reliance on foreign technology and to the mobilizational character of the Soviet bureaucracy? Did the regime make any serious efforts to overcome them? And how did it manage to generate sophisticated military technology when it was incapable of similar achievements in most non-military spheres?

The Role of the USSR Academy of Sciences

At the outset of the industrialization drive, the Academy became a major focus of the effort to build up the scientific potential necessary for a broad future changeover to indigenous technologies. Between 1927 and 1937, the total number of persons working in Academy institutions grew from 1,018 to 7,090. By 1940 the figure had more than doubled again, to 16,335.[186] Though subject to exaggeration because of inflation, figures for the Academy's budget suggest that its material resources grew even faster than its personnel. The budget rose from 3 million rubles in 1928 to 28 million in 1934, and over 175 million in 1940.[187] This enormous expansion of the Academy was not accompanied by a growth of research in the other customary home for basic science, the universities. During the 1920s the Bolsheviks had preferred to concentrate research in the Academy, partly to limit the professoriat's opportunities for subverting students,[188] and this tendency became even more pronounced in the 1930s, when the governmental demands for the rapid creation of trained manpower by higher educational establishments reached an extraordinary pitch. Some observers warned against removing research from the orbit of the universities,[189] but the warnings had little effect. According to one estimate, expenditures from the state budget on R & D in higher educational establishments came to only about 4 million rubles in 1935.[190] Most basic research, as well as a good deal of applied work, thus became the province of the Academy.

As it expanded the Academy's resources, the party leadership pushed to make the Academy's political and research roles compatible with its own vision of the USSR's development. Between 1928 and 1931 the attack on "bourgeois specialists" swept over the Academy, subordinating it to governmental and party controls.[191] Meanwhile, political leaders also pressed the Academy for more tangible payoffs from its research. In addition to the Academy's work in exploring untapped natural resources, which may have constituted its greatest contribution to the industrialization effort,[192] the regime sought to reorient other Academy research programs to economic needs. The government-drafted charter adopted by the Academy in 1930 put new weight on economically useful research; soon after came a major increase in contractual financing of Academy research by economic organs, along with a noticeable shift in the disciplines of newly elected members toward technological fields.[193] Taken together, these were important changes. Whereas in the 1920s the Academy's reputation as an "unreconstructed" remnant of the Tsarist era had left its future in doubt, by the early 1930s its new political orthodoxy and research priorities raised the prospect that it might assume a central role in the development of socialist science. The prospect was enhanced by the party's formal commitment to the idea of a single institution for the planning of all Soviet research.[194]

The administrative leaders of the Academy energetically sought to capitalize on this opportunity. V. Volgin, then the Permanent Secretary of the Academy, was in the forefront of the campaign to expand the Academy's functions. Volgin argued that the Academy was becoming the "directing center of all scientific work in the Union [sic]." Although he acknowledged that the Academy could not execute all research undertaken in the USSR, he stressed the need for closer ties between the Academy and the main scientific institutes in industry and in other governmental agencies. In his view, some of the applied research institutes then controlled by governmental departments should be directly subordinated to the Academy, and the scientific investigations of those remaining in the jurisdiction of other organs should be harmonized with the Academy's research plans.[195] It soon became clear that the Academy intended to exercise decisive authority in the type of harmonization suggested. At the end of 1932 its General Meeting urged the government to obligate all governmental departments to obtain the Academy's authorization before assigning basic lines of research to their major institutes.[196]

The political authorities never followed up their commitment to create a single center for science planning. But they did continue to express dissatisfaction with the existing arrangements for coordinating scientific activities,[197] thereby affording the Academy's leaders a continuing opportunity to emphasize its special virtues in research administration. In early 1935 Volgin, remarking that recent organizational changes were "merely . . . preliminary measures" paving the way for the Academy to become "the real directing center" of Soviet scientific thought, noted that this raised the question of the Academy's relations "with the whole network of scientific institutions" in the USSR.[198] Another Academy official, P. I. Kagan, contended that the industrial commissariats lacked the competence to select the most promising projects and distribute them among institutes. In his view, the most suitable organs for this task were the various scientific groups of the Academy, supplemented by representatives of the institutes that did not belong to it.[199] The Academy's representatives were equally assertive in competing for the material resources and scientific manpower allocated by the regime. In early 1936, for example, G. Krzhizhanovskii, a Vice-President of the Academy, commented that after the massive expansion of the national research network in previous years the problem of the network's quality was becoming "more and more acute." He acknowledged a general need to improve the state's research programs by paring them and bringing together the best scientific facilities and personnel. "But," he added, "in setting the whole network . . . in order it will be necessary to direct special efforts to the strengthening of the basic citadel of this network—the Academy of Sciences."[200] Appeals of this kind evoked a positive response from the political leadership; by 1937 the Academy's manpower had grown 54 percent beyond the level initially targeted for that year by the Second Five-Year Plan.[201]

The Academy and the Push for Indigenous Technology

Although efforts had been under way since the late 1920s to strengthen the tie between the Academy's research program and economic practice, they intensified in the mid-1930s. In large part the new push was an attempt to adjust institutionally to the regime's growing commitment to indigenous technological advance sustained by domestic research. To this end, the proportion of the Academy's funding channeled through contracts rather than budgetary payments was reduced from 57 percent in 1932 to 14 percent in 1937, by one calculation.[202] The apparent motive for this change was to reduce the disruptive impact that the narrow demands of economic agencies were having on the profile of the Academy's research; too many demands for short-term results could prevent the generation of the new fundamental knowledge on which indigenous technological progress depended in the long run. At the same time, however, it was important to find an institutional link to ensure that the Academy's theoretical discoveries would ultimately be applied to industrial needs.

Part of the regime's answer was to create such a link within the Academy. The 1930 charter had maintained the earlier organization of the Academy into two subdivisions, one for mathematical and natural sciences, the other for social sciences. By 1935 the idea of setting up a third division, for the technological sciences, was being canvassed. Evidently this idea was a logical extension of the Technological Council, which had been established within the Academy in late 1934 to co-ordinate the transfer and industrial application of the Academy's discoveries.[203]

The proposal to create a Division of Technological Sciences sparked a serious conflict over the Academy's proper purpose. In part the conflict stemmed from the political leadership's desire for more concrete benefits from research. The initiative for creation of the Division reportedly came from Ordzhonikidze, who was at the time both Commissar of Heavy Industry and a member of the Politburo.[204] Within the Academy the idea received some support, especially from the newer academicians who were trained in engineering and wanted more applied research done in academic institutes.[205] But there was strong resistance as well. According to the President of the Academy, "a long struggle" preceded the decision to set up the Division. "Many persons" opposed the decision on the grounds that the research projects envisioned were "simply not a matter of science but of engineering, and for that there should be an engineering academy."[206] Circumstantial evidence suggests that pressure from the political leadership was necessary to overcome these defenders of the Academy as an institution devoted strictly to basic research. In March 1935 the Academy's General Meeting discussed in detail the draft of a new charter submitted to it by a commission of academicians; after making amendments the General Meeting confirmed the draft and submitted it for governmental approval.[207] Eight months later, however, the Academy found it necessary to confirm a draft charter once more in order to submit it to the government.[208] This indicates that some aspect of the earlier draft had been found wanting by the political authorities and that its reworking had required a new vote of the Academy membership. Because the creation of the Division of Technological Sciences was the most significant organizational change to appear in the new charter, it is a plausible guess that the Division was the bone of contention that delayed government approval.[209] At any rate, the new charter was confirmed by the Council of People's Commissars in November 1935.[210]

The argument over applied research within the Academy did not dissolve once the Division of Technological Sciences was formed. It still had to be decided whether the Division was to serve primarily as a clearinghouse connecting the Academy with applied research done elsewhere, or whether the Division itself was to do extensive applied work. In 1935 the former conception enjoyed substantial support, at least among the Academy leaders. The Division started by operating mainly through interdisciplinary commissions rather than institutes,[211] and its Academic Secretary warned against excessive growth. He had repeatedly been subjected to arguments for establishing new institutes in the Division, he said, but *"it is not necessary to create numerous institutes, especially with a branch profile* [that is, with an applied orientation]. We propose to create a very limited number of new research institutions in areas where scientific work either is not set up at all, or is unsatisfactorily developed in branch [commissariat] institutes. In this connection, we will orient the plans of our future institutes toward profound theoretical projects."[212] In 1937 the Division still contained only two full-fledged institutes, although some of its sixteen commissions may have been conducting research on their own.[213] Top Academy officials thus promised greater responsiveness to immediate industrial requirements but also defended a high priority for exploratory research.[214]

If the creation of the Division of Technological Sciences represented an attempt to move toward greater reliance on indigenous technology, it also symptomized the difficulties of making this transition. Late in the decade there was increasing evidence that, although it was paying more attention to applied research, the Academy was encountering major difficulties in transferring its scientific results to industrial production. In 1938 the Academic Secretary of the Division remarked that while it had achieved "some successes" in strengthening its ties with industry, this problem still demanded a great deal of systematic effort.[215] A short time later the Academy President ordered all institutes to submit monthly reports detailing steps to introduce their research results into economic practice.[216] Simultaneously, the Academy Presidium emphasized that institutes had "a direct obligation" to see to the practical utilization of research findings, and it instructed them to report immediately any difficulties they encountered in doing so.[217]

These measures were the harbingers of a political storm. In May 1938 the Council of People's Commissars met to discuss the Academy's work and heavily criticized it. The criticism was provoked partly by the controversy then raging in scientific and political circles over the relation between agrobiology and theoretical genetics. In addition to officials from the USSR Academy, Trofim Lysenko, President of the Lenin Academy of Agricultural Sciences and the chief proponent of agrobiology, participated in the Council meeting.[218] But the Council's criticism was also provoked by the sporadic transfer of scientific results from the USSR Academy to the industrial sector. The Council complained that "even the Technological Division," which at the time consisted exclusively of units working in areas with industrial applications,[219] "has still not achieved the connection of its activity with the practical work of socialist construction." Refusing for the time being to confirm the Academy's 1938 research plan, the Council ordered the Division to "establish supervision" over the economic application of major research findings. It also ordered an enlargement of the Academy's membership and decreed a shake-up in the ranks of the Presidium and all the divisional bureaus.[220] A week later, Stalin proposed a public toast favoring the sort of science that "does not allow its old and recognized leaders to enclose themselves in the shell of priests of science, the shell of monopolists of science," and that supplemented old cadres with new.[221] Viewed in context, the speech gives the impression that Stalin was utilizing the same techniques in scientific matters that he employed to discipline and purge other bureaucracies. Young researchers' ambitions for upward mobility were to be used to wring better technological results from established scientists and institutions.[222]

This clash greatly strengthened the hand of those academicians who opposed the Academy leadership's effort to limit the amount of applied research done in the Technological Division. Shortly after the Council meeting, the Academy Presidium gathered to discuss the problems of innovation. While it took pains to point out that these problems were the fault of industry as well as the Academy, it was obviously under pressure to remedy them. Thus the Presidium decreed that the Academy research plan should be reformulated to give more coverage to fuel production, power engineering and other applied fields, and it instructed the Technological Division to present "a concrete plan for a basic reconstruction" of its operations within ten days.[223] In the following six months four new institutes were organized in the Technological Division, and another was set up in 1939, making the Division one of the Academy's largest research components.[224] The Presidium also stepped up its contacts with the commissariats and its efforts to promote new industrial processes.[225]

These measures, however, did not satisfy the political leadership. At a meeting in April 1939, the Council of People's Commissars reviewed the 1939 research plan submitted by the Academy. Though it approved the plan, the Council subjected it to sharp criticism, inserted several changes, and again emphasized the need to improve the Technological Division's work.[226] The Council also ordered the Academy Presidium to identify finished research that would be economically beneficial; and it appointed a special commission led by S. V. Kaftanov, Chairman of the State Committee for Higher Schools, to review the Academy's suggestions and shepherd them through the process of innovation.[227] This commission selected fifty-one projects for introduction into the economy.[228]

Two related motives underlay this campaign to link the Academy's research more closely to practice. One was the gradual change of technological strategy during these years. The authorities were increasingly committed to deriving original technologies from Soviet basic research, and this presupposed a stronger tie between industrial engineering and the Academy's investigations. A second, more specific motive was military. As the international situation deteriorated in the late 1930s, the regime's efforts to develop sophisticated weaponry were stepped up from their already formidable pace, and the Academy was called upon to help. Precisely how much of its research was channeled into military projects is not clear, but the amount appears to have been considerable. At the end of 1938 the General Meeting referred to the threat of war as one reason for making a special effort to include economic and defense projects in the 1939 research plan.[229] A year later an Academy official stated that "many" Academy institutions were conducting defense research and remarked that the Academy "must orient itself still more to problems connected with defense topics." About the same time, the Academy set up an internal Special Department to facilitate administrative contacts with defense organizations.[230] Soviet historians have reported that the Academy worked on "approximately 200 projects" assigned by military commissariats in "the prewar years."[231] It therefore seems likely that one of the concerns of the Kaftanov commission was to expedite Academy research projects with military applications.

The exertions of this commission must have had positive effects, since the Academy Presidium voiced appreciation for its efforts and asked that its work be continued.[232] In doing so, the Presidium asserted that practical application of the projects in question had been "delayed for a series of years by causes which, for the most part, did not depend on the USSR Academy of Sciences."[233] The Presidium obviously felt that much of the responsibility for the inadequacies in industrial innovation belonged to the commissariats, and it saw the commission as an instrument for prodding the commissariats into greater activity.

The crucial question was not whether the commission's endeavors were useful, but whether they were useful enough. A report in early 1940 by a Vice-President of the Academy came too soon after the fact to offer conclusive evidence, but his prognosis was not optimistic. He noted that the intervention of the top leadership had strengthened the ties between the Academy and the commissariats, but he added: "We still have not found the means to overcome the gap which exists in our country between the stage of laboratory research . . . and the factory installation with output in tons."[234] During the next two years, despite some successful innovations in fields such as metallurgy, problems arose in developing the Academy's findings on polymer chemistry, catalytic cracking of petroleum, and automation.[235] Cases of this kind led one commentator to conclude that "there still is no organic connection between theoretical research projects and their practical utilization in industry," even though some improvements had occurred.[236]

Conditions inside the Academy helped impede such a connection. Citing the problems of automation and telemechanics as an illustration, the 1941 General Meeting called attention to "the inadequacy of the organic tie between scientific institutions belonging to the same Division, and the almost complete absence of this connection between institutions belonging to different Divisions."[237] Aside from the administrative failings of the Presidium, two other factors probably hurt cooperation among Academy institutes. One was the scope of scientific secrecy. Among several "weak places" in the Soviet research effort, Academy President Komarov singled out "a very important question—about the adequate dissemination of scientific and technological information concerning everything new in both Soviet and in foreign literature." R & D specialists, he added, "must be fully assured" of such information.[238] Since evidence given below suggests that widespread secrecy was inhibiting technological innovation, it is not unreasonable to believe that Komarov was alluding to this issue, at least in part. A second factor that complicated the generation of useful applied research within the Academy was the lack of competition in some fields. One member charged that "healthy efforts at criticism are often extinguished [and] hushed up," and the General Meeting noted the "weak development of scientific criticism in all the work of the Academy."[239] No doubt the Presidium's attempt to eliminate research overlap by enforcing the rigorous specialization of institutes intensified this problem.

Although some obstacles to innovation were rooted in the Academy's structure, others extended far beyond its reach. The Academy Presidium was correct in insisting that the industrial bureaucracy bore a large measure of responsibility for the problems which appeared during the 1930s. The commissariats had accepted a broad expansion of the Academy's research role not because they were eager for better R & D administration, but because most research was a secondary matter outside their immediate concerns.

Innovation and the Industrial Bureaucracy

During the First Five-Year Plan, the campaign for rapid growth based on Western technology left the industrial bureaucracy little incentive to generate innovations of its own. Under the intense pressures of the industrialization drive, even the use of foreign technology was difficult. The application of know-how garnered abroad presented the industrial bureaucracy with demanding tasks, which it did not always perform satisfactorily.[240] But foreign borrowing was nevertheless far easier than creating entirely new technologies. Laboring under the political leadership's relentless demands for rapid technological change, few industrial administrators were willing to risk the added uncertainties raised by heavy reliance on domestic R & D.[241] This attitude was one factor that shaped the structure and operation of the industrial research sector.

The official policy toward industrial research combined expanding material support with tremendous pressure for short-term results. During the First Five-Year Plan the resources available to research institutions within industry increased dramatically. The number of research institutes and their affiliates grew from 30 in 1928 to 205 in 1931, and from 1930 to 1932 expenditures on research projects in industry more than tripled.[242] At the same time, the regime's quest for Western know-how increased the proportion of these institutions engaged in adapting foreign technology,[243] and the demand for expeditious responses to the needs of the enterprises assimilating this technology grew. One way of sensitizing industrial R & D units to this task was to make them more dependent on enterprises and other operating agencies for financing. Between fiscal 1925–26 and 1928–29 the proportion of such nonbudgetary income in industrial research spending rose from 23 to 50 percent.[244]

The desire to orient research to the short-term demands of the transfer process was also reflected in debates over the organization of industrial R & D. By 1928, pressures were mounting for the central scientific research board of the Supreme Council of National Economy (VSNKh) to relinquish control over many of its research institutions to a lower level of the industrial bureaucracy. Proponents claimed that this step would compel industrial research establishments to provide more direct aid to enterprises.[245] Not everyone agreed. Bukharin, for one, resisted on the grounds that it would cause the institutes to cut back on creative research, with the result that industry would lose rather than gain.[246] Nevertheless, a large number of institutes were shifted from central to lower jurisdictions in 1929 and 1930.[247]

This move was probably motivated by a real desire of Stalin's supporters in industry to strengthen the links between research and production. But it also had the political benefit of dispersing authority over some of the technical specialists who had adopted a critical attitude toward the First Five-Year Plan and who otherwise would have fallen under the direct command of Bukharin in his capacity as head of the VSNKh sector administering scientific research.[248] The political aspect of the dispute is shown by the fact that on two separate occasions Lazar Kaganovich, a close ally of Stalin, charged that Bukharin had neglected the phenomenon of "wrecking" among researchers in making recommendations on the organization of industrial research.[249] The obvious implication was that Bukharin was protecting VSNKh researchers from the purge of "bourgeois specialists" then under way. Stalinist anxieties about technical specialists as a possible source of support for Bukharin had clearly been aroused,[250] and when the next show trial occurred in November 1930, representatives of the Scientific Research Sector (NIS) and its research establishments were prominently featured.[251] While the trial was going on, a bid was made to strip the NIS of control over virtually all scientific institutes.[252] Whatever temporary success it enjoyed, the longer-term results of this gambit were mixed. As of August 1931 twenty, or roughly a quarter, of VSNKh's institutes remained under NIS control.[253] Bukharin's position thus seems to have won at least partial vindication, although within two years he changed his opinion on this issue.[254]

After 1931 the debate over industrial research expanded to include the question of scale as well as organization. The recent expansion of research within VSNKh had been extremely rapid, and some deterioration of research quality had certainly resulted[255]—worsened, no doubt, by the imprisonment of many qualified scientists and engineers during the attack on "wreckers." At the same time, the industrial commissariats were becoming aware that as long as foreign borrowing remained the main technological strategy, further growth of their R & D programs was not essential to their primary economic mission. The tasks of wholesale technological borrowing and massive economic construction during the First Five-Year Plan had subjected the commissariats to acute strain. A curb on their research effort offered the welcome prospect of a reduction in the relentless demands on their limited means.[256] This must have been one of the motives of industrial officials like Ordzhonikidze in seeking to extend the Academy's responsibility for industrial research and working to limit the commissariats' role.

The Commissariat of Heavy Industry, where most industrial research was conducted in these years,[257] illustrates the commissariats' tendency after 1931 to restrict the amount of R & D done under their auspices. In mid-1932 the Commissariat issued a decree limiting further increases in the number of its institutes and calling on its science board to weed out "feeble" institutes from the network. Five months later, pursuant to an order of the Council of People's Commissars, it issued a further decision reorganizing the research network. As a result, the number of institutes in the Commissariat declined from 162 to 136.[258] Given the suddenness of the earlier expansion of the research program, these cutbacks had some justification as steps to improve its quality. But by early 1934 an additional large reduction of industrial research was under consideration.

Attempting to ward it off, Bukharin, who had served as head of the Commissariat's Scientific Research Sector since 1929, published a defense of the projects being conducted under the Sector's supervision. He began by stressing that no other agency—neither any commissariat nor the Academy of Sciences—possessed "such a significant and high-quality experimental base, as [does] our heavy industry," and he pointed out that steps had already been taken during the preceding year to slim the branch's research network and improve its quality.[259] To cut the program further, he argued, would constitute a narrow preoccupation with practical results that would allow imperialist enemies to outflank the USSR.[260] Bukharin made an especially vigorous effort to emphasize the value of the research program for regular production. "A whole series of institutes," he said, "have already occupied *a definitive place in the process of material production*."[261] The USSR had entered a stage "when a *theoretical orientation* in questions of technology . . . is directly and immediately becoming the lever of *practical action*."[262]

Events soon suggested that Bukharin's appeal had not received a sympathetic hearing from the leaders of the Commissariat. In early 1934 the Council of People's Commissars passed a major resolution demanding large reductions in government expenditures.[263] In implementing it, the Commissariat of Heavy Industry imposed cuts on the Scientific Research Sector that were far more drastic than those applied to the main administrations for industrial production. The sector was ordered to cut its expenditures on salaries and administration by more than 17 million rubles—though, notably, military projects, together with studies of natural resources, were exempted.[264] This amounted to a reduction of about 15 percent from the sector's manpower and administration expenditures for 1933.[265] None of the decreases imposed on individual production administrations (glavki) reached the absolute size of this one, and in proportion to their total budgets the decreases were radically smaller.[266] By a rough calculation, the total cuts scheduled for the glavki amounted to no more than 2 percent of their wage funds for 1933.[267] The limitations on the Scientific Research Sector were thus quite disproportionate, and they signalled a setback for Bukharin's conception of research as the guiding light of industrial practice.

Just how real a defeat this was for Bukharin's views is difficult to say, however. According to a later report on the actual research expenditures of the Commissariat during 1934, the final total exceeded the 1933 level by some 20 percent.[268] It is conceivable that the increase was due to a major expansion of military R & D, and that other programs favored by Bukharin were in fact cut back. But it may also be that the cut was initially decreed partly to satisfy Stalin's personal animus toward Bukharin, and that with the aid of more moderate political leaders such as Ordzhonikidze, Bukharin succeeded in reversing the implementation of the order.[269] We have already seen that Bukharin had been able to limit the drastic reorganization of industrial R & D proposed in 1931, and he may have enjoyed a similar success over R & D funding. In other respects he was making a modest political comeback, having become chief editor of *Izvestiia* in 1934.

The immediate injury to the Commissariat's research program may thus have been more qualitative than quantitative. Even if total research expenditures remained high, the proportion channeled directly to major research institutes for applied projects of longer-term character decreased. Between 1934 and 1935 these direct allocations from the Commissariat's budget dropped by about 18 percent, whereas funds channeled to institutes via contracts with production organizations increased by about 46 percent.[270] Because the amount of contract research was now between five and six times as large as research funded through budgetary grants, the research programs of the Commissariat were liable to be powerfully influenced, and probably distorted, by the short-term demands of enterprises.

The evidence, while incomplete, suggests that during the next three years the resources devoted by the Commissariat to research not only shifted focus but may also have declined in scope. Between 1933 and 1935 the number of researchers and auxiliary personnel in industry dropped by about 25 percent, and at the end of 1938 the number of researchers working in industry and construction remained below the 1933 level.[271] During these years at least some scientific fields suffered severe reductions in funding as well as personnel. By 1937 investments in industrial chemical research, for example, had been cut to 5 percent of the level originally projected in the Second Five-Year Plan.[272] The stabilization and possible decline in the volume of industrial resources devoted to research in this period contrasts strikingly with the Academy system, where manpower and resources were growing very rapidly, and it helps to explain why the Academy's Department of Technological Sciences was increasingly called on to help speed indigenous technological development. In a state bureaucracy built primarily on mobilizational lines, the Academy seemed the best place to generate new technology.

Given the pressures of the Commissariat of Heavy Industry on slack resources and its persistent bias toward foreign technological ideas, this was no easy assignment. In 1936 Academy Permanent Secretary Gorbunov took note of the "enormous gulf" that still existed between the Academy and industry. In seeking an explanation, he accepted some blame for the Academy but put the bulk of it on industry's shoulders. "It is one thing," he said, "when discussion concerns the purchase of some American patent which has won itself full recognition in capitalist industry." But when a new idea "is still emerging from a [Soviet] laboratory or even a semi-industrial installation, it is still necessary to travel a very difficult road to obtain its broad recognition and application."[273] Gorbunov plainly felt industry's preoccupation with the rapid absorption of proven foreign technology was preventing it from adjusting to the requirements of indigenous innovation. Doubtless this preoccupation helps explain why at a 1936 Academy meeting "not a few curious examples" were adduced of instances in which the research findings of Soviet physics institutes had been applied abroad and then introduced into Soviet industry "significantly later, in the form of foreign experience."[274]

The attempt to generate more indigenous technology was complicated by the exceedingly parsimonious attitude of the industrial apparatus toward development facilities. The capacity of applied research institutes to assimilate the findings of basic research and create novel technology depended on the availability of experimental plants where the institutes could test their ideas in industrial conditions. At a 1936 conference on research in heavy industry several institute directors emphasized that the strategy of foreign technological borrowing could be discarded only when such facilities were provided.[275] These, however, were the resources which the Commissariat and its glavki, constantly preoccupied with the current economic plan, were loath to furnish. Of ninety-nine institutes listed in a 1935 directory of industrial research, eleven had had test facilities in 1928; seven years later, only four more institutes had them.[276] Even those that did were often caught in a losing struggle with the Commissariat over their actual use.[277] Nor did the industrial bureaucracy show any inclination to provide more of the facilities. In May 1937, for example, the chemistry section of Gosplan offered revised estimates on capital investment. The funds originally earmarked by the Second Five-Year Plan for investment in regular chemical plants were reduced to about 95 percent of their old level. Investment in plants for the testing of new processes and products was slashed to 54 percent of the original figure.[278] Along with curbs on increases in industrial research, this decrease indicates that, under the pressures of economic mobilization, the industrial bureaucracy found it expedient to pare R & D programs whose reduction would not immediately affect the volume of economic output.

If the Commissariat's research institutes had trouble creating novel technology, product- and plant-design organizations faced pressures that discouraged them from drawing on such technology even when it was available. Pressed to draft blueprints rapidly and cheaply, they were wary of accepting the risks of development, especially when reliable Western technology could be copied instead. The machine-tool industry provides an example. Near the end of the First Five-Year Plan the Research Institute for Machine Tools and Instruments was nearly dissolved by branch officials intent on foreign technology and short of money; only the intervention of the party hierarchy prevented this.[279] In 1933 the Institute was combined with a design bureau and a test plant to form a new organization, ENIMS for short, which was given primary responsibility for creating advanced Soviet machine tools.[280] The Commissariat's emphasis on copying foreign technology, however, weakened the new institution's attention to original R & D. In 1935 ENIMS was criticized by two engineers, who asserted that it was "fatally detached" from scientific research and was only imitating foreign designs. While the writers' criticism of the particular foreign designs selected is difficult to evaluate,[281] their article did point up some consequences of the stress on foreign borrowing and rapid industrial expansion. ENIMS, they charged, possessed none of the laboratories and equipment needed to conduct independent investigations in the field of machine tools, even though the organization was supposed to do such research.[282] Allowing for some hyperbole in the accusation, it still indicates that the connection between domestic applied research and industrial design was tenuous.

The link was also weak in the institutes planning new factories. The tremendous pace at which these institutes were required to create technical plans was one of the major sources of the construction delays and errors about which the top state organs frequently reproved the commissariats.[283] The natural response of the commissariats to such criticism was to press their subordinate design bodies to make maximum use of previous blueprints.[284] In conditions where the alternative was often to begin construction without any technical plans at all, this step was a logical one. But it did not encourage the institutes to stay abreast of the latest research and to improve their plant designs, although an effort to facilitate such improvements was made in the mid-1930s by amalgamating some applied research establishments with plant design institutes.[285] Nor did the resource limitations and administrative rules under which the institutes worked provide an incentive to innovate. In early 1934 the Commissariat of Heavy Industry imposed strict norms on the overhead expenditures of all its design organizations. In 1936, when the Council of People's Commissars decried the "impermissibly high cost of planning" construction, the Commissariat cut these norms by 15 percent.[286] The same preoccupation with cost-cutting underlay the premium system enacted at the end of the decade, which gave the institutes a strong incentive to shirk on quality and avoid the unpredictable expenses of new technology.[287] Together these conditions restricted the institutes' receptivity to new technological ideas. Often only detailed administrative orders from above could spur the institutes to pursue ideas for novel methods of industrial construction and production. This arrangement depended on the administrative organs' ability to know as much or more than their own specialized technological agencies, and its effectiveness was therefore limited.[288]

The pressures of the industrialization drive also made it difficult to elicit voluntary participation by industrial plants in the process of innovation. Though anxious to receive technical advice on urgent matters pertaining to current production, the enterprises did not ordinarily pursue major improvements of their manufacturing methods or of their output. Several institute directors at the August 1936 meeting on heavy industrial R & D complained that they were overburdened with "small change" requests from enterprises.[289] The enterprises, preoccupied with meeting the demanding short-term requirements of their production plans, played a relatively passive role in the whole process of innovation. In the words of a Western scholar, "once enterprises were built, they were subject to rare changes in process technology and to only occasional changes in product specifications. . . . The chief function of the enterprise in this period was to master technology assigned from above rather than to decide technology."[290]

This arrangement was feasible partly because most technological innovation during the 1930s was channeled through major construction or remodeling projects small enough in number to be administered by the central industrial organs. In 1937, 80 percent of all industrial output was being manufactured by plants built or basically remodeled after 1928.[291] The arrangement had some serious drawbacks, however. Testing and perfecting new technology was difficult in central laboratories that had no direct connection with industrial plants, particularly when special development facilities were in short supply. The blueprints and drawings of the central R & D organizations often contained "an enormous number of voluntary and involuntary errors."[292] This situation frequently forced industrial plants to rework the designs on an ad hoc basis, and it strengthened the bias of enterprises against major innovations. It also limited the factories' ability to adapt their products to the purchaser's exact technological needs.[293]

Beginning in 1932, the regime attempted to enlarge the enterprises' R & D capacities and link their laboratories to central research units.[294] But despite some progress, including the establishment of a few large factory labs, the typical factory's R & D program remained weak and isolated. A conference on factory labs in 1935 complained of "the bad organization of information and exchange of experience among scientific research institutes, factory laboratories, [and] design organizations, in connection with which plants, even in the system of one main administration, often do not know about the problems being solved in the institute or at other similar plants."[295] One writer explained this problem by referring to "quite numerous cases" in which R & D personnel intentionally withheld their findings from other researchers.[296] The information flow within the industrial bureaucracy, in short, was not conducive to technological innovation.

The Search for Administrative Remedies

The persistence of these impediments to innovation provoked a push to decentralize industrial R & D. In July 1935 Ordzhonikidze issued a decree which stated that the current organization of product-design work could not ensure further improvement of machine-building output. Ordering that most manpower of the central bureaus for machine design be transferred to enterprise design bureaus, the measure called for the designers remaining in the central bureaus to serve only as review panels for blueprints produced at the plants. It also set up sizable salary bonuses for designers who did especially good work.[297]

Ordzhonikidze's order was followed by a similar but more cautious move to decentralize control over the Commissariat's research institutes. Not long after the order A. A. Armand, acting head of the Scientific Research Sector, published a discussion of the need for a "deep reorganization of the whole system" of industrial R & D.[298] In his opinion it was necessary to "reorganize as fast as possible scientific-technological service to industry, developing and strengthening factory laboratories." This reorganization would include changes "at the expense of the network of branch institutes" for applied research.[299] Though Armand was quick to aver that the existence of branch institutes was not threatened, he was clearly suggesting a basic alteration of the system for administering applied research. In February 1936 the Commissariat tentatively backed this idea when it ordered the Scientific Research Sector to examine the research network within three months and, in cases where the work of the institutes duplicated that of factory labs, to transfer the institutes' personnel and equipment to the plants.[300] The idea also received some support from the central party organs. In May 1936 the party Orgburo recommended to the Commissariat that some tasks normally done at the research institutes be transferred to the plants.[301]

Nevertheless, these efforts to decentralize industrial R & D failed. The issuance of Ordzhonikidze's 1935 order on machine building had been preceded by what one proponent termed "fierce disputes," with considerable opposition coming from the central design bureaus,[302] and this resistance continued after the promulgation of the decree. Most glavk officials who discussed the subject of machine design studiously ignored the existence of the order or made comments that were obliquely negative.[303] The bureaucratic resistance was strong enough to be condemned by the Commissariat newspaper as bordering on sabotage, and by another proponent as "efforts to frustrate the order of the People's Commissar of Heavy Industry."[304] Three years later, calls were still being heard for changes of the type embodied in the original order, indicating that its implementation had been blocked.[305]

The same fate befell the effort to decentralize the activities of the Commissariat's research institutes. By the time the Commissariat held its August 1936 conference on R & D, Armand had reversed his position on this question.[306] A decree on R & D published shortly afterward over Ordzhonikidze's signature put the main stress on improved management and financing of projects by the glavki; administrative devolution was not mentioned.[307] The failure of this attempt to decentralize was due in significant measure to political opposition from the research institutes that felt their existence to be in jeopardy. A year earlier the institutes had reportedly resisted increasing the size of factory laboratories because they feared the loss of their research monopoly and feared even more that they might ultimately be abolished.[308] The 1936 conference reflected the same attitude. Several institute directors spoke against decentralization, and the one or two who supported it did so only equivocally.[309] As in the central design bureaus, the weight of opinion in the institutes was clearly on the side of centralization. Although the institute directors complained bitterly about mismanagement of research by the glavki, on this matter their interests and those of the glavki coincided; probably they felt that the demands of enterprise managers would be even less compatible with good applied research than those of higher-level bureaucrats. Moreover, by mid-1936 the opposition to any far-reaching decentralization had picked up more support within the central party apparatus. K. Ia. Bauman, head of the Central Committee's Department of Scientific-Technological Inventions and Discoveries, made a formal bow to the idea of decentralization but insisted that "the basic mass" of the applied research institutes should remain attached to the glavki.[310] Resistance of this kind was probably due partially to the changed political atmosphere as Bukharin came under attack, and as the great purge approached. In these circumstances, talk of decentralizing R & D—a step Bukharin had advocated—became politically dangerous.[311]

If the efforts to decentralize industrial R & D failed partly for political reasons, they also foundered because there was no incentive for industrial managers to support research. The idea of giving greater authority to the enterprises was based on the assumption that they wished to pursue serious research and to make basic changes in their technology; it was the lack of scientific manpower and equipment which prevented them from doing so.[312] In reality, the mobilizational character of the industrial bureaucracy gave them other preoccupations. The overriding concern of plant managers was to meet short-term production plans and to locate the scarce supplies necessary for this purpose.[313] The constant pressure of the plans destroyed most of the incentive to innovate, particularly because the tautness of consumer enterprises' production schedules and the lack of market competition made it possible to dispose of all output, no matter what its quality. As a rule, only cost-saving manufacturing methods that could be introduced without disrupting production interested the enterprise staff. However, in cases where they decided to risk major innovations, the shortage of assured supplies and information about possible new inputs presented further barriers.[314]

For all these reasons, enterprises frequently failed to maintain laboratories and used existing laboratory facilities only as a quality-control service.[315] For the most part, factories did not voluntarily improve their product designs, and they engaged in widespread efforts to exaggerate their achievements by reporting the number of new models being produced without saying how many of each model were being manufactured.[316] When enterprises were alloted equipment and engineers for development projects, they channeled them into ordinary production work instead.[317] According to a 1939 government resolution, the experimental shops of a majority of machine-tool plants had been diverted to current production.[318]

In such conditions the decentralization of responsibility for R & D threatened to eliminate these activities rather than improve their results. Until the bureaucratic structure within which the factory operated was basically altered to create a strong interest in innovation at this level, R & D could not be decentralized successfully. Effective R & D decentralization presupposed a relaxation of high-pressure planning and the abolition of many central controls over the enterprises. In the late 1930s there was no room for changes of this kind. Faced with the enterprises' persistent tendencies to neglect technical specifications and quality standards, the Stalinist leadership redoubled its emphasis on centralization. The negative side of this response consisted of a vigorous effort to keep authority over technological decisions at the top of the industrial bureaucracy.[319] The positive side consisted of an attempt to extend the central economic plan to include R & D activities.[320]

Nevertheless, evidence of serious impediments to innovation continued to appear. In early 1941, G. M. Malenkov complained that because many managers underestimated the significance of new technology, valuable inventions lay unused for long periods in research institutes. The commissariats' technical councils, he added, had become "auxiliary organs" unconcerned with innovation, and the design of new types of production took an unjustifiably long time. As a result, frequently "the series output of a new product begins when that product is already far from advanced."[321]

There were at least two other impediments to innovation that Malenkov did not mention. One was the poor circulation of information among research establishments and enterprises, which must have worsened after accusations of espionage and divulgence of state secrets became commonplace in the purges of the late 1930s. In 1940 a group of authors lamented "the lack of well organized technological-economic information" to aid the process of innovation.[322] The problem was described with unusual frankness in the journal of the Commissariat of Aircraft Industry. Asserting that the exchange of technological information must be "decisively" improved, the journal observed that "very often our designer solves problems which have already been solved in an adjacent designing bureau." To illustrate the "ridiculous" results to which this could lead, the journal cited a case of three design bureaus located at the same Moscow aviation plant. Working in secret from each other, all three bureaus had developed "the same unsuccessful scheme," even though the first bureau had already tried and rejected it. Concluded the journal: "It is said that such isolation is necessary, that it is caused by secrecy. But too much secrecy is ruining our work."[323]

A second impediment to innovation that Malenkov failed to mention was the purges. Ever since the Shakhty and Industrial Party trials, specialists had been subject to the charge of "wrecking" through their technical decisions,[324] but in the late 1930s such allegations became more frequent. The charges made against R & D personnel were extremely diverse. They included the unnecessary delay of R & D projects, the expeditious application of "pseudoscientific" findings which were damaging, and the application of new ideas without adequate preliminary testing.[325] In short, any conceivable difficulty that might arise in the process of innovation was open to charges of wrecking. Extensive arrests of specialists on such charges seriously disrupted innovation and bred timidity in the specialists who remained at liberty. One index of the specialists' caution is that the number of inventions registered with the government fell from an annual volume of 7,000 in 1934 to a total of only 3,902 for two and a half years in the late 1930s when the purges were at their height.[326] Although we lack statistics on the number of inventions actually introduced into production, it must have declined as well. To protect themselves, specialists frequently declared themselves in favor of innovation in principle but sought to avoid responsibility for introducing any innovation in fact. As one emigré commented years afterward, a final decision to innovate could be made "only when there are so many people involved that in the case of a failure nobody would be able to trace the responsible ones."[327]

In an effort to limit the disruption of innovation by the purges, the political authorities set up institutes staffed by arrested specialists under the control of the secret police. Such establishments appear to have been in existence by 1930.[328] Leonid Ramzin, the star defendant in the Industrial Party trial, completed the work on his novel once-through boiler in such a prison, and incarcerated aircraft designers are also reported to have been working on research projects by 1931.[329] Fragmentary evidence suggests that in the mid-1930s the scope of these institutes may have diminished, as the relative moderates within the Politburo gained the upper hand over Stalin in questions of economic policy and obtained the release of some imprisoned specialists.[330] But the onset of the great purge sharply expanded the network. According to a samizdat account by a former prisoner of one of the institutes, in 1938–40 they contained a total of about three hundred aircraft designers.[331] The creation of such establishments lessened but by no means obviated the technological damage done by the purges. Not all imprisoned specialists found their way to such institutes, and it was sometimes difficult to transfer the institutes' findings to nonpolice enterprises, which feared to be associated with the work of convicted wreckers.[332]

The Organization of Military R & D

Although the Stalinist system showed many deficiencies in creating and diffusing new technology, there was one area where it transcended such difficulties and achieved significant technological successes: the arms industry. Several factors distinguished weapons research from other types of industrial R & D. The first was a plentiful supply of slack resources, including test facilities. At the end of the 1920s the bureau of military inventions within the Scientific-Technological Department of VSNKh was "by far the largest" of any of the Department's subdivisions,[333] and the preference accorded to military research projects increased during the 1930s. During the First Five-Year Plan, the regime embarked on a major program to re-equip its military establishment with new weapons while keeping the level of military manpower roughly stable. A large part of the military budget thus went into weapons procurement, and it is logical to conclude that a substantial portion of this amount was channeled into military research. The priority of military R & D is shown by the fact that when the Commissariat of Heavy Industry decreed blanket cutbacks in industrial research in 1934, military projects were explicitly exempted.[334] The same priority was reflected in the distribution of development installations. The aircraft industry, for example, stood out from other branches because of the volume of such facilities at its disposal.[335] As for the actual production of military goods, the output of all factories was regularly divided into three descending categories of quality—for military use, for export, and for nonmilitary use.[336] This system of allocation cushioned arms enterprises against the chronic shortages that afflicted the rest of industry, and it assured them of higher-grade inputs than were available to plants producing nonmilitary goods.

A second factor that distinguished military from nonmilitary R & D was the close scrutiny it received from top leaders, who were far more inclined to intervene to overcome administrative bottlenecks in this area than in others. In 1934, for example, M. Tukhachevskii, the Red Army's Head of Ordnance, wrote Politburo member Kirov to request that he expedite the production in his Leningrad domain of supplies necessary for research on the radiolocation of aircraft.[337] The other ranking members of the Politburo were also closely involved in weapons development. In June 1935, for example, Stalin and Ordzhonikidze, along with Defense Commissar Voroshilov, inspected models of the country's artillery; soon afterward the Politburo devoted a session to this question and changed the previous policy for artillery development.[338] Stalin in particular took a close interest in all aspects of weapons innovation. One military memoirist reports that he "knew well dozens of plant directors, party organizers, [and] chief engineers" in the defense branches, and regularly met with them to demand that goals for new weaponry be achieved.[339] A retired defense industrialist recalls that Stalin devoted his attention to the aircraft industry "every day." When difficulties arose in starting up production of a new aircraft machine gun, Stalin called together the designers, plant officials, and military consumers and arbitrated the disagreements among them.[340] Similarly, in 1936 the Politburo reportedly ordered Ordzhonikidze to devote his full attention to supervising the defense sectors of the Commissariat of Heavy Industry, freeing him temporarily from overseeing the Commissariat's nonmilitary sectors.[341]

Paradoxically, this kind of close supervision by top leaders gave talented military designers more professional autonomy than their nonmilitary counterparts possessed. Because the arrangement allowed a discontented designer or an unhappy military procurement officer to appeal decisions by mid-level officials to the top leadership, it sensitized these administrative officials to the importance of innovation. In view of the leadership's inclination to regard outdated weapon designs as "wrecking," few administrators could afford to ignore the ideas or needs of a talented designer. In 1936, for example, the aircraft designer A. S. Iakovlev became embroiled in a disagreement over a series of new light airplanes with the main administration for aircraft industry, to which his design bureau was subordinate. Iakovlev appealed directly to the Deputy Chairman of the Council of People's Commissars and won the dispute.[342] It is undoubtedly true that top-level intervention also contributed to some mistaken choices in weapons policy during the 1930s.[343] But on balance it was better than the existing Soviet alternatives. In the Stalinist system, leaders were more likely to acquire some familiarity with the major issues in fields of technology they thought crucial. They were also more likely to demand that administrators and ideologists heed the views of qualified specialists in such fields. Better informed and more concerned about these areas, the leaders were less likely to endorse erroneous technical ideas or to bestow authority on incompetents like Lysenko.

A third factor, the unusually powerful position of the military "consumer," reinforced the pressure for technological advance in armaments. This power derived largely from the political leaders' overriding desire for weapons innovation. But it also derived from the fact that the military, unlike other organizational consumers of industrial output, did not have its own production targets to meet. It was free to reject inferior goods without underfulfilling the high production plans that made most industrial consumers desperate to obtain whatever supplies were available in a seller's market. Thus, although the military needed large quantities of weapons in the long run, it could afford to demand high quality as well. Institutionally, this demand was transmitted to the arms industry via military representatives who were involved in all stages of weapons development[344] and compared the effectiveness of Soviet and foreign designs.[345] At arms factories the military maintained a network of independent representatives whose sole responsibility was to exercise quality control. Because these officers were not paid by the enterprise, they were free of the intraplant pressures to sacrifice quality for plan fulfillment that undercut similar personnel in nonmilitary sectors.[346]

A final factor that differentiated military from civilian R & D was parallel research and technological competition among weapons designers. Although limited, the evidence suggests that this practice became a standard feature of military R & D later than the other features, as the party chiefs concluded that large budgetary allocations and top-level supervision were not closing the military gap between the USSR and its rivals fast enough. In the aircraft industry the trend at the start of the decade was toward concentrating design work in a single large organization, but this change did not speed up innovation, and in 1931 the regime broke the organization into smaller units. Since these units were supposed to specialize in different kinds of planes,[347] however, it seems unlikely that systematic competition was introduced at this time. Evidence that prototype competitions were still not widespread in 1936 comes from the recommendation of a conference on high-speed aviation that such competitions should be arranged more frequently.[348] By the late 1930s, however, design competitions had become a regular feature of the aircraft development program. The poor performance of Soviet planes against German aircraft in the Spanish Civil

War produced a major shake-up in aviation R & D and contributed to the decision to promote competitive design.[349] From this time onward the Soviet leadership was unwilling to entrust the design of a certain category of military hardware to a single R & D establishment. It began to encourage overlapping work by researchers and designers on the same problem, and it also began to conduct regular design competitions that were decided on the basis of the comparative performance of prototypes. Around 1940, for example, the regime mounted a major competition for the design of a new generation of Soviet fighters. Roughly a dozen separate design organizations were enlisted in the competition, which resulted in the selection of three new fighter models for series production.[350] Similar competitions were conducted in other branches of the arms industry.[351] This direct confrontation and argument between experts with different points of view provided the political leadership with a clearer understanding of possible alternatives. It also gave the experts a strong incentive to push the pace of technological change as rapidly as possible.

Conclusion

As the industrialization drive got under way, the political elite vigorously asserted that the USSR was superior to Western capitalist systems not only morally but also economically. Spokesmen underscored the Soviet system's tremendous capacity to mobilize resources, contrasting this capacity with capitalism's inability to deploy the great productive forces at its disposal. In addition they soon began to claim a similar discrepancy between the capacities of the two systems to promote technological innovation. Mobilization and innovation were thus viewed as complementary rather than contradictory tasks. By the mid-1930s some writers started to warn that innovation was continuing to occur under capitalist systems. These warnings assumed special significance as the international situation deteriorated, and the political leadership undoubtedly took them seriously, especially where military technology was concerned. But this awareness of continuing Western achievements did not undermine the elite's belief in the ultimate technological superiority of the USSR. At the end of the decade Soviet industrial growth rates still far exceeded those of most Western countries, which seemed to be entering a new phase of the economic crisis that had exploded in 1929. Thus it was plausible to assert that in the longer run the USSR's superior powers of mobilization would allow it to outdistance the West technologically and in per-capita output. To "overtake and surpass" was still a credible goal, and only a major military defeat or a dramatic slowdown in Soviet economic growth could undermine the legitimacy that the political elite drew from pursuing it.

The technological strategy by which the regime sought to surpass the West was complex and not entirely consistent. Because the ultimate purpose of the industrialization drive was technological independence, the leaders supported a massive expansion of the research establishment. At the same time, because they were in a hurry, they demanded that science and industry concentrate at first on the wholesale importation of foreign technology. As the decade progressed, the leadership strengthened its emphasis on technological independence from the West, and it gradually shifted technological borrowing toward non-negotiable and impersonal modes of transfer. These avenues of technology transfer were more compatible with the fear of foreign subversion fostered by the Stalinist regime, but they were not as effective as other kinds of contacts. The severity of the restrictions on foreign contacts in the late 1930s meant that industry received less foreign technology than during the First Five-Year Plan, and the technology was less up-to-date than in the period when foreign contacts were more extensive. A further complication was that the acute nationalism of the late 1930s encouraged many specialists to neglect the opportunities for technological borrowing which were still available.

Why, despite these impediments and the regime's commitment to technological self-sufficiency, did officials continue to try to borrow some technology from the West? The answer is that creating advanced indigenous technologies was even harder than borrowing foreign ones. In spite of the large expenditures on Soviet science, the links between research and production remained weak. The biases of the mobilizational bureaucracy against slack resources, against professional autonomy and lateral cooperation among specialists, and against the diffusion of technological information all militated against indigenous innovation. As a result, relations between the Academy and the industrial commissariats were afflicted by delay and lack of coordination; so were relations between research and production units within the commissariats. On a limited part of the R & D front, military weaponry, the political leadership succeeded in insulating innovation from many of these organizational obstructions. But it did not manage to foster flexible, innovative behavior in the bureaucratic apparatus as a whole.

The first decade of the industrialization drive had clearly established some of the preconditions for a general meshing of Soviet science and industry. Chief among these were the creation of a large supply of trained manpower and the establishment of a network of research institutions necessary to sustain autonomous technological advance. Along with the massive expansion of industry, these major achievements marked an important step away from the kind of foreign technological dependence that had characterized the Tsarist era and the first decade of Soviet rule. With the benefit of hindsight, we can see that these measures had not overcome many obstacles which were liable to hinder the USSR's further movement toward rapid indigenous innovation in a wide range of economic sectors. In the short term, however, the system faced a different test: whether it could survive a crushing assault from the greatest military leviathan the world had yet seen.

War and High Stalinism,
1941–1953

For the Soviet Union, World War II was cataclysmic. In four years the war killed at least twenty million Soviet citizens and destroyed an enormous amount of physical property. Soviet suffering and sacrifice were extraordinary. Faced with the Nazi onslaught, the USSR mounted a prodigious military and economic effort, narrowly escaped destruction, and finally overpowered the German armies. In some respects this grueling victory helped to legitimize the Stalinist regime by identifying it more closely with Russian nationalism and by vindicating many of its earlier industrial policies.[1] But the war also left the country economically prostrate and poorly prepared to meet potential threats from other powers in the postwar years.

Socialism and Capitalism: Doubt Raised, Doubt Suppressed

The war had ambiguous implications for official Soviet thinking about the nature of the political and technological challenge from the West. The Nazi attack seemed to verify the inherited Stalinist theory of an aggressive imperialism bent on the destruction of socialism. But, too, the anti-Nazi alliance muted British and American hostility toward the USSR and elicited a great deal of Anglo-American material support for the Soviet war effort. Which tendency would prevail after the war? Should the Soviet elite anticipate a renewal of "capitalist encirclement" led by the United States and Great Britain, or could it expect that the policies of these countries toward the USSR would be permanently softened by the wartime partnership? From the Soviet standpoint these were hard political questions, and the difficulty of answering them was compounded by uncertainty about the economic implications of the war.

On the one hand, the defeat of Nazi Germany helped buttress the Stalinist notion that centralized socialism was economically superior to capitalism. On the other hand, the war gave the Soviet elite tangible proof of the great economic and technological dynamism of the United States. The influx of American Lend-Lease supplies after 1942 must have made many party members aware that while the Soviet economy was being wracked by the German invasion, U.S. output was expanding at a tremendous rate.[2] Was the U.S. economic surge only a momentary anomaly in the downward spiral begun by the Great Depression, or did it mark a fundamental change in the long-term prospects of capitalism? Did capitalism's capacities for technological innovation remain the same as before the war, or had they improved in comparison to Soviet capabilities?

These political and economic questions were critically important for Soviet policy choices. They bore directly on the significance that should be attached to further Soviet technological progress, including progress in military weaponry, and on the possibility of maintaining economic ties with the West after the war.

Soviet Conceptions of Capitalism during the War

Soviet wartime commentators interpreted the political dynamics of capitalism in two quite different ways. Stalin gave the cue for this approach by suggesting that there were basic differences between states belonging to the Axis and the capitalist members of the Allied camp. Nazi Germany had a "reactionary, black-hundred essence" that prompted it to oppress the working class and to engage in predatory territorial expansion. England and the United States, on the other hand, offered "elementary democratic liberties" to the working class, were not plutocracies, and were linked to the USSR by a common program of opposition to Nazi goals.[3] During the war the Soviet leaders virtually banned general discussions of capitalism in terms of the Leninist theory of imperialism,[4] and as late as the fall of 1944 they implied that postwar relations with the West might not be worsened by the capitalist character of the United States and Great Britain. In November Stalin declared that "the foundation of the alliance between the USSR, Great Britain and the U.S.A. lies not in chance and passing considerations but in vitally important and long-term interests."[5] Although Stalin and his associates obviously had expediential motives for such statements, their distinction between the political characteristics of Anglo-American capitalism and the Axis capitalist regimes was still the baseline from which postwar evaluation of the capitalist world would have to begin. As such, it was a distinction of some consequence.

Soviet commentators observed similar distinctions in their wartime discussions of capitalist economic performance. Articles about Germany emphasized the superiority of the Soviet economic system. The unique strengths of socialism, it was said, were shown by the unprecedentedly rapid Soviet industrialization before the war. During the war the socialist system had demonstrated "indisputable advantages" over capitalism in mobilizing all available resources for the war effort, and socialism posed no barriers to the rapid diffusion of advanced technology.[6] In time of war, the writers granted, capitalist states did undertake greater governmental regulation of their economies, but measures of this kind did not eliminate constant collisions between private monopoly interests and state interests. Such collisions were "a great obstacle" to a country's mobilization of resources and sharply retarded the domestic diffusion of advanced technology.[7] Expositions of this sort satisfied the party's demand that economists and ideologists highlight the approaching economic exhaustion of the Axis countries and the steady worsening of their workers' standard of living.[8]

Predictably, Soviet wartime discussions presented a much more favorable picture of the economic achievements of the United States and United Kingdom. Unlike discussions of Germany and Italy, these commentaries played down capitalist impediments to economic mobilization and technological change. While noting the reluctance of some capitalist corporations to join the war effort, they also pointed out cases in which the bourgeois state had eliminated monopolies that threatened the effort, and they avoided any assertion that socialism was economically superior to capitalism.[9] One article, for instance, stressed that the economic resources at the disposal of the Western Allies far exceeded those of the Axis. The American economy, "capable of enormous development," was already growing at "unprecedented tempos" and undertaking a gigantic expansion of military production.[10] No less than accounts of Axis economic troubles, these upbeat discussions of American economic power helped sustain the faith of leaders and citizens under siege.

The Debate Begins

Although useful as wartime propaganda, this dualist depiction of capitalist politics and economics shed little light on the policies the USSR should adopt after the war, and by early 1945 a dispute had begun over which half of the wartime image of capitalism was actually correct. Many public participants in this debate were academics, but it had more than academic significance. The debate bore on the dangers the West might pose to the USSR after the war, on the amount of additional effort that would be required to overtake the West technologically, and on the advisability of seeking wider economic relations with the West to aid Soviet reconstruction.

Soviet policymakers apparently helped trigger this academic dispute. In August 1945 G. F. Aleksandrov, head of the Central Committee's Directorate of Agitation and Propaganda, acclaimed the Bolshevik willingness "to replace outdated Marxist ideas with new ones that correspond to changed historical circumstances" and asserted that a great virtue of Soviet social science was its willingness to attack "the sharpest and most burning issues of the contemporary period."[11] Aleksandrov went on to complain that Soviet economists were doing too little research on important questions such as the war economy of capitalism, the transition of capitalist countries to a peacetime footing, and "the postwar economic ties of the USSR with foreign countries." All these issues required "deep and immediate development."[12] The most controversial book published in the scholarly debate was evidently prepared in response to this summons,[13] and the academic disputants were in close contact with some policymakers.[14] The timing of events also suggests a connection between the academic debate and policy choices. The earliest signs of scholarly disagreement over the nature of capitalism surfaced about the time the USSR made its first official request for an American loan of six billion dollars for postwar reconstruction, and just a few months before the government commissioned the writing of a new five-year plan.[15]

The debate focused on the role of the state under capitalism and socialism. The traditionalists asserted that the role of the state was fundamentally different in the two systems and rejected the notion that the heightened economic role of the capitalist state in wartime indicated any basic political change in capitalism or in the state's ability to control the anarchy of the market. In their view, the bourgeois state was still dominated by a small group of monopoly capitalists who selfishly used it to pursue only profits for themselves. Thus, disrupted by class divisions and economic crosscurrents, capitalist systems had been unable to mobilize for war as effectively as had the Soviet Union.[16] Moreover, predicted the traditionalists, the capitalist transition from war to peace would immediately cause unmanageable market dislocations that would create "a situation critical in the highest degree."[17] The Western powers, in other words, would probably slide back into an economic depression after the war.[18] This prognosis, when taken together with standard Leninist notions of the dynamics of capitalism, also suggested that the economic wellsprings of imperialist aggression remained as strong as ever. Even before 1945, some adherents of this school revived the general theme of capitalist encirclement and warned that its importance should not be forgotten.[19] A number of scholarly advocates of the traditionalist view, which was doubtless held by many members of the Soviet elite, were associated with the Academy's Institute of Economics.[20]

Other commentators, especially those who had studied the wartime economies of the Western Allies at the Academy's Institute of the World Economy and World Politics, contested the traditionalist outlook.[21] Evgenii Varga, Director of the Institute, claimed that under the impact of the war the bourgeois state had *"attained decisive significance"* in the capitalist economy and had fundamentally altered the economy's nature.[22] The state, representing the interests of the bourgeoisie as a whole rather than only the giant monopolists, had curbed monopoly interests for the sake of prosecuting the war, and in the postwar United States and Great Britain there was widespread agreement on the need for government management of the economy and thorough social reform.[23] While Varga noted difficulties in governmental efforts to regulate the capitalist economy, most of his exposition highlighted the efficacy of such market regulation. Observing that capitalist industry had grown tremendously during the war, especially in the United States, he pointed out that state action had reduced many of the obstacles which had impeded Western technological progress during the 1930s.[24] Despite the gloomy prognosis for the West presented in the body of his book on capitalism, the introduction, by stating that serious capitalist economic difficulties really lay a decade in the future, conveyed the impression that capitalism might actually be quite durable economically.[25] Meanwhile researchers from Varga's institute hinted that mass unemployment could be avoided in postwar America and suggested that the burden of postwar demobilization might not be shifted entirely onto the shoulders of the American workers.[26] The overall effect was to portray the West as being more economically dynamic and more politically moderate than the traditionalists felt it to be.

Policy Implications and Leadership Attitudes

The dispute between the traditionalists and nontraditionalists had major implications for Soviet relations with the West.[27] After abruptly interrupting Lend-Lease shipments at the close of the European war, the American government showed its determination to offer postwar economic aid only in return for substantial Soviet foreign policy concessions in Eastern Europe.[28] Nevertheless, the persons who leaned toward the nontraditionalist view evidently hoped for postwar economic and technological cooperation between the United States, the USSR, and Eastern Europe. Arguing that a desire for foreign trade had created a vested American interest in a durable peace, one writer from Varga's institute emphasized that some capitalist circles in the United States were pushing for a modified form of Lend-Lease for the postwar period.[29] In the same vein, the institute's journal published a note that itemized the vast scale of America's past Lend-Lease assistance to the USSR and omitted any reference to the sudden interruption of the program in the summer of 1945.[30] Varga predicted that the attitude of capitalist countries toward the USSR "will not be the same as in the prewar period." The democratic forces in all countries, he said, would strive for cooperation with the USSR and would oppose a resurgence of fascism and aggression "in any form whatsoever." Therefore, although some reactionary forces still existed, foreign governments "will not lightly decide" to attack the Soviet Union.[31]

Proceeding from these premises, Varga intimated that Soviet plans for reconstruction and technological development should include large-scale cooperation with the West, rather than only competition. The pace of the economic recovery of Europe, he said, would depend on how willing the United States, Canada, and Great Britain were to supply producer goods on credit to the region.[32] The obvious implication was that such assistance was desirable. It is true that Varga did not touch explicitly on the question of aid for the USSR, and that he stated that Eastern Europe would receive assistance "in the first place" from the Soviet Union.[33] But when he discussed the pivotal role of American credit and capital goods for reconstruction, he spoke of Europe as a whole, not just of Western Europe, and he still classified the East European countries as capitalist countries, albeit with some special redeeming qualities. He was clearly suggesting a postwar world that would not be divided into economic blocs; and in view of his emphasis on the West's conciliatory policies toward the USSR, it is plausible to conclude that he favored the vigorous pursuit of Western economic relations for his own country as well as for Eastern Europe.

Traditionalist thinkers, however, emphasized the aggressiveness of imperialism, the danger of American economic penetration of Europe, and the virtues of Soviet economic self-sufficiency. As early as 1944, traditionalist writers began to downplay the wartime significance of American Lend-Lease aid and to underscore the importance of self-sufficiency.[34] In 1946 another writer of this school quoted Stalin's militant prewar declaration that the USSR must rapidly close the economic gap with the West or be destroyed by external enemies. By stressing that before the war the West had either refused to make any loans to the Soviet Union or had attached "clearly unacceptable, fettering conditions" to loan offers, he seemed to rule out the possibility of obtaining Western aid in the postwar era.[35] Shortly afterward an editorial in the party's main journal claimed that the reactionary forces which had spawned Hitlerism not only persisted in the imperialist world but were becoming more active, and that the United States had launched an unprecedented effort at economic expansion intended to bring the countries of Europe under American domination, rather than give them genuine assistance.[36]

As this sparring continued, the top leaders pondered several policy questions whose answers depended partly on the kind of political and economic behavior the USSR could expect from the West in the coming years. How rapidly should the regime push the advancement of science, technology, and heavy industry? A related choice was how vigorously to pursue the development and manufacture of new military weapons. And should the USSR seek extensive ties with the West to hasten its economic recovery? These issues were probably the focus of top-level deliberations in late 1945 and early 1946. On 8 August 1945, two days after the United States dropped the first atomic bomb on Japan, the political chiefs authorized the compilation of a new five-year plan.[37] In December 1945 the Politburo began to hold regular biweekly meetings, and as work on the five-year plan neared completion, devoted several sessions to a comprehensive review of Soviet military policy. The review considered changes in the international situation, recent developments in science and technology, the weapons programs of foreign powers, and the condition of the domestic economy.[38] In February the U.S. government, breaking a long silence, expressed its willingness to renew diplomatic negotiations on a postwar loan to the

USSR.[39] In March the Soviet authorities announced the general outlines of the new five-year plan.

Although incomplete and sometimes inconsistent, the evidence indicates that the leaders disagreed over these postwar choices.[40] In speeches during this period, they offered noticeably different interpretations of the international situation. The backdrop for most of the speeches was a ceremonial address in January 1946 by Aleksandrov, the administrative overseer of party propaganda. Arguing that inherited Marxist theory should not be regarded as "complete and inviolable," Aleksandrov asserted the need for a strong state to protect the country against external pressures but carefully avoided any reference to capitalist encirclement. Instead he painted a relatively benign picture of the USSR's current international environment. Although the nation lacked a full guarantee against attack, he said, the war had brought an "unprecedented destruction" of reactionary forces abroad. Consequently the prospects for building communism in the USSR had improved "not only from the point of view of internal, but also external conditions."[41]

Some Politburo members did not accept such a sanguine view. Lazar Kaganovich acknowledged that the defeat of Germany and Japan meant that the Soviet strategic situation was no longer the same as before the war. "However," he quickly added, "we must remember that our country continues to exist in capitalist encirclement." This fact, he said, ruled out "good-heartedness" and "self-satisfaction"—failings which Kaganovich apparently thought some of his fellow leaders were exhibiting.[42] In the same vein, L. P. Beriia cautioned that although the fascist governments had been destroyed by the war, fascist forces persisted in many countries; and one of his supporters warned that some foreign countries were creating conditions favorable to a resurgence of fascism.[43]

These variations in appraising the international environment were linked to discrepancies in the leaders' domestic priorities. Kaganovich, expounding the persistence of capitalist encirclement, pointedly recalled Stalin's 1931 statement that the Soviet regime, if it slowed its rate of industrialization, would be defeated by its foreign enemies. Under such conditions, he asserted, the party and the people would generously support and strengthen the armed forces.[44] Molotov, while he did not speak directly about encirclement, argued even more strongly for the same priorities. The USSR, he observed, was returning to the task of rapidly overtaking the West in per-capita industrial production, and technological progress should receive "paramount attention" in new economic plans. Citing atomic weapons as an example of the application of science to practical ends, he promised that the USSR "will have atomic energy, and much else."[45] Military technology was obviously uppermost in his mind, since he stressed that the country's "great obligations" to meet the needs of the military should not be forgotten for a minute.[46]

Other leaders, however, took issue with these priorities. Although the main party journal had recently censured "some comrades" who thought it was now possible to relax rather than to step up the accumulation of military and economic power, a few leaders came close to espousing this position.[47] In keeping with his upbeat view of international affairs, Aleksandrov made no mention of increasing Soviet military power or forcing heavy industrial growth.[48] A. A. Andreev, a Politburo member with special responsibility for agriculture, ignored the international scene in favor of domestic needs. Highlighting the sacrifice in economic welfare caused by wartime military requirements, he argued that it was now necessary to reorganize the economy "more rapidly" and to raise the living standard of the population. Andreev accented the importance of building up "all branches of our economy" and said nothing about the preeminent place of heavy industry. Although he praised the military for its wartime accomplishments, he pointedly omitted any mention of the need to strengthen it in the future.[49] Andrei Zhdanov, another Politburo member, adopted a less daring version of this position. He referred ambiguously to the existence of capitalist encirclement before the war, and endorsed the goal of increasing Soviet military strength. But he refrained from stating that encirclement still existed, and he put unusual emphasis on the need to convert industry rapidly from military to consumer needs. Some people, he added, failed to grasp that this conversion required changes not only in light industrial plants, but in the heavy industrial plants supplying their equipment. Whereas Kaganovich had asserted that Soviet citizens would willingly support major new investments in heavy industry and defense, Zhdanov stressed that the population would respond with gratitude to heightened spending on consumer needs.[50] Obviously there were significant tensions among Stalin's deputies over resource allocations, tensions that persisted after the announcement of the new five-year plan.[51]

Stalin himself chose the middle ground in these leadership debates. In a speech in February 1946, he refrained from repeating his prewar formula describing capitalist encirclement. Instead he still numbered the United States and Great Britain among the "peace-loving states" of the wartime antifascist coalition. Moreover, he remarked that during the next five-year plan the regime would give special attention to the production of consumer goods and rebuilding devastated regions of the country.[52] Other parts of the speech, however, suggested that these statements were tactical accommodations rather than basic changes in Stalin's policy preferences. Explaining that the past war had resulted from economic struggles within the capitalist world, he indicated that wars would recur because the capitalist states were incapable of adjusting their economic conflicts peacefully. He also asserted that the priority of heavy industry during the 1930s had been a key condition for the Soviet victory in World War II, and that the party had been justified in purging those persons who had opposed preferential investment in heavy industry before the war. Promising extensive new support for scientific research, Stalin called on Soviet scientists "not only to overtake but to surpass in the near future the achievements of science beyond the borders of our country." In the longer run, he added, the regime would strive for a "new powerful surge" in the economy, with special attention to heavy industry. Only this policy could ensure against "all sorts of accidents," he said, in an obvious reference to possible foreign military attacks.[53]

While evidence of disagreement within the Soviet leadership over economic and military priorities is clear, what we know of leadership attitudes toward economic relations with the West remains sketchy. Most leaders did not discuss this issue publicly during 1946, and we can do little more than speculate. Shortly after the United States offered to renew loan negotiations, one leader, Voznesenskii, seemed to dismiss the idea. The USSR, he said, would develop economic relations with foreign countries, but only within the limits of the time-tested line of preserving its technological and economic independence. Official policy was to ensure the steady growth of productive capacity "on the basis of domestic resources."[54] His words seemed to rule out a major loan and capital imports from the United States. But the USSR did show an interest in these possibilities by sending representatives to the March 1946 meetings that organized the World Bank and the International Monetary Fund.[55] Further, in late 1946, after voicing skepticism that the United States and Great Britain either wanted or were able to encircle the USSR, Stalin expressed interest in an American loan and spoke favorably of enlarging the volume of trade between the United States and USSR.[56] Given Stalin's deeply suspicious nature, it seems improbable that he expected the USSR's long-term policies to evolve along these lines. But he did hope to follow such policies in the short term.

The policies that resulted from these internal discussions were initially rather differentiated. The leadership, having inaugurated a crash A-bomb program in August 1945, made a dramatic additional commitment of resources to R & D early in 1946, and military projects received the lion's share of this new allocation. Simultaneously, however, the leadership sharply cut the wartime level of military manpower and gave preferential treatment to consumer industries over the producer-goods branches.[57] It also continued to explore the possibilities for large-scale Western economic aid, and it concluded new technical-assistance contracts with American corporations. On balance, the elite appeared to be preparing for the eventuality of an arms race with the United States while seeking to alleviate the terrible domestic hardships caused by the war and to keep alive the chance of outside economic aid for reconstruction.

Intensified Debate and the Swing toward Traditionalist Policies

By 1947, however, Stalin and his deputies could no longer concentrate on preserving their options; the course of events required that they choose. In March the hardening American attitude toward the USSR was expressed in the proclamation of the Truman Doctrine. In May and June came a great watershed—the American offer to launch a massive program of economic assistance to Europe under the Marshall Plan.[58] The terms of the offer left open the possibility that the United States might accept aid requests from the Soviet Union and its East European client-states, as well as from Western Europe. In deciding how to respond, the political leadership had to weigh the intentions of the Western powers, particularly the United States, and the political costs of developing such economic ties. Would the lack of outside aid be a crucial economic setback, or should the regime rely strictly on domestic resources, confident that its enduring superiority would allow it rapidly to overtake the West in any case?

As decisions on these questions became unavoidable, the academic debate between the traditionalists and nontraditionalists broke into the open. The nontraditionalists accused traditionalists of underestimating recent changes in capitalism, of failing to come to grips with the postwar economic problems of the USSR, and, possibly, of attaching excessive

importance to the accumulation of Soviet military power.[59] The traditionalists rebutted at a gathering of specialists convoked in May 1947
to discuss Varga's book on capitalism.[60] While a number of participants
treated the book as a serious contribution to scholarship, almost all
took issue with its line of argument. Many disputed Varga's view of
the economic role of the bourgeois state, and several rebuked him for
exaggerating the economic dynamism and prospects of capitalism.[61] A
second matter of controversy was Varga's prediction that capitalist
domestic politics would be benign and that the policies of the West
toward the USSR would be conciliatory. In point of fact, said one critic,
there was a serious possibility of the rise of new fascist regimes and
of a capitalist attack on the Soviet Union.[62] Finally, Varga was vigorously
criticized for treating the regimes of Eastern Europe as still capitalist
and for suggesting that American aid was a benefit without which East
European economic recovery would be impossible. In actuality, maintained the traditionalists, the new Eastern European regimes were no
longer capitalist, and this shift had greatly bolstered the world forces
of socialism. In Eastern Europe the state was playing a new economic
role that allowed it to achieve rapid economic recovery without the
benefit of subversive economic ties with America.[63] By ignoring these
realities, the nontraditionalists were injuring the domestic legitimacy
of the Soviet regime and its links to its East European clients.[64]

Faced with this barrage of criticism, Varga gave surprisingly little
ground, and on some points he even hardened his position. He flatly
refused to modify his depiction of the economic role of the bourgeois
state. Instead, he emphasized the scope of capitalist economic regulation
more strongly than ever before, and on this matter he received qualified
support from I. A. Trakhtenberg, another member of the Academy who
worked at his institute.[65] Concerning the foreign policy goals of the
capitalist world, Varga maintained a stubborn silence at the May sessions. He did not reiterate his belief that capitalist attitudes toward the
USSR would be different from those of the 1930s, but neither did he
recant. The only point on which he admitted any error was his treatment
of the nature of the "new democracies" of Eastern Europe, and even
this admission he gave grudgingly.[66]

While traditionalists and nontraditionalists vied on the level of social
theory, the Soviet leaders pondered possible Soviet and East European
participation in the Marshall Plan. Whatever their misgivings, the leaders
still had enough interest in the possibility that they sent Molotov to a
June meeting in Paris, where he explored the details of the American

offer with some care.[67] Soviet indecision was indicated by the fact that the USSR gave its East European clients no clear signals on the matter up until the very last minute, and thereby revealed embarrassing differences of opinion between itself and some of those regimes.[68] The Soviet leaders probably hesitated, as Adam Ulam has pointed out, because by requesting aid under the Plan they might have gained either large strategic or large economic benefits. American attitudes toward the USSR were sharply divided, and it was quite possible that a Soviet request would cause the U.S. Congress to reject the entire Plan in order to deny the USSR assistance. In this case Western Europe would lose all aid as well, and the prospects for political stability and American influence in the region would be gravely damaged. If, on the other hand, the Congress accepted the Marshall Plan with Soviet participation, the USSR would obtain a large infusion on new technology for its badly damaged economy. Either way, the USSR was likely to benefit.[69]

In the end, however, the Soviet leadership condemned the Marshall Plan. A central reason for this decision was the traditionalist distrust of Western capitalist regimes. The basis on which Molotov elected to leave the Paris meetings was that the requirements for sharing economic data under the Plan were unacceptable and violated Soviet law.[70] Probably this objection stemmed from a fear that participation in the Plan would reveal the full extent of the economic damage the USSR had suffered in the war, thereby encouraging the United States to adopt a more assertive policy toward the Soviet Union. The Soviet leadership also feared that participation by its East European clients would ensnare them in the powerful economic attractions of Western Europe and especially the United States. Before the war, Eastern Europe had been an integral part of the European economy, with only limited ties to the USSR, and by 1947 there was a growing possibility the old pattern would be reestablished.[71] Since the United States had already tried to use its economic power for diplomatic ends, this prospect must have worried the Soviet leadership, even though Eastern Europe was under Soviet military control. Coupled with these foreign-policy considerations were anxieties about the domestic political effects of closer contacts with the West, no matter how benign Western intentions might be. For some time the regime had been showing renewed concern about unorthodox intellectual tendencies among Soviet citizens, and extensive economic intercourse of the kind envisioned in the Marshall Plan would encourage such tendencies. The Soviet regime wanted American technology, but it was unwilling to risk any erosion of its domestic legitimacy in return, particularly when the traditionalist outlook suggested that the USSR could still overtake the industrial West in the longer run without external aid.

Rather than accept Marshall Plan aid, the Soviet Union chose to rely primarily on its own resources and strove to consolidate its ties with Eastern Europe. In addition to signing new trade agreements with several East European countries, it convened a September meeting of Communist parties that proclaimed the world was divided into two camps and set up the Cominform as a vehicle for international Communist cooperation.[72] Andrei Zhdanov, who had shifted his stance dramatically since early 1946, set the prevailing tone at the meeting with a highly traditionalist speech. He asserted that the Communist camp must resist imperialism on all fronts, including the economic and ideological ones.[73] Citing the lesson of appeasing Hitler at Munich, he warned that concessions to imperialism would only strengthen its aggressiveness.[74] The main thing to avoid, he admonished, was underestimating the strength of the working class in relation to imperialism. The Marshall Plan was intended to enslave European nations to foreign capital, and the "basic and decisive condition" for economic reconstruction in Europe was self-reliance.[75] If the countries of Europe united in refusing the fettering terms proferred under the Marshall Plan, it was possible that the United States, which was threatened by economic crisis and badly needed to export capital, "may be forced to retreat."[76]

It is not entirely clear from Zhdanov's wording whether he meant that the United States might retreat from Europe in a broad strategic sense, or that it might retreat on the aid terms and offer credits with no political preconditions. Judging by the tenor of the speech, it is hard to believe that Zhdanov thought the latter was even a remote possibility, and the ambiguity of his remark may have been intended as a sop to those in Eastern Europe, and possibly inside the USSR, who still felt some interest in the Marshall Plan. At any rate, Zhdanov asserted that the Soviet system and its East European clients were sufficiently dynamic that American aid was not essential for their recovery and rapid further development. The final resolution of the September conference endorsed these traditionalist views.[77] Presumably this same outlook underlay the proposal of Voznesenskii, a close ally of Zhdanov, to formulate a new twenty-year plan for overtaking the capitalist West in per-capita production. The proposal received party and governmental approval in August 1947.[78] The regime's reliance on the mobilization of domestic resources was also reflected in the fact that in 1947, for the first time since 1944, the output of producer goods grew more rapidly than the volume of consumer goods.[79]

Even at the founding meeting of the Cominform, however, there was an echo of disagreement within the top leadership. Malenkov's address to the September gathering had a tone noticeably different from Zhdanov's. True, Malenkov warned of American expansionism and endorsed building up the USSR's military power and strengthening its economic independence.[80] But he simultaneously stressed the Soviet desire for good relations with the capitalist world. The USSR, he said, accepted the inevitability of a long period of coexistence with the West, and it supported "loyal good-neighborly relations with all those states" that showed a desire to cooperate with it.[81] The tenor of this remark was distinctly different from Zhdanov's diatribes against imperialism. Moreover, Malenkov acknowledged that the unavailability of external economic aid would slow Soviet economic growth. He observed that the Soviet inability to obtain foreign equipment which might have been imported under "more normal international conditions" was one of several economic problems that "cannot fail to manifest themselves in a slowing of the tempos of our development."[82] Malenkov did not suggest that this situation could be reversed, but neither did he maintain that the Soviet Union and Eastern Europe could develop equally well without Western economic ties. Here, too, was a significant discrepancy between his speech and Zhdanov's, and it suggests disagreement within the leadership over the nature of capitalism and the capacity of the Soviet Union to develop rapidly without the benefit of capitalist technology. Whereas Zhdanov was putting increasing emphasis on this capacity, Malenkov seemed increasingly skeptical.

These policy differences must have contributed to the bureaucratic infighting during this period. After the September meeting, the publication of Malenkov's speech was held up for a full month after Zhdanov's appeared—a sure sign that someone had tried unsuccessfully to suppress it.[83] Leadership disagreements may also have emboldened Varga and his associates to publish in December 1947 a stenographic record of the formal debates over capitalism that had transpired seven months earlier. The stenographic record gave the most favorable public account to date of Varga's position in the debates, and its appearance drew fire from his traditionalist opponents.[84]

Nevertheless, events at the close of 1947 demonstrated that the non-traditionalist view of capitalism and socialism was losing out to the traditionalist view on all levels. Voznesenskii published a book denouncing nontraditionalist ideas about capitalism as "nonsense" propounded by pseudo-Marxist theoreticians. Voznesenskii underscored the differences between socialism and capitalism, arguing that "a new devastating economic crisis and chronic unemployment" were imminent in the West.[85] The United States was moving toward a new war, and the Soviet Union still faced capitalist encirclement, making the ability to meet all economic needs with domestic output a key condition of Soviet military strength.[86] About the same time that the book appeared, the Council of Ministers, in response to a request from Voznesenskii, effectively abolished Varga's institute and journal by fusing them with the Institute of Economics under traditionalist academic leadership. The amalgamated institute was put under Voznesenskii's supervision.[87] To judge by sketchy evidence, Voznesenskii may have felt some affinity for nontraditionalist views during the war.[88] But now, like Zhdanov, he was espousing a rigorously traditionalist line.

This was a judicious change of stance, since it matched Stalin's shifting policy predilections. Indeed, the moves against nontraditionalist academics were made on Stalin's initiative. Until the second half of 1947, while the debate raged over capitalism and possible Western economic ties, Stalin refused to make any categorical statements on the dispute. The few Olympian remarks he offered tended toward the traditionalist outlook but did not foreclose the issue.[89] Once he had decided to reject the Marshall Plan, however, he gave the traditionalists a clear signal to press their attack. Voznesenskii had completed the draft of his book around the end of 1946—soon after Varga's book appeared—and had circulated it to other Politburo members for comment. None had replied, including Stalin, whose approval was essential.[90] At last, in September 1947, the month of the founding of the Cominform, Stalin returned the carefully edited manuscript to Voznesenskii with a recommendation that it be published, along with additions and corrections "insignificant in their volume"[91] but possibly weighty in their content. No doubt Voznesenskii and others construed this, correctly, as an authorization to move against the nontraditionalist school of thought in general, and against Varga's institute in particular. After toying with the alternatives, Stalin, in a characteristically circuitous manner, had renewed his support for the traditionalist outlook he had first articulated in the 1930s.

The Suppression of Nontraditionalist Ideas

With the publication of Voznesenskii's book, the traditionalist outlook gained virtually unchallenged dominance in public discussions of capitalism and Soviet economic development. Party commentators emphasized capitalist encirclement and the importance of building up Soviet military might and economic self-sufficiency.[92] The USSR would soon surpass America "in all fields" of R & D and in economic output.[93] Except for a few fields of military research, most areas of Western science and technology were stagnating.[94] In June 1948 *Bol'shevik* topped off this traditionalist campaign by announcing that Voznesenskii's book had won a Stalin Prize and condemning Varga's defunct institute for using "bourgeois methodology" to analyze the capitalist economy.[95] This was the first time the charge of ideological deviation had been raised against nontraditionalist thinkers by a party editorial rather than by individual opponents, and the distinction was significant. Five days after giving its imprimatur to the traditionalist contrast between the strengths of socialism and the weaknesses of capitalism, the party leadership launched the Berlin blockade.

Surprisingly, however, not all nontraditionalist thinkers succumbed during 1948. Deprived of direct access to the media and under intense pressure to recant, several did repudiate their past positions in favor of the traditionalist perspective. But a few leading scholars did not. Varga, in particular, continued to advocate his view of the West in academic meetings. Emphasizing that the socialist and capitalist camps influenced each other in important ways, he reportedly discussed Western military policies in terms of "defense" rather than profit-motivated aggression, and he expressed strong doubt that a future war was inevitable, including an imperialist war against the USSR.[96] At any other time such a statement, implying that Western actions might be a response to Soviet policies rather than the result of inherent capitalist aggressiveness, would have been extremely unusual. At the height of the Berlin blockade it was astonishing. Varga also argued that the Soviet and capitalist economies should not be analyzed separately, but through closer comparative study.[97] The implication appeared to be that the economic superiority of the Soviet system was not self-evident. Varga's colleague Trakhtenberg likewise resisted pressure to repudiate his earlier views of the West.[98] The tenacity of these few nontraditionalist holdouts probably reflected hidden support for them somewhere in the top leadership, but I know of no published evidence that this was so.

At any rate, events in early 1949 made the position of the holdouts untenable. Provoked by the Berlin blockade, the United States moved toward the North Atlantic alliance with Western Europe and tightened its economic embargo against the Soviet bloc.[99] Partly in response, the

USSR organized the Council of Economic Mutual Assistance (CEMA) with itself and its East European clients as members.[100] Perhaps this increase in international tension helped precipitate the shake-up of some top leaders in March 1949 and contributed to the postponement of the party congress that, to judge by indirect evidence, had been scheduled to occur during the year.[101] The shake-up of the leadership seemed momentarily to offer the nontraditionalists a reprieve when Voznesenskii, the most outspoken Politburo critic of their ideas, fell from power as a result of high-level intrigues and conflicts over economic policy that may have been sharpened by the worsening international situation.[102] But although Voznesenskii was gone, Stalin was not, and Varga was soon forced to make a full public recantation. He confessed that he had committed "a whole chain of errors of a reformist tendency" which together amounted to "a departure from a Leninist-Stalinist evaluation of modern imperialism." Stating that he had exaggerated the degree of concord between the USSR and its imperialist allies during the war, he drew a parallel between Nazi Germany and the contemporary United States.[103] He had erred, he said, in suggesting that the bourgeois state could act independently of the capitalist oligarchy to direct economic development, and his treatment of the status of Eastern Europe and the relative power of the Eastern and Western blocs had been wrong.[104]

Lingering Doubts

Varga's recantation marked a major public victory for the traditionalist view. Not until after Stalin's death would Soviet commentators resume the public discussion of the issues raised by the nontraditionalists and attempt, however gingerly, to arrive at answers on the basis of empirical observation. In this sense the recantation was a watershed. It signified the public rejection of all doubts about the veracity of the traditionalist view of the world and the Soviet path of technological development. For the time being, any possibility of a significant change in official Soviet thought and policy on these matters had disappeared.

At the same time, it would be mistaken to conclude that the doubts were completely laid to rest, and that every member of the political elite now accepted the traditionalist outlook unqualifiedly. Between 1949 and 1953 the members of the Politburo continued to emit subtle signals of disagreement about precisely how the traditionalist perspective should be interpreted and applied.[105] As we shall see, there are solid reasons to believe that Stalin was determined not to let such differences alter the prevailing course of Soviet foreign and domestic policy; and he may have tolerated disagreements of this kind primarily to keep his lieutenants politically divided. Nevertheless, these nuances in interpreting traditionalist ideas were not devoid of policy significance, and their persistence sheds considerable light on the major changes in official Soviet thought that occurred after Stalin's death.

Speeches by Stalin's lieutenants during the fall of 1949 reflected such differences. In discussing the economic performance of the West and the USSR, Mikoian and Malenkov struck a rather subdued note. Mikoian referred only perfunctorily to the existence of a Western economic crisis and did not dwell on the economic superiority of the Soviet system.[106] Malenkov prophesied an approaching Western economic crisis, but he also warned against exaggerating Soviet economic and cultural achievements and implied that "ill-starred leaders" had raised the "danger of stagnation" in the USSR.[107] Molotov and Suslov, on the other hand, drew a sharper contrast between Soviet economic achievements and the difficulties of the imperialist countries, underscored the speed of the Soviet postwar recovery, and emphasized that the country was making rapid progress in science and new technology.[108]

Significant nuances also separated the leaders' depictions of the international environment. In a major speech, Malenkov described Soviet borders as being so secure that he indirectly called into question the reality of capitalist encirclement. The one time Malenkov mentioned the term was when he described American policy as having been directed toward that end—but he immediately noted that the recent Communist triumph in China marked a major victory for anti-imperialist forces in Asia.[109] Mikoian took a similar tack. He noted that before World War II the USSR had been "hemmed in by the 'cordon sanitaire' of capitalist encirclement," but that with the creation of the neighboring people's democracies the Soviet Union had "destroyed the enemy 'cordon sanitaire.'" Referring then to the Communist victory in China, Mikoian left the unspoken but palpable impression that capitalist encirclement had been destroyed as well.[110]

These attempts to mute the themes of capitalist encirclement and aggressiveness did not meet with the approval of other leaders. Suslov, a rising member of the party Secretariat, argued vehemently that the imperialists were preparing for a new war, a contention seconded by Molotov.[111] Moreover, Lazar Kaganovich and Nikolai Bulganin, both Politburo members, signaled that the doctrine of capitalist encirclement should not be tampered with. Using the same new phrase, both re-marked that Stalin's doctrine of the socialist state in conditions of en-circlement was "final and complete."[112] Their terminology implied that the prewar concept of capitalist encirclement was no longer universally regarded as the last word in doctrine, and that someone within the Soviet elite wished to modify the doctrine by advocating its "further development."

Kaganovich and Bulganin were probably trying to keep Malenkov and Mikoian from making a persuasive case for a relaxation of tensions with the West now that the Soviet Union had lifted the Berlin blockade. However, Malenkov, who had previously denounced "pedants" for treating inherited Marxist theory as unalterable, reiterated that the USSR was pursuing a program that included cooperation with other great powers and "a broadening of trade-economic ties."[113] A more striking bid for greater trade with the United States came from Mikoian. After implying that capitalist encirclement was a thing of the past, Mikoian mentioned the term, but tried to change its meaning. In a clever gloss, he asserted that the term included two opposing capitalist tendencies: the tendency to act aggressively toward the USSR and the tendency to seek markets and necessary goods in the USSR. Mikoian, who had been Minister of Foreign Trade until a few months before, maintained that the state monopoly of foreign trade both facilitated the capitalist tendency toward trade and protected the Soviet Union against external economic blows. In other words, national-security arguments should not get in the way of trade expansion. Moreover, trade with the West "gives additional material resources for accelerating the rate of devel-opment of the Soviet economy."[114]

The sparring over the doctrine of encirclement also bore on the ques-tion of how much effort the Soviet Union should put into heavy industry and military technology. There was probably already some internal tension over this question in 1949, when the formation of NATO spurred the acceleration of military R & D and presumably prompted Bulganin to cite the doctrine as an argument for strengthening the armed forces.[115] But the tension increased as a result of two events. One, announced in March 1950, was the start of work on a new five-year plan for 1951–5.[116] The other was the American response to the outbreak of war in Korea. This response, which probably took the Soviet leadership by surprise, included a drastic increase in American military expen-ditures, and it accented the need for stepped-up growth of the Soviet military budget in order not to lose ground in the arms race with the

United States.[117] The available evidence suggests that the new requirements were not easily met. Near the end of 1950 the USSR compelled its CEMA partners to revise their five-year plans to help meet the needs of the Communist combatants in Korea, and it did not publish its own five-year plan until August 1952, over a year and a half late.[118] Furthermore, three months after work on the Soviet plan began, a major academic conference on the theme of the transition to communism revealed clear differences over the importance of heavy industry and technological progress versus popular consumption.[119] It is a plausible assumption that these delays and disagreements reflected the difficulty of settling on the large military increases the regime made in 1950–51.[120]

Apparently the differences carried over into the next year, when the regime again raised its defense expenditures sharply.[121] Even as the traditionalist view held unchallenged sway in the media, hints of disagreement continued to surface. At the end of August 1951, two months before the Central Committee began extensive closed discussions of the draft of a political economy textbook intended to codify official thought on such matters,[122] Bol'shevik published an important article reaffirming that capitalist encirclement remained "the fundamental factor determining the international situation of the Soviet Union." Changes since World War II did not signify that encirclement had ceased to exist and that the danger of foreign intervention had disappeared.[123] "Some comrades," said the author pointedly, had mistakenly perceived the triumph of the people's democracies and the receding influence of capitalism as the end of encirclement; but this was "entirely incorrect."[124] Also sounding the theme of encirclement, a second article in the same issue criticized the tendency of some economists to treat the transition to communism solely in terms of distribution, rather than also as a question of developing the necessary material-technical base.[125] The implicit message was that increased consumption must wait on the further development of industrial capacity.[126]

Warnings of this kind may have been aimed partly at those who still dared to question the doctrine of encirclement in the secrecy of closed academic meetings,[127] but they were also probably intended for some leaders who wanted, if not to abandon the doctrine entirely, then to reduce its reach. A few months after Bol'shevik cautioned against discounting the reality of capitalist encirclement, Lavrentii Beriia, a Politburo member, delivered a speech that appeared to do precisely that. Beriia did mention the term, as one might expect after Bol'shevik's recent warning. But appealing for authority to a 1927 statement of

Stalin in favor of exchanging raw materials for Western equipment, he also made a bid for wider commercial ties with the West. By comparison with 1927, he said, "we now have incomparably larger possibilities for business ties. . . . We are not against a significant broadening of business collaboration with the U.S.A., England and France and other bourgeois countries on the basis of mutual benefit."[128] This was a noteworthy statement, not only because there had been a very large increase in Soviet-Western trade in the years immediately after 1927, but because just a month before Beriia's address the United States had tightened the Western embargo of the Soviet bloc by passing the Battle Act.[129] Shortly afterward *Bol'shevik* took a line quite different from Beriia's. Stalin's doctrine of the socialist state in conditions of capitalist encirclement, the editorial remarked meaningfully, was "final and complete." Quoting Lenin's vitriolic description of American imperialism as "the most reactionary and naked, the most bloody and rabid imperialism," the editorial noted that Lenin had underscored America's constant attempts to interfere in the affairs of other peoples "both by means of direct military intervention and under the guise of 'assistance.'" Lenin's view of American imperialism was particularly relevant at present, when the United States was preparing to instigate war against the Soviet Union, the editorial concluded.[130]

What explains this discrepancy? One explanation is that Beriia, favoring a softer policy toward the West on strictly diplomatic and economic grounds, was willing to support such a policy even at the risk of offending Stalin. This possibility is consistent with evidence that in mid-1953 Beriia advocated concessions to the West on the German question. An alternative explanation is that Beriia was trying to protect himself against Stalin's domestic intrigues by playing down the sense of external danger. In 1951 Stalin was beginning to lay the groundwork for a purge of Beriia and other leaders on charges of collusion with foreign imperialists.[131] Beriia's more temperate depiction of the West implicitly undermined the rationale for such a purge, and he may have hoped it would hamper Stalin's designs against him. An expansion of Western economic ties may have seemed attractive because it fitted this defensive purpose, as well as for the technological benefits it promised. The *Bol'shevik* editorial, on the other hand, seemed to embody the views of leaders who did not fear a purge and who felt that expanded Western economic ties were undesirable.

The most important of these leaders was Stalin himself. As discussed earlier, Stalin had questioned his own prewar notion of capitalist encirclement in 1946. But in mid-1950, several months after Malenkov and Mikoian had implied that the idea might be outdated, Stalin served notice that it was still correct. Implicitly rejecting the argument that Communist victories in China and Eastern Europe had ended the capitalist encirclement of the USSR, he asserted that the doctrine was valid "for the period of victory of socialism in one or several countries."[132] He later gave a fuller exposition of the traditionalist outlook in his "contributions" to the discussion of the draft text on political economy. These assorted writings, completed at various points between February and September 1952, were subsequently published under the title *Economic Problems of Socialism in the U.S.S.R.*

The ideas in this tract dovetailed neatly with Stalin's conception of capitalist encirclement. Stalin depicted capitalism as a system in which the giant monopolies use the state to oppress the workers and militarize the economy. In keeping with this image of the West, he maintained that war remained inevitable and that Soviet heavy industry must be further developed on a preferential basis.[133] He also underscored that socialism possessed far greater technological dynamism than did the West. Modifying dicta pronounced by Lenin and himself before 1930, Stalin rejected the notion that capitalism had achieved stable markets and was growing more rapidly than in the past. Whereas the USSR was sustaining "uninterrupted and stormy growth" with continuous technological innovation, capitalism was restricted to "low rates of growth," recurring economic crises, and periodic interruptions in technological development. Stalin added that there were "two parallel world markets," the capitalist and the socialist. While he acknowledged that the Western embargo had influenced the formation of the socialist economic bloc (CEMA), he praised the socialist market and argued that the economic bifurcation of the world had harmed the West rather than the socialist camp, whose members benefited from sophisticated Soviet technical assistance the West could not match.[134] In this view, the economic sources of imperialist aggression remained as powerful as ever, and the USSR need not risk a significant expansion of technological ties with the West.

For all his megalomania, Stalin did not publish this work primarily to satisfy his intellectual pretensions; it had a concrete political purpose. Although some material in the book had been completed several months earlier, it was held back and published on the eve of the 19th Party Congress in October 1952. The book's contents show that Stalin intended it as a decisive rebuttal to the deviant thinking of some economists, and the circumstances of its publication suggest that he may have had in mind the thinking of some of his Politburo associates as well. According to the recollections of Khrushchev, before the 19th Congress Stalin "went so far as to propose that the speakers at the . . . Congress address themselves to the theoretical questions he had raised" in his book.[135] Doubtless Stalin wanted his lieutenants to offer an unequivocal public endorsement of his ideas, not a dispassionate analysis of them. He could and did exact such fealty from academic commentators; a month after *Economic Problems* appeared, Varga, in a symbolic touch which was more than coincidental, was once more made to recant his doubts about the inevitability of war.[136] But it was probably harder for Stalin to obtain the wholehearted support of his political lieutenants for his outlook, especially when they knew that he was preparing to purge some of them as imperialist agents. In February 1953 *Pravda* printed an article which again condemned "certain propagandists," "certain would-be theoreticians," and "dogmatists" for suggesting that creation of the socialist camp had eliminated capitalist encirclement and the imperialist threat. The phrasing clearly implied that the objectionable view was held not only in academic but in party circles.[137]

Thus, between 1941 and 1953 official Soviet thought went through some important variations, as the elite debated Western economic prospects and intentions toward the USSR, the desirability of commercial ties with the capitalist world, and the amount of effort that should go into developing Soviet military and industrial power. The shifting balance of opinion had a profound impact on the regime's technological strategy and its policies toward indigenous innovation.

Technological Strategy

R & D Spending and Wartime Attitudes toward Foreign Technology

During the war R & D, like every other aspect of Soviet national life, experienced acute new economic strains. The 1941 plan, approved before the German attack, had called for an R & D outlay of 1,651 million rubles.[138] No figures have been published for 1942 and 1943, but there must have been a severe cutback, since in 1944, when the most disruptive effects of the invasion were already past, the regime set aside only 1,300 million rubles for R & D.[139] Even in the elite USSR Academy the number of researchers dipped by 20 percent between 1941 and 1943.[140]

Although the war forced a temporary reduction of the R & D effort, it strengthened the political elite's conviction that research was an essential ingredient of national power, and by 1945 R & D spending was again climbing sharply. The figure budgeted for R & D in 1945 rose to 2 billion rubles. In 1946, when Stalin called for Soviet scientists rapidly to surpass the achievements of foreign research, it shot up to 6.3 billion.[141] The 1946 increase embodied a determination not to be left behind in the quickening international race for technological preeminence, and its enormous size must have been influenced by the American use of the atomic bomb against Japan; as we shall see, the actions of the Soviet leadership after the war demonstrated a special concern about military R & D projects. After 1946, R & D spending grew at a slower but still impressive pace. Total expenditures reached roughly 9 billion rubles in 1950, and about 11 billion in 1953.[142] Thanks to the postwar increase in material support, the number of researchers employed in research institutions climbed from 26,400 in 1940 to 70,500 in 1950, and the number of institutes increased from 786 in 1941 to 914 in 1945 and 1,157 in 1951.[143]

The exigencies of war greatly increased the Soviet desire for foreign technology and the willingness of the other Allies to provide it. Soon after the Nazi attack on the USSR, the United States offered Lend-Lease shipments, first on credit and then as outright grants.[144] The direct demands of the war on the U.S. economy meant that many of the earliest Soviet requests under Lend-Lease were not met fully, and occasionally were not met at all. But by 1943 the American shipments were assuming very large proportions. The value of American aid during the war came to more than eleven billion dollars, including a substantial quantity of weapons and more than one billion dollars' worth of industrial machinery and equipment.[145] The utility of such inputs, of course, depended on the Soviet capacity to mobilize and ration the vast stocks of additional resources needed to prosecute the war. But American shipments of equipment and pivotal industrial materials did play a vital role in sustaining the Soviet war effort.[146]

Although the war increased Soviet reliance on foreign equipment, it did not produce a significant relaxation of the domestic obstacles to other forms of technology transfer from the Western Allies. During the war the Soviet leadership remained wary of high-level contacts that might reveal information about the general condition of the Soviet economy. "No proper information about Russia's internal economic situation was ever forthcoming. Russian requests under lend-lease were simply presented as lists of equipment, and the American government had to make its own decisions on the justification for each request. These decisions were usually made in an atmosphere of the deepest suspicion and distrust."[147] Contacts between Soviet and American specialists at lower levels were also very limited. During the war the Soviet government signed a small number of technical-assistance contracts with American firms, but these agreements were exceptions to the rule.[148] The sources that I was able to examine mention few specific instances in which American specialists accompanied Lend-Lease equipment to help with its installation.[149] When such experts went, they received a mixed reception.[150] The shortage of contacts in the USSR may have been mitigated, but only slightly, by the access to American plants that was reportedly granted to Soviet engineers attached to the Soviet Purchasing Commission in the United States.[151] No doubt some of the limits on contacts among specialists were due to wartime travel restrictions and to an understandable element of wartime secrecy. But some of them were due to the ingrained official suspicion of personal contacts between Soviet citizens and foreigners of any stripe.

Nevertheless, the spirit of the anti-Nazi alliance emboldened some Soviet scientists and engineers to push for closer dealings with their foreign counterparts. By doing so, these specialists clearly challenged the practices established in the late 1930s, when party watchdogs had beaten down such requests in the name of a separate "socialist" and "native" science. In 1943 Petr Kapitsa, the eminent physicist who had been prevented from returning to a joint research project in Britain in 1934, lobbied openly for a change from the scientific isolation of the 1930s. Alluding to the barriers that had cut off the flow of foreign scientific visitors before the war, he reminded the Academy's Presidium that some foreign scientists had wanted to visit the USSR but had been unable to do so because Soviet links with foreign science had been severed. These links should be reestablished in the future, he said, since they were a normal requirement for research and since "all world science constitutes a single indivisible whole."[152] In the same spirit, the Academy's Department of Technological Sciences called on the Presidium to petition the government for research trips to the United States and Great Britain.[153] At the Academy's anniversary celebration in June 1945, Kapitsa denied that there could be separate national sciences and promised the visiting foreign scientific emissaries that the USSR would expand its exchange of scientists and publications with the outside world.[154] An engineering spokesman also noted the importance of exchanging technological information with R & D establishments abroad.[155]

Postwar Attitudes toward Foreign Technology

Even in the celebratory atmosphere at the end of the European war, the political leadership received such arguments with deep ambivalence. At the Academy's jubilee, Molotov toasted closer cooperation with foreign science, and Stalin reportedly promised a maximum effort to obtain foreign scientific literature (though not personal contacts) for Academy researchers.[156] Such statements should not be discounted as meaningless. Even the most traditionalist leaders must have favored foreign connections that could help the USSR draw abreast of the United States in vital fields like atomic research. But the leaders were also very reluctant to allow the extensive contacts needed to promote technology transfers along a wide front. They still feared the impact of foreign ties and travel on the loyalty of Soviet citizens, such as the large number of soldiers exposed to Europe during the war, and they cannot have been reassured by events such as the demand of leading social scientists for more personal meetings and book exchanges with foreign social researchers.[157] Before the close of the war the regime had begun to tighten domestic ideological controls, and oblique evidence suggests that Stalin may have intended to downplay the achievements of foreign science and technology as part of this process.[158] At any rate Molotov bore witness to the elite's concern for domestic political legitimacy in late 1945, when he rejected the notion of broadening the general exchange of ideas and information between the USSR and the West. Machines, he asserted, were the only fruits of Western society in which the Soviet Union had an unequivocal interest.[159]

The regime's policy toward foreign technology was squeezed between these conflicting technological and ideological needs. As indicated above, during 1946 the political authorities were still cautiously sounding out the possibility of a large American loan for extensive capital imports. Meanwhile a spokesman who had supervised wartime R & D asserted that the USSR should study and use all the major recent discoveries in Britain and America—an injunction that was reflected in the Soviet pursuit of new technical-assistance agreements with U.S. firms.[160] The Soviet appetite for such agreements had grown noticeably in the last eighteen months of the war, when the government opened negotiations on roughly twenty-five new technical-aid contracts with U.S. firms.[161] In late 1945 and early 1946 at least nine further contracts were tentatively negotiated, including several in electronics, one in radar, and possibly one for aid in producing a machine cannon. By the end of 1946 the USSR reportedly had some fifty current technical-aid contracts with American companies.[162] Simultaneously, however, the regime blocked a large number of invitations for Soviet scientists and engineers to visit the United States, and a prominent official of the political police warned that all foreign engineers and businessmen in the USSR were potential spies.[163] One sign of the barriers to wider contacts was the fate of V. V. Parin, Secretary General of the Soviet Academy of Medical Sciences, who visited the United States in late 1946. During his visit Parin emphasized the internationalism of science and called for fuller Soviet-American scientific exchanges. Soon after his return to the Soviet Union, he and the Minister of Health were removed from their posts, and he disappeared from public view.[164]

The conflict between political and technological needs was less pronounced in Soviet dealings with the vanquished Axis powers, whose resources were often free for the taking. While in Manchuria, Soviet forces shipped home about two billion dollars' worth of Japanese industrial equipment.[165] In Germany, the 60,000 or 70,000 dismantlers accompanying the Red Army removed a large amount of machinery and technical documentation, albeit often with great waste.[166] In the zone of Germany under their control, the Soviet forces also set up German-staffed R & D centers that continued to function until October 1946, when as many as 20,000 to 40,000 German scientists and technicians were deported to work in the USSR and help train Soviet specialists.[167] This kind of technology transfer, which could obviously be applied only to defeated enemies, was appealing because it allowed the regime to maintain its barriers against day-to-day interchange with the outside world. But even here the Soviet authorities were leery of ideological contamination. The political police played a central role in procuring German technology, and they ordered Soviet specialists in Germany not to send home any Western literature or use German newspapers to wrap equipment destined for the USSR.[168]

Although fragmentary, the evidence suggests that the party chiefs made a special effort to acquire militarily useful technology and to protect such transfers from the political barriers that hampered technological borrowing in other sectors. According to an aeronautical engineer who served as an agent in gathering information on science and technology in occupied Germany, the Soviet leaders displayed a pronounced military bias. They were intensely curious about German research with military potential but rebuffed attempts to report on research and technology with purely civilian uses.[169] Similarly, the small number of specialists who were permitted to travel to the United States in 1946 were primarily in fields like radio engineering and machine tools,[170] which promised some military benefit. On occasion the Soviet press also gave unusually open accounts of Western developments in military fields such as rockets and radar (though not atomic energy). In early 1946 the main armed forces newspaper reported on a U.S. conference that discussed German ideas for using winged or multistage rockets to deliver bombs over long distances.[171] Even more striking was the public attention given to Western developments in radar. In 1946, as part of a broad drive to improve Soviet radar, the regime set up a special information center to disseminate the relevant foreign and domestic literature.[172] One of the center's organizers was the author of an unusually detailed discussion of Western radar published a few months later.[173] Such extensive discussions of Western technological achievements were exceptional; there were no comparable articles, say, on Western tractors or rolling mills. It thus appears that the leadership was more interested in providing adequate information on foreign developments in military technology than in other sectors.

Decisions about copying a particular piece of Western military technology were often made at the political apex and were less influenced by the increasingly anti-Western atmosphere than were decisions made lower in the hierarchy. Even as party leaders glorified native technology in their general pronouncements, they did not hesitate to emulate Western military technology they privately deemed superior. On occasion, of course, decisions at the top went against technological borrowing. Thus at the end of 1945, after a wide-ranging debate about whether to copy the Messerschmidt 262 jet fighter or expand R & D on domestically designed jet airframes, the leadership rejected the idea of copying and sacked the industrial commissar who had proposed it.[174] But this decision hinged on which course of action would produce the most sophisticated aircraft, not on which was ideologically correct.[175] Simultaneously the government chose to copy captured German engines as the power plants for the first generation of Soviet jet fighters, and in 1947–48 it authorized the reverse-engineering of jet engines purchased from Britain. Soviet versions of these engines were incorporated into several new fighters and bombers.[176] The leadership also elected to model a new heavy bomber, the Tu-4, on American B-29s that had made emergency landings in Siberia during the war. According to an emigré designer who worked in military electronics during this period, Stalin ordered that the new bomber be a detailed copy of the American plane. Soviet specialists were threatened with punishment for introducing any changes in the design or the materials—a condition they satisfied only with considerable difficulty.[177] After the new machines were unveiled, the press touted them as an indigenous achievement superior to American weapons.[178] On balance, Western research and designs made a major contribution to Soviet advances in aircraft and radar technology after the war.[179]

The top leadership, however, gave such close attention and encouragement to technology transfers mostly in the military sphere. In other sectors having lower priority, industrial researchers and administrators had to take their cues from the leaders' general pronouncements about domestic versus foreign technology, and these remarks were increasingly nativist. By 1947 Molotov was attacking "groveling and servility" before Western culture and was calling for "merciless criticism" of all such attitudes.[180] To political officials and specialists who had lived through the Great Terror before the war, the theme was all too familiar, and it drastically reduced the flow of foreign technology into the country. This is not to deny that the Western embargo seriously impeded the movement of technology from West to East.[181] But the political atmosphere inside the USSR obstructed the transfer of much technology, especially technology in the form of documents and specialists, which remained available. Although the regime had slightly lowered the domestic barriers to technological borrowing during the war, it had never dismantled them, and now these barriers were raised higher than ever.

The xenophobic line did not triumph without some resistance. In 1948 B. M. Kedrov, the editor of *Voprosy filosofii* (*Problems of Philosophy*) who had worked in the Central Committee apparatus during the 1930s, registered his disagreement. As a specialist on the philosophy of natural science, Kedrov warned against carrying the idea of separate Western and Soviet scientific traditions too far. He granted that the principle of party-mindedness required avoiding apologetics for bourgeois science, but argued that it also required avoiding an indiscriminate rejection of the positive achievements of contemporary Western research. Lenin, Kedrov said, had distinguished the negative idealistic tendencies of bourgeois science from its rich factual findings, and Leninism had nothing in common with vulgar criticism that failed to make this distinction. By drawing a parallel between the current situation and the one faced by Lenin in the 1920s, Kedrov was clearly arguing that a more receptive attitude toward Western R & D was in order.[182] This was part of his objection to "an incorrect understanding of the task of struggling against servility" that often entailed efforts to create the appearance of ideological errors where none existed.[183]

Although this dispute was about philosophy, it was about policy as well.[184] Kedrov's position was backed by G. F. Aleksandrov, and in a surprisingly concrete way. After being removed from his Central Committee propaganda post in 1947, Aleksandrov had been made Director of the Academy's Institute of Philosophy, which sponsored the journal edited by Kedrov. In early 1949, in a provocative gloss on a letter written by Lenin in 1922, Aleksandrov made it clear that he opposed the prevailing Soviet policy on Western science and technology. *Pravda* printed Lenin's letter, previously unpublished, on the January 1949 anniversary of Lenin's death. Addressed to Stalin and headed by a request that it be circulated to all Politburo members, the letter stated that the Soviet radio network should be expanded. Its final paragraph remarked that the English had recently invented a means of sending radiotelegrams secretly. "If it were possible [for us] to buy this invention," Lenin concluded, "then radiotelephone and radiotelegraph communications would assume still more enormous significance for military affairs."[185] Soon after the letter appeared in *Pravda*, Aleksandrov followed up with a forceful statement on the importance of foreign science and technology. He began by quoting another document in which Lenin attacked a Soviet government agency for ignoring a series of technological discoveries. The quotation demanded that the agency should either be awakened or that its scholarly good-for-nothings should be thrown out of office. It was necessary, Lenin had said, to establish clearly who would bear responsibility for familiarizing the Soviet Union with European and American technology "on time, practically, and not formalistically." In particular, it was necessary to have "one copy of *all* the most important, *newest* foreign machines, in order to learn from them and to teach." Immediately after quoting these words, Aleksandrov artfully embarked on a discussion of Lenin's interest in radio for the purpose of detecting enemy air squadrons at long distances. He summed up by saying that the Lenin letter recently published in *Pravda* "has enormous scientific and theoretical interest on this question."[186]

Aleksandrov's article contained an obvious implication that someone deserved to be punished for failing to obtain valuable Western technology and a definite hint that some of this technology was connected with radar. In view of his earlier political role and comparatively conciliatory attitude toward the West, it may be that his article was intertwined with the high-level political disagreements analyzed above. Perhaps it was part of a veiled discussion of the impact of the Western strategic embargo and the advisability of aggravating the embargo through a hard-line foreign policy. It may also have meant that the top leadership was having difficulty acquiring some foreign military technology it desired because many channels of technology transfer were being choked off by the USSR's own anti-Western political campaign. What is certain is that controversy over the proper policy toward Western science and technology was still brewing among secondary members of the elite in early 1949.

Despite their outspokenness, moderates like Aleksandrov and Kedrov were losing the argument. Rebuked by other philosophers for underestimating the uniqueness of Soviet science, Kedrov lost the editorship of *Voprosy filosofii* about the time the journal published a highly chauvinistic assault on "cosmopolitanism" in all spheres of culture, science included.[187] *Bol'shevik* backed these criticisms by stating that Kedrov's ideas about the unity and internationalism of science served the enemies of the USSR.[188] A few months later *Pravda* castigated both Kedrov and Aleksandrov for showing excessive sympathy for Western values.[189] This accent on the distinctiveness of Soviet science was probably carried furthest by the Lysenkoites in their final assault on recalcitrant biologists, who tended to invoke the authority of world science against Lysenkoism.[190] But there were strong efforts to extend the campaign to other fields such as chemistry, where critical scientists likewise tried to resist ideological incursions by suggesting that the world prestige of Soviet science would suffer.[191] In early 1949 the Academy of Sciences devoted several ceremonial meetings to the praise of "native science," in an atmosphere which indicated unmistakably that Soviet accomplishments had to be shown to be superior, whether or not they were in fact.[192] Soviet writers began to manifest extraordinary chauvinism in claiming international priority for the past achievements of Russian science and engineering.[193]

The official obsession with "native science" spilled over into the ordinary work of scientists and engineers, further attenuating their connections with foreign R & D. By 1949 *Pravda* was writing virtually as if borrowing Western scientific ideas were equivalent to advocating foreign political doctrines.[194] Meanwhile the Academy *Vestnik*, spurred by the party campaign, inveighed against the uncritical use of "bourgeois sources." Emphasizing that it rejected not just the general theories of bourgeois science, the journal challenged the reliability of Western experimental data as well, thereby casting aside earlier efforts, such as those of Kedrov, to justify using Western research findings.[195] In this atmosphere, scientists and engineers were attacked for merely citing foreign journals and consequently "neglecting" Soviet research. For instance, reviewers castigated the author of a book on water resistance to ship hulls for devoting excessive attention to foreign scientists and scientific literature; the authors of texts on physics and mining were criticized on similar grounds.[196] Having been instructed to survey recent advances in Blooming machines, a group of design engineers was attacked for devoting too much attention to American accomplishments.[197] In yet another case, two medical researchers had their doctoral degrees revoked for utilizing "tendentious bourgeois statistics."[198]

Political onslaughts of this kind weakened the USSR's already tenuous links with technological developments abroad. Soviet foreign trade, the most politically antiseptic form of technology transfer, remained extremely limited—including trade with the USSR's new socialist clients.[199] The impact on other varieties of technology transfer was still more pronounced. In 1950 a total of "about 200 foreign scholars" visited the USSR, and the number of Soviet scientists traveling abroad also remained exceedingly small.[200] Despite the formation of the CEMA, even travel between the USSR and other bloc countries was sharply limited, probably because of the Stalinist fear that Titoism or other heresies might infiltrate the USSR from the newly communized Eastern Europe. From 1948 through 1952 only 750 Soviet technical specialists made professional trips to these countries, and only 775 of their specialists visited the USSR.[201] On the average, this represents an annual exchange of specialists less than a tenth as large as that reached by the USSR and its socialist client-states at the end of the 1950s.[202]

Similar impediments affected the access of Soviet R & D personnel to foreign scientific and technological literature. As indicated previously, incoming publications were screened by the secret police, who winnowed out material regarded as subversive. Moreover, the higher authorities were rather reluctant to spend scarce hard currency to obtain foreign technical writings.[203] Exacerbated by the hysterical nativism of the late 1940s, such obstacles sharply restricted the amount of literature reaching Soviet scientists and engineers. The available data on the receipt of foreign publications are not comprehensive. But to judge by the figures for the Academy, the amount of specialized literature received in 1950 was well below the prewar nadir reached in 1940 and about four times below the figure for the peak prewar year of 1936.[204] In the last two years of Stalin's reign, the regime began to draw back from the extremes of this policy.[205] But for virtually the whole period of high Stalinism, Soviet access to foreign scientific and technological literature was painfully limited.

Although we cannot prove it, the political police probably took on a larger role in acquiring foreign technology as other avenues of transfer were squeezed off by domestic politics and the Western embargo. A police official reportedly headed a secret purchasing agency set up in Berne to acquire Western equipment whose export to the USSR was prohibited under the embargo.[206] And the Main Economic Administration of the Ministry of Internal Affairs, one of the two principal security agencies at the time, was responsible for "certain technical studies abroad." Disguised as diplomats, engineers and economists from the Ministry were attached to Soviet embassies to study the industries and inventions of the host country.[207] One benefit of this arrangement was that it helped minimize exposure of less reliable Soviet specialists to the outside world. Another was that although most materials acquired via this channel were obtained legally,[208] it established a foreign network of police specialists to assist with technological espionage, especially in the military sphere. But there is no reason to think the efforts of the police compensated for the many other channels of technology transfer the Stalinist regime had closed off. The principal mission of the police was to isolate Soviet citizens from the outside world, not to foster closer connections between the two, and the basic impulse of the police agencies was to withhold foreign information rather than to disseminate it.

Thus, although the exigencies of the struggle against Nazi Germany had sharpened the Soviet appetite for Western equipment and technical-assistance contracts, the onset of the Cold War and a new ideological crackdown at home ultimately produced an unprecedented measure of Soviet isolation from Western sources of technology. Meanwhile, the forces that shaped the regime's technological strategy along these lines were also affecting its policies toward indigenous innovation.

The Domestic Organization of Innovation

World War II influenced the organization of R & D less deeply in the USSR than in many other belligerent countries, because the regime had already taken numerous steps to link research to its economic and defense goals.[209] Nevertheless, the war created a pressing need for more research with immediate military value, and Soviet efforts to meet that need merit our attention. We must also examine the adjustments made in R & D programs as the euphoria of victory and the prospect of further infusions of U.S. technology gave way to new tensions with the West. How vigorously did the authorities strive to overtake America's advancing military technology, and were they able to reconcile this aim with the goal of raising the technological level of the nonmilitary sectors of the economy?

Wartime Changes in R & D

Although Soviet R & D was already strongly slanted toward military technology in the late 1930s, the Nazi attack triggered an urgent new push to harness research to military requirements. In the first days of the war the regime established an all-powerful State Defense Committee to direct the mobilization of the economy.[210] The State Committee relied primarily on the existing party and state administrative apparatus to enforce its policies,[211] but it did set up new staff agencies in some fields, among them R & D. In July 1941 the Committee appointed S. V. Kaftanov, the leading Soviet official for higher education and a top-level R & D troubleshooter in the late 1930s, as its plenipotentiary for science; and it created a Scientific-Technological Council to assist him in coordinating war-related research.[212] The Council's initial field of responsibility was R & D in chemistry. By October 1941, when the Presidium of the Academy passed a resolution approving its work and asking that its responsibilities be expanded, it was already responsible for chemistry and physics.[213] Later its terms of reference were widened further to include geology and other fields.[214] Kaftanov's eight assistants headed its administrative sections, which screened proposals from specialists and establishments in individual scientific fields, passing along to the full Council those which seemed useful. The Council, in turn, reported on the proposals to the State Defense Committee, the Central Committee staff, and the Council of Ministers.[215] Aside from his Moscow assistants, Kaftanov also had official representatives based in the cities where substantial amounts of R & D were being done.[216]

Faced with the advancing Nazi juggernaut, no one challenged the principle that war needs should receive precedence in setting new R & D priorities. In July and August 1941, while the regime started a massive evacuation of factories and research facilities to the east, the Academy reshuffled its programs to concentrate on work with military and strategic economic applications, and Kaftanov's Council hastily approved the revised plan.[217] Further, the Academy established a central Projects Commission (*Tematicheskaia komissiia*) to plan its war-related R & D and maintain contact with the military. It also set up separate commissions to develop antitank measures, locate new supplies of natural resources, and assist industrial production in the Urals.[218] But general acceptance of the overriding importance of military R & D did not resolve the issue of precisely what kind of research would contribute to the war effort. How quickly should the research yield practical results? How closely should it be tied to weapons production? The Academy was besieged with aid requests from many organizations, and each request raised these questions afresh.

The search for answers sparked serious disagreements among leading Academy officials. At a meeting of the Academy Presidium late in 1941, a spokesman for the Bureau of the Department of Technological Sciences cautioned implicitly against being carried overboard by the demand for military relevance. The Bureau, he said, felt that some of the most important research projects begun before the war should be continued even though their completion would require an extended period. The Department believed that it should concentrate on defense projects of broad intellectual scope, rather than fragmented short-term undertakings, and that some minor projects should be transferred to ministerial research bodies. Another speaker seconded this view, contending that economically valuable research was in danger of being lost.[219] But other academicians vigorously disagreed. One warned that such a policy would weaken urgent military R & D programs and that less critical projects must be postponed. Asserting that science should contribute more to the war, another member remarked sarcastically that "some people," although considering themselves antifascists, thought the enemy could be defeated without their help.[220] Similar disputes broke out in the Departments of Chemical and Biological Sciences. As one academician remarked, there was "quite a large divergence" in the way various Academy members interpreted the tasks of science in time of war.[221] The most striking thing about these arguments is that almost all the disputants quoted above were or soon became members of the Projects Commission, the Academy's principal executive organ for planning and coordinating its war-related research.[222] The disagreements thus existed among the scientists who were directly responsible for guiding the Academy's contribution to the war effort.

Perhaps in response to such debates, the political authorities intervened to strengthen the links between the Academy and the military. At the beginning of 1942 Central Committee officials convened a meeting with representatives of the Academy and the scientific departments of the People's Commissariat of Defense to discuss the introduction of Academy research into production. After the meeting agreed on ways to strengthen the ties between the institutes and defense organizations, the Central Committee participants served notice that they would monitor the fulfillment of these arrangements on a day-to-day basis.[223] Moreover, in late March Stalin notified the Academy president of measures adopted by the Council of People's Commissars to improve the leadership of the Academy, and he emphasized that the Academy must meet the country's growing military requirements.[224] While the content of the Council's resolution has not been revealed, the Academy did modify its research plans in response and set up a new commission of scientists and military specialists to work on naval R & D. In addition, it installed an Academy member from the military as its Academician-Secretary, an administrative post that allowed him to wield authority over secret Academy research.[225] The Presidium members also met separately with the Defense Commissariat's chief of R & D and military supply. The Commissariat then issued a decree specifying which defense agencies the Academy should aid, and the procedures that should govern these contacts.[226]

Under such prodding, a far-reaching shift of the national R & D effort toward military needs occurred during 1942. A majority of Academy researchers set to work on war-related tasks as various as improving the aerodynamics of airplanes, optimizing artillery fire-patterns, designing bombs, producing military fuels, discovering new stocks of natural resources, and adapting evacuated enterprises to local raw materials. The list could be extended almost indefinitely.[227] It typifies the large swing toward military needs that took place in the R & D establishment as a whole.[228]

The demands of the war caused not only the cancellation of much basic research, but the postponement of weapons projects that could not produce quick results. In 1941 the regime set aside systematic research on atomic energy because it felt that any military result would come too late to influence the outcome of the war.[229] The same pattern held in the aircraft industry. Stalin reportedly said early in the war that the time was not right to work on experimental airplanes, and designers spent most of their energy improving the planes already in series production. Prewar exploratory work on jet engines was suspended, and experiments with a rocket-propelled plane received very low priority during the conflict.[230] The pressures of the front had a similar effect on other weapons sectors, such as the tank industry.[231] The demand for quantitative increases in weapons was so great that designers were forced to make most technological changes in ways that would not slow the assembly lines.

Nonetheless, the technological performance of the arms sectors was extremely impressive. Designers and managers made important qualitative improvements in Soviet weapons, and they showed great ingenuity in devising substitutes for scarce inputs. The bureaucratic apparatus manifested its formidable mobilizational capacity by rapidly relocating strategic industrial plants and by drastically increasing military production, which rose from 26 to 64 percent of total industrial output between 1940 and 1942.[232] The tremendous quantitative increase in weapons production was aided by major improvements in the manufacturing methods of the weapons branches, which experienced a sharp increase in productivity during the war.[233]

Once the tide of war began to turn against the Germans, the regime gradually relaxed its demand that all R & D yield swift results. At the end of 1943 the party leadership authorized Gosplan, the Academy, and the commissariats to develop an economic plan for the postwar period.[234] Thus encouraged, the Academy officials started to emphasize that theoretical research should be expanded, partly because of the ultimate military benefits it promised, and they launched a systematic inquiry into which lines of research should be promoted over the long term.[235] This planning exercise was supervised by officials of the Central Committee Secretariat, and the resulting document was approved in October 1944 by the Academy's General Meeting. Among the topics singled out for special attention were atomic energy, radar, jet and rocket technology, electronics and semiconductors, calculating devices, and the theory of combustion.[236] Meanwhile Gosplan sought to incorporate emerging technological trends into its general economic plans. Early in 1943 Gosplan's Council of Scientific-Technological Assessment received a contingent of new members from the Academy. The Council's new head was A. A. Baikov, an Academy Vice-President, and twenty of its twenty-six members were also from the Academy.[237] In August 1944 the government gave the Council the power to oversee the commissariats' plans for technological innovation, and the Council presented Gosplan with a study outlining the anticipated directions of technological change in the economy through 1947. When the government set about formulating a five-year plan for 1946–50, it incorporated the Council's proposals into the new document.[238]

As the time horizon for R & D was extended, several programs for creating novel weapons gathered momentum. In March 1943 the Central Committee initiated a comprehensive discussion of the steps needed to accelerate the development of Soviet radar, and in July a Council on Radar was created under the aegis of the State Defense Committee. Malenkov chaired the new body, which was administered by Deputy Chairman A. I. Berg, a radio specialist who was simultaneously appointed Deputy Commissar of the electrotechnical industry.[239] The Council included a Central Committee official, the commissars of the defense industries, planners, military officers, and top scientists and engineers.[240] Its duties were to coordinate R & D on radar, disseminate information, train radar specialists and expand the output of radar sets.[241] To increase output the Council sought the aid of several commissariats of military industry, but Berg encountered considerable difficulty because these agencies, hard-pressed to meet their own arms production targets, resisted sharing their facilities.[242] This may be why he complained in his diary early in 1944 that Malenkov was giving the Council no support, with the result that progress in radar was being seriously delayed.[243] At any rate, by the start of 1945 Malenkov was lecturing the defense industry commissars on the need to give maximum backing to the development of radar.[244] In 1944 and 1945 the output of radar sets increased dramatically.[245]

The authorities began to look further into the future of other weapons as well. In February 1944 the State Defense Committee, prompted by reports of research on jet engines in Germany and the other Allied countries, set up a special institute to experiment with jet and rocket engines and instructed the Commissariat of Aircraft Industry to present concrete proposals for developing jet planes.[246] Work on jets at first grew rather slowly. The only available Soviet power plant was comparatively backward, and the improvement of propeller-driven aircraft continued to occupy most Soviet designers.[247] In December 1944 the political authorities resolved to increase the scale of aircraft R & D, which must have facilitated work on jets by the designers Iakovlev and Mikoian. Probably even more important was the capture of more powerful German jet engines, which furnished the propulsion technology used in the first successful Soviet jet flights in 1946.[248]

At roughly the same time the USSR increased R & D on jets, it renewed research on the much more problematic project of developing a nuclear bomb. In this case, too, intelligence reports about German and American efforts played a role in stimulating Soviet research.[249] At the end of 1942 the State Defense Committee ordered the establishment of a special Academy laboratory to work on the nuclear problem. Gradually the project gained momentum, until at the end of 1944 about one hundred researchers were employed at the laboratory.[250] But neither the jet nor the A-bomb project was pursued on a crash basis. On balance, it seems that the Soviet authorities began to force the pace of R & D on jets and atomic weapons only after the defeat of Germany, as we shall see below.

Postwar Changes in the Organization of Research and Innovation

At the close of the war the regime faced two large technological tasks. In the nonarmaments sectors of industry, most of which had been neglected during the conflict, the level of technology badly needed improvement. Meanwhile military technology had to be developed at a rate that would protect the USSR's interests against possible pressures from its erstwhile American ally. Both these goals required expanded R & D and rapid technological innovation. But it was not clear how easily the two aims could be reconciled.

One way the authorities tried to step up innovation was to create separate technological plans that paralleled plans for regular production—a method that had been initiated in the late 1930s but disrupted by the war. In 1946 Gosplan began to formulate five-year, annual, and quarterly technological plans for the nation.[251] In 1947 a separate agency, Gostekhnika, was set up alongside the overburdened Gosplan to handle such plans.[252] A few months later the government ordered the creation of a special system for collecting statistics on new technology, since the existing arrangements for reporting on innovations were found to contain serious inadequacies.[253] More comprehensive planning and more accurate information were thus key elements in the regime's approach to its postwar technological tasks.

Even more important was the decision to channel a very large quantity of scarce resources into R & D spending and high salaries for R & D personnel. We have already noted that aggregate R & D expenditures shot up between 1945 and 1946, and continued to grow steadily thereafter. A large share of the increase went into preferential salary raises and bonuses for scientists and engineers. In March 1946 the Council of Ministers (as the Council of People's Commissars was now called) adopted a resolution entitled "On Increasing the Salaries of Scientific Workers and Improving their Material Living Conditions."[254] While the resolution's contents have never been published, its general effect is quite clear. According to one historian of Soviet science, in 1946 the average salary for scientists doubled or tripled. This step raised their standard of living well above the prewar level, at a time when the living standard of the general population was still depressed far below its prewar level.[255] In addition, between 1945 and 1950 the government quadrupled the number of lucrative Stalin Prizes awarded annually for exceptional achievements in science and technology.[256] As for enterprise incentives, the regime instituted a special bonus for plants designing and producing experimental prototypes of new technology, to be paid even if the enterprise's overall target for regular production was not met.[257]

Coercion had not been abandoned as a method of spurring specialists to strive for technological excellence, but a change of emphasis had occurred. The system of prison institutes founded before the war was maintained intact by the secret police. According to a recent estimate by insiders critical of the regime, about 15 percent of major scientific projects and at least 50 percent of atomic energy research were conducted in such establishments in the late 1940s.[258] The system, though, does not seem to have received massive new intakes of Soviet specialists, as it did in the 1930s. Prison institutes housed the foreign scientists forcibly removed from Germany, and they may have obtained some new researchers from the occasional arrests of Soviet specialists after the war. But such arrests appear to have been relatively isolated cases.[259] On balance, the regime was clearly moving toward a greater use of positive inducements to spur technological advance.

As they adopted these measures, the Soviet leaders faced a question of R & D priorities. Now that the Axis powers had been defeated, should the national R & D effort be shifted from military to civilian requirements, and if so, how much? The forced production of heavy industrial goods, primarily weapons, had created unprecedented imbalances between light and heavy industrial output; and within heavy industry, between military and nonmilitary articles. In 1944–47 the leaders took steps to reduce these imbalances, although for the 1940s as a whole the growth of heavy industry still greatly exceeded increases in light industry.[260] But it remained to be seen whether this adjustment of production priorities would be accompanied by a serious effort to upgrade the technological level of the nonmilitary sectors, including the parts of heavy industry not directly involved in military production.

Judging by the fragmentary evidence, a limited shift in research and innovation toward civilian requirements did take place. Postwar discussions of the Academy's plans and of the country's technological future stressed that researchers had a responsibility to aid the economic reconstruction of liberated territories and to help restore prewar levels of output. There was also considerable stress on applying wartime manufacturing techniques to civilian industry.[261] During the war, labor productivity had grown much faster in the arms sectors than in other branches, partly as a result of new mass-production methods; and commentators occasionally suggested that these methods were now being applied to civilian output.[262]

Despite some adjustment favoring civilian branches of technology, however, it appears that military-related projects continued to enjoy a very strong priority over other R & D undertakings. Out of the Politburo review of military policy in early 1946 came a "broad program" for reorganizing and strengthening the armed forces[263] with a heavy emphasis on military R & D. Authoritative discussions of science and technology at this time laid special stress on atomic energy, radar, and jet propulsion. In March 1946 Voznesenskii put particular weight on these fields in describing the new five-year plan.[264] An article on the plan in the main party journal highlighted them as "the latest and greatest achievements of science and technology in our century,"[265] and Kaftanov cited them to underscore his point that the technical qualifications of specialists must be raised.[266]

A very large share of the vast postwar increase in R & D spending was undoubtedly channeled into military programs. In August 1945, after the destruction of Hiroshima, Stalin ordered an all-out effort to create a Soviet atomic bomb, and a Scientific-Technical Council was established to set policy for this undertaking. A rapid expansion of institutes, factories, and technical schools ensued.[267] At the end of 1945 the top leadership, having monitored a heated debate among aircraft specialists about the danger of falling behind in aircraft technology, rejected a proposal to copy German jet fighters wholesale and resolved to increase sharply the number of design organizations working on jets. At least thirteen separate design groups were put to work on jet engines and airframes.[268] The regime also established a Scientific Council to guide the development of military rockets, and in mid-1946 it created a new R & D network to design and manufacture such weapons.[269] In June of that year the Council of Ministers set up an Academy of Artillery Sciences under the Ministry of Armed Forces. Headed by a member of the USSR Academy of Sciences with the rank of Lieutenant General, the new institution comprised a wide range of scientists, military officers, and plant engineers; its administrative structure included departments for rocket weapons and radar instruments.[270] Meanwhile the Council on Radar took the lead in formulating an ambitious general plan for the rapid development of radar. Adopted by the government in July 1946, the general plan called for a series of new institutes, design bureaus, technical schools, and industrial plants.[271] Steps were also taken to tighten the links of the Academy of Sciences with weapons research.[272] The testimony of several Soviet specialists offers further evidence that the party leaders provided abundant material and administrative backing to researchers working in these weapons programs.[273]

In addition to material support, the party chiefs gave close attention to military projects and frequently relied on competitions in order to obtain the best results from weapons specialists. In March 1947, at Stalin's suggestion, the regime stepped up work on long-range rockets and set up a special commission to oversee this effort.[274] The decision stimulated a widespread desire among R & D groups to build a transatlantic rocket. In the fall at least two groups were at work on different approaches to the problem—although at the end of 1947 the group working on the less promising approach was forcibly disbanded.[275] For military aircraft design, competition continued to be the norm. In one case, a fly-off of bomber prototypes led Stalin to choose a design by Iliushin over one by Tupolev. In a later competition, another Tupolev design bested one by Iliushin.[276]

The few efforts to encourage civilian spin-offs from military R & D programs apparently achieved little success. Early in 1946 the Academy's Academician-Secretary, underscoring the "great danger" that wartime research would remain inaccessible to broad scientific circles, urged that ways be found to avoid this danger.[277] About the same time, the Academy appointed a commission to identify materials that no longer needed to be held in the secret repository of the Academy Library and to formulate rules of operation for the repository in the future. While the commission's make-up suggests that it focused mainly on social studies, its terms of reference did not exclude the natural sciences and technology.[278] In addition, a brief relaxation in the censorship of publications in nuclear physics occurred at the end of the war.[279] By 1947, however, the pendulum had swung decisively in the opposite direction. In June of that year the regime adopted two decrees which greatly increased the scope of legal secrecy in science and technology. Abandoning the distinction drawn by previous Soviet laws between military and nonmilitary technical information, the decrees classed as state secrets research projects and inventions "in all fields of science, technology and the national economy, until they are finally completed and authorized to be published."[280] An authoritative exegesis of the new decrees asserted that divulging such information was just as dangerous now as during the war.[281] In this political atmosphere the prospects for increasing the exchange of information, especially between military and civilian R & D, were extremely poor.

Compounding the shortage of information was the low priority attached to civilian innovation at every administrative level. One writer complained that "many" enterprises that had mass-produced military goods during the war were lagging in applying mass-production methods to civilian articles like agricultural machines, and he pointed out that such transfers of manufacturing processes were necessary to raise the productivity of civilian machine building.[282] Product innovation met similar difficulties. According to a Soviet history published in the early 1960s, Gosplan and the industrial ministries delayed increasing the number of designers at civilian engineering plants from which designers had been removed to do military work at the start of the war, and this slowed the creation of new civilian products.[283] In 1946 the Academy President remarked that innovation sometimes occurred with "uncommon slowness," and that in most cases the cause was the "very great inertia of some of our plants and main administrations." Calling the struggle for innovation an urgent task, he asserted that industry could do much more to solve the problem than could scientists.[284] This did not keep *Pravda* from criticizing "extremely serious inadequacies" in the Academy's applied research, or from advocating tighter party control over scientific institutions as an answer to the problem.[285] But much of the responsibility did lie with the industrial ministries. A review of plant design in ferrous metallurgy, for example, concluded that the agencies planning new factories were making little use of the latest domestic and foreign research and were neglecting economic criteria in design decisions. Among the causes cited were the agencies' lack of familiarity with enterprise conditions, their poor connections with scientific institutes, and their inadequate attention to recent technical publications.[286]

A New Push for Accelerated Innovation

After decreasing in 1947 and 1948, political friction over the rate of domestic technological advance rose again sharply in 1949. In March the authorities inaugurated a campaign for closer cooperation between science and industry by announcing public meetings and pledges to improve innovation in Leningrad. By May the campaign had expanded to Moscow and other cities.[287] It may have been launched by Voznesenskii's enemies partly to build a case of economic malfeasance against him and other Zhdanovites who hailed from Leningrad. But in any case, it raised the substantive issues of the scope of military R & D and the quality of innovation in the civilian industrial sectors.

The issue of the scope of military R & D surfaced during the compilation of the Academy's research plan for 1949. The Academy Presidium presented a draft plan to the Council of Ministers in January. The Council approved the draft in principle, but ordered the Academy to make changes in keeping with its recommendations. Ten more weeks were required for the Academy to make the necessary revisions and obtain the Council's final approval.[288] Obviously referring to the Berlin crisis and the formation of NATO in April, the official who described this process remarked that the conflict between socialism and capitalism had recently sharpened as the imperialists sought to ignite a third world war. The Academy's research plan, he said, could not fail to reflect international trends, and the final version had been formulated to take account "of the international situation which took shape at the start of the current year."[289] His words suggested that the Council of Ministers had ordered the Academy to put more effort into military-related projects than the draft plan had called for.[290] The confirmation of the revised plan, incidentally, occurred almost simultaneously with Varga's recantation of his previous views about the moderate intentions of the West toward the USSR.

The shift toward still greater emphasis on military projects coincided with signs of top-level dissatisfaction with the Academy's contribution to technological innovation. On 7 April 1949, the day the Academy

Presidium approved the final research plan, the Council of Ministers ordered the Academy to alter its Charter to create a Scholarly Secretariat within the Presidium. The duties of the Secretariat were to ensure fulfillment of government research targets and to handle personnel selection.[291] Within a week the Presidium met with ministerial representatives to discuss the pace of innovation, heard a report critical of the Academy's role in that process, and set up the Secretariat. At about the same time it removed N. G. Bruevich from the post of Academician-Secretary, where for seven years he had exercised major influence over the direction of Academy research, particularly military-related research.[292] Although the new Secretariat was headed by another Academy member, it evidently gave the central party apparatus greater day-to-day influence over research policy; one member of the Scholarly Secretariat was Iu. A. Zhdanov, head of the Central Committee's Department of Science.[293]

Following these events, the regime took vigorous steps to make Academy personnel more responsive to research targets handed down as special assignments from the Council of Ministers. As the head of the new Scholarly Secretariat observed, some physics, chemistry, and technological institutes had delayed completing projects ordered by the government, and this was "completely impermissible."[294] Aside from a new Academy inspectorate created to monitor plan fulfillment, the corrective measures included an extensive shake-up of research administrators within the Academy. Changes were made in the membership of the Presidium and all eight departmental bureaus, in the Academy's Council for the Study of Productive Forces, in the administrative leadership of at least six institutes, and in the scientific councils of nineteen institutes. The editorial boards of twenty-four scientific journals were also reorganized.[295] There can be little doubt that a shake-up of this magnitude was ordered by the political authorities.

Although the authorities were probably concerned about several varieties of technology, they expressed particular displeasure about calculating and mathematical machines—the forerunners of modern computers. On April 6 the Council of Ministers adopted a decree that condemned the "exceptional lag" in developing and producing such equipment. Because of this lag, it said, the handling of economic data was being mechanized too slowly. Moreover, the shortage of computational equipment had hampered research in several fields essential to the creation of new technology: atomic physics, jet and rocket propulsion, ballistics, electronics, and gas dynamics.[296] The latter complaint may have contributed to Bruevich's removal from his administrative post in the Academy, since he was closely connected with defense projects in these fields and was the sole Presidium member with expertise in the theory of computational equipment.[297] At any rate, the decree instructed the Academy to formulate specifications for new computational machines to be used in research and to set up training courses for the designers of such machines. It also created a new ministerial institute in this sector. Ordering the series production of several new machine models, the edict obligated various economic agencies to introduce calculating equipment into their operations.[298]

The public campaign for better technology evoked franker discussions of the impediments to innovation. S. I. Vavilov, the President of the Academy, remarked that there were a great many cases in which scientific findings were not being applied, and that these cases occurred in "the most diverse areas of science and technology." Scientific institutions and industrial enterprises, he stated, often took opposing, hostile positions on proposed innovations. Departmental barriers not only separated the institutes of the Academy from ministerial institutes, but isolated the organizations in each adminstrative hierarchy from one another. To overcome conflicts between researchers and enterprises it frequently became necessary to resort to the "complicated means of special commissions or special resolutions." Adamantly asserting that such measures were ineffective, Vavilov suggested the creation of more research units within enterprises as a way of bridging the gap between many branch institutes and the factories.[299]

Meanwhile, the press demanded that ministerial officials pay more attention to innovation. *Izvestiia* editorialized against ministries that produced only small runs of new products—presumably because this allowed them to report fulfillment of their targets for new technology—rather than manufacture the large quantities the economy needed.[300] To put teeth in such complaints, the newspaper tried to make an example of the Ministry of Oil Industry. Detailed articles and an editorial accused the Ministry of technological inertia and laid responsibility at the feet of the Minister and his deputies.[301] Strikingly, however, *Izvestiia* was ultimately forced to retract its charges, saying that they had been based on an "incorrect generalization" of "particular facts."[302] The incident shows that the ministries were sometimes strong enough to resist vigorous public pressure for better technological performance. In these circumstances the level of cooperation between researchers and industrialists was unlikely to improve without major administrative reforms.

Stalin and his minions, however, were uninterested in major reforms. Their only significant change in R & D administration—the abolition of Gostekhnika in 1951—was a negative one. Otherwise, the leaders simply increased their demands for more rapid innovation. Faced with mutual complaints between the Academy and the ministries,[303] they heightened their stress on "criticism and self-criticism" as a source of progress in science and technology. Late in 1951 the leaders began to tout this slogan as a way of combating the research monopolies allegedly held by some scientists.[304] They also emphasized that the "unlimited" opportunities for promoting young scientific cadres could speed up scientific development.[305] In all likelihood this accent on upward mobility for young specialists was connected to Stalin's simultaneous preparations for a general political purge to be fueled by the ambitions of junior officials.[306] Even more than in 1938, Stalin and his supporters seemed to be falling back on the politicization of relations among R & D specialists as a means of generating more rapid innovation.[307]

The slogan of criticism and self-criticism, however, could easily produce bad rather than good technological results. It was most likely to produce good results when embodied in institutional competition among rival R & D specialists—the kind of competition that still existed in the arms sectors during Stalin's last years. Moreover, whether conducted between institutions or strictly between individuals, disputes over questions of technology could yield beneficial consequences only where the criteria of success remained primarily technical, not political.[308] It is conceivable that the new emphasis on criticism and self-criticism may have stimulated a marginally greater amount of innovation in some fields.[309] But the slogan was also a powerful club with which ideologists and semicompetent specialists beat down valid intellectual arguments raised by researchers in fields such as chemistry and genetics.[310] On balance, the campaign probably produced many more destructive "innovations" from hacks than real achievements from genuine experts in R & D. However tempting such slogans were, the overall performance of the Soviet system in domestic R & D, particularly in the nonmilitary spheres, could be improved only through institutional change. And serious attempts at institutional change would have to wait until Stalin closed his eyes for the last time.

Conclusion

The debate between traditionalists and nontraditionalists during the 1940s was not simply a dispute between "propagandists" and "realists," although most of the propagandists were on the traditionalist side. It reflected genuine uncertainty within the elite about what to expect, politically and economically, from the capitalist world after the war.

The public victory of the traditionalist outlook had several causes. One was the ingrained feeling of many elite members that the capitalist world would experience the same kind of debilitating economic crisis and would give rise to the same kind of aggressive behavior as in the 1930s. Many persons thought that the Soviet Union, although it had suffered disproportionate damage in the war, had previously demonstrated its superior dynamism and would do so again. Moreover, they feared a Western inclination to capitalize on the Soviet weaknesses caused by the war. In the short term, they felt, the regime should concentrate on internal political consolidation, eschewing policies that might undermine its domestic legitimacy or make it susceptible to pressures from abroad.

The triumph of traditionalist ideas deeply affected the USSR's technological strategy and technological priorities. Although at the close of the war the regime initially showed a strong interest in obtaining Western technology, it also undertook a massive expansion of its own R & D effort and restricted the human contacts necessary for extensive technology transfers from the West. Abetted by Western restrictions on the movement of technology, this xenophobic policy cut off Soviet specialists from virtually all contact with foreign R & D. As for technological priorities, in the mid-1940s the party attempted to meet some elementary civilian needs by enlarging the output of consumer goods faster than producer goods. But this relatively short-lived policy never involved a basic change in the R & D system to upgrade civilian technology relative to military technology. After the war the regime first embraced high expenditures on military research, then renewed the priority of heavy industry, and finally boosted spending on stocks of military equipment.

Looking back at Stalin's last decade, the members of the elite had some causes for satisfaction with their system's technological performance. The USSR had, after all, restored and expanded industrial production with remarkable speed after the war. The rates of industrial growth were extremely high by international standards, reflecting the system's capacity to mobilize large quantities of investment capital and labor. The regime had created an atomic bomb, as well as other types of sophisticated new military technology. On the other hand, many industrial sectors were still comparatively backward, and the popular standard of living was extremely low. The capitalist West had not slumped into a second Great Depression and was growing at a fairly steady pace. Moreover, thanks to the tense international situation, the USSR had very little access to the fruits of the bourgeoning R & D programs in the West. This was the situation that confronted Stalin's successors.

4

The Khrushchev Years, 1953–1964

Socialism and Capitalism: Stirrings of Change

After 1953, official Soviet thought entered a new phase. No longer restrained by their fear of Stalin, the new leaders reevaluated the international competition with capitalism and gradually concluded that the threat from the West was no longer as great as the traditionalist outlook assumed. While some leaders manifestly believed this much more firmly than did others, the overall departure from the traditionalist view on this question was marked. The new leaders also modified the inherited picture of Western economic performance, but less sharply. Although they became more realistic about current Soviet technological deficiencies, they retained the traditionalist conviction that the USSR was rapidly overtaking the West technologically.

The general result was a consensus on the feasibility of surpassing the West, coupled with recurring differences over the proper means of doing so. A number of leaders felt the regime should assign continuing priority to the advancement of heavy industrial and military technology, whereas others felt this was no longer the best formula for winning the economic competition with the West.[1] Disagreements also surfaced over the renewed acquisition of foreign technology. Some officials believed that extensive reliance on Western technology for backward industries was a way to leapfrog the Western powers, but others regarded this as a recipe for permanent lags and diplomatic vulnerability.

Debating the Imperialist Threat

The new leaders felt that Stalin's foreign policy had contributed to an undesirably high level of international tension, and most of them (though possibly not all) supported the decision to reduce the tension by negotiating an end to the Korean War.[2] But they differed over how much further a relaxation of tension could be carried. Malenkov highlighted the "serious successes" recently achieved in lowering international tensions and avoiding a new world war and argued that it would be "a crime against humanity" if these changes gave way to a renewal of hostility.[3] He avoided any direct suggestion that the changed internal dynamics of capitalist regimes had eliminated their warlike impulses, but he offered an analysis that had some of the same policy implications as this nontraditionalist argument. The USSR, he pointed out, had built up its industrial might; it now had the hydrogen bomb, and its military strength was clear. Relying on that basis, the socialist bloc could oblige the capitalist world to follow a policy of peaceful coexistence. Malenkov claimed that all disputes between the United States and the USSR could be solved by negotiation, and he called for the expansion of Soviet-American trade and "real progress" in arms control.[4] A similar argument was advanced by Mikoian, who like Malenkov had inclined toward nontraditionalist views of the international environment in Stalin's last years.[5]

Klement Voroshilov, however, took the traditionalist position that despite the USSR's many economic and diplomatic successes, "we live in encirclement all the same."[6] Khrushchev, too, declared that the country was still in capitalist encirclement, and Kaganovich argued that the failings of capitalism were increasing the West's disposition to resort to war.[7] The controversial nature of these assertions is indicated by the fact that nontraditionalist opponents sought, sometimes successfully, to excise them from the central press.[8] Meanwhile, traditionalist commentaries asserted pointedly that the worsening crisis of capitalism and the growing aggressiveness of American imperialism were phenomena not only of the past but of the present and should be discussed in this light.[9] Traditionalist advocates also condemned as subjectivism the notion that the internal laws of capitalism could somehow be "paralyzed," thereby avoiding the phenomena (such as war) to which those laws gave rise.[10] This was precisely the assumption on which Malenkov had based his contention that the increased power of the USSR had given the imperialist states a new motive to accept peaceful coexistence.

The debate over the accuracy of the traditionalist view of the West was linked to disagreements over policies for domestic research and production. The new leaders may well have agreed on the need to make some adjustments in the harsh economic priorities that Stalin had enforced, but they differed sharply over the scope of such changes.[11]

Malenkov and his associates wished to restrain the growth of the Soviet military budget and emphasized the sufficiency of the military means the country already had at its disposal.[12] They also pushed for a reduction in the priority enjoyed by heavy industry. Malenkov granted that in the 1930s any slackening of the stress on heavy industry would have exposed the USSR to imperialist attack. But since that time, he claimed, the USSR had built up so much heavy industrial capacity that it could now shift its priorities and promote a "steep rise" in light industrial output.[13] Some signs indicated that Malenkov and his supporters were trying to revise the traditionalist interpretation of the "laws of socialism" that helped undergird the top priority of heavy industry, just as they were obliquely questioning the traditionalist conception of the laws of capitalism.[14] Meanwhile, other commentators backed Malenkov's economic preferences by asserting that the low level of manufacturing technology in many branches of light industry could no longer be tolerated. It was necessary to introduce into these sectors advanced manufacturing methods, especially those methods of mass production that had already been applied in heavy industry,[15] and to devote more R & D to the needs of light industry.

Adherents of the traditionalist view of international politics disputed these policy suggestions. Khrushchev, Bulganin, Molotov, and Kaganovich argued that Soviet military power should be increased.[16] Emphasizing that heavy industry was the foundation of military might, Khrushchev maintained that the high priority accorded this sector could not be diminished.[17] Others asserted that proposals for a fundamental change in investment priorities would leave the country helpless before the imperialist world.[18] More specifically, they assailed the Malenkov program on the grounds that it would inhibit the progress of Soviet military research and innovation. Arguing that complacency about military power would be "an irreparable error," Bulganin stated that "the most important thing in military affairs is the uninterrupted perfection of the armed forces. This is especially true of aviation, where technological progress is very rapid. . . . We cannot assume that the imperialists expend enormous material and financial resources on armaments only to frighten us."[19] Dmitrii Shepilov sharpened this case. In the face of capitalist encirclement, a change of priorities to light industry would mean that "we surrender the advantage of forcing forward the development of heavy industry, machine construction, energy, chemical industry, electronics, jet technology, guidance systems, and so forth, to the imperialist world. . . . It is hard to imagine a more antiscientific, rotten theory, which could disarm our people more."[20] Both statements implied that Malenkov's policies would deny the USSR the long-range bombers and missiles which traditionalist leaders evidently felt were necessary to make Soviet nuclear weapons a real deterrent against the United States.[21]

Although at odds over the traditionalist image of imperialist foreign policy, none of the leaders seemed to doubt the traditionalist belief that the USSR could surpass Western technological levels quickly. The radical superiority of Soviet growth rates and the declining technological dynamism of capitalism remained fixtures of official discourse, and surpassing the West continued to be touted as the country's central economic task.[22] Yet the issue of overtaking Western technology did play a tactical role in this phase of the political infighting among Stalin's successors. Early in 1955 *Pravda* spoke out against administrators who "continue for many years to be lost in admiration for obsolescent technical achievements, undismayed by the fact that these achievements have long been surpassed." Such administrators showed little interest in foreign technological accomplishments, and their complacency "conceals great dangers and does harm to the Soviet state."[23] At a Central Committee plenum six months later, Bulganin, echoing this charge, censured "certain comrades" who thought that because Soviet industry was developing rapidly and had met the demands of World War II, "we will not have to make any special effort in the future in order to ensure our retaining the leading place in world technical progress. The error and harmfulness of such an argument are obvious . . . We cannot and do not have the right to forget that technology in capitalist countries is not standing still."[24]

Bulganin and the other politicians who struck this sober note must have known that the growth of industrial labor productivity had slowed somewhat from the exceptionally high levels of the preceding years, and they apparently concluded that the extreme Stalinist emphasis on the superiority of Soviet technology had contributed to this decline by fostering complacency within the scientific and industrial establishments.[25] As we shall see presently, such technological inertia was a real problem. But Bulganin and others seized on this issue largely because it constituted a useful weapon in the struggle against Malenkov and his associates. Timed to coincide with the final anti-Malenkov offensive, the campaign against technological laggards was given added bite by equating the prospect of further technological progress with priority for heavy industry—a priority which Malenkov had disputed—

and by launching a blistering attack on economists who supported his views.[26] Bulganin was seeking to brand Malenkov as one of the "certain comrades" who favored the dangerous idea that the regime could relax its drive to surpass the West technologically. For two years Malenkov, as Chairman of the Council of Ministers, had exercised direct supervision over the industrial ministries. By concentrating political fire on the ministries' technological shortcomings, the campaign strengthened the case against Malenkov and offered a pretext for weeding out many of his allies from the ministerial apparatus.[27] Moreover, the campaign enjoyed the support of First Secretary Khrushchev and other party officials because it allowed the party bureaucracy to claim greater control over day-to-day industrial administration.[28] From Khrushchev's standpoint, the attack was an important step in an ongoing struggle to build up the party's administrative leverage over the economy.

This political coalition defeated the bid by Malenkov and his associates to make basic changes in the traditionalist view of international politics and in traditionalist priorities. As the opposition mounted, Malenkov gave ground. Granting that one should not overestimate the relaxation of international tension, he called for increased Soviet military power and tempered his commitment to consumer goods. Nevertheless, the differences over these questions came to a head in early 1955.[29] In February Malenkov was compelled to resign as Chairman of the Council of Ministers, although he retained his seat on the party Presidium.[30] The exponents of the traditionalist outlook thus ensured continued high priority for military technology and heavy industry. That they championed this outlook against Malenkov partly for opportunistic reasons is indicated by the fact that some of them claimed to discern a relaxation of international tension soon after he resigned, and by the regime's substantial reduction of military manpower in 1955 and 1956.[31] But these belated concessions did not infringe the traditionalist priority of heavy industry and military weaponry that Malenkov's critics had fought to protect. Soviet military expenditures remained on a high level, and according to Western sources, the Soviet missile-development program was drastically speeded up in 1955–1956.[32]

The Selective Modification of the Traditionalist Outlook

Once Malenkov's political eclipse seemed assured, some of his opponents began to embrace certain of his ideas. Khrushchev was a key figure in this process of doctrinal evolution. Under his prodding, the leaders reluctantly revised their image of the West's intentions toward the USSR, and they began a reassessment of Western economic performance. This shift of outlook did not affect Soviet technological priorities immediately. In particular, given the example of Malenkov's recent defeat, the preeminent importance of heavy industry and military power went uncontested for several years. But the change in official thought did gradually impinge on Soviet decisions concerning Western technology, heavy industry, and military weaponry.

The 20th Party Congress in 1956 displayed both change and continuity in the official outlook. There Khrushchev adopted a new attitude toward the West. The Leninist doctrine that imperialist wars are inevitable, he proclaimed, had been absolutely correct in an earlier era. But since 1945 world politics had changed fundamentally, and international war was no longer a "fatal inevitability." This new circumstance, he explained in an analysis taken directly from Malenkov, was due primarily to the growing power of the socialist camp, which served to counterbalance the warlike tendencies that continued to be rooted in the economies of capitalist systems.[33] Khrushchev also redefined the notion of peaceful coexistence between the socialist and capitalist worlds. In place of the Stalinist idea of coexistence as a factual description of a current military standoff created by impersonal historical forces, he interpreted the slogan as a normative Soviet policy aimed actively at increasing the scope of cooperation with the West.[34] Prominent among the foreign-policy goals he listed were the relaxation of international tensions and the extensive expansion of trade and scientific relations with Western Europe and the United States.

At the 20th Congress, however, this attitude was not allowed to affect the official commitment to the top priority of heavy industry and increased military power.[35] One reason is that not all of Malenkov's old opponents believed the international scene had suddenly changed as much as Khrushchev now contended. Molotov, in particular, registered a strong reservation. He accepted that the international forces favoring peace had increased, thereby creating a chance of preventing war. But under the guise of elaborating on Khrushchev's discussion, he flatly restated Lenin's idea that imperialism invariably gives rise to war. Far from dismissing this notion as outdated, Molotov implied that war would cease to be inevitable only when imperialism had been destroyed by revolution, and he warned that "we should not minimize the danger of war nor . . . surrender to complacency as if it were possible to convince the imperialists with nice speeches and pacific plans."

Whereas Khrushchev had stated that the antiwar forces within imperialist regimes had grown enormously stronger, Molotov remarked that the "decisive role" in such regimes still belonged to social groups willing to go to war to preserve their privileged position.[36]

Subsequent commentaries showed that these variations signified political disagreement. A gloss on the decisions of the Congress remarked critically that "some propagandists" were having trouble understanding how the denial of the inevitability of war was compatible with the Leninist doctrine of imperialism.[37] Adopting an opposing attitude, a military writer complained that "some of our propagandists give a one-sided interpretation of the question of wars in the contemporary epoch." These propagandists were paying too much attention to the possibility of preventing wars and were treating the idea that new wars might break out as "insignificant." However: "What is basic for Soviet military ideology is . . . that even now the economic basis for war exists and that the imperialists will seek to unleash war. Our chief attention should be devoted to that aspect of the question."[38] Obviously, the movement away from the siege mentality of traditionalist thought was difficult for some persons to accept.

The 20th Congress's review of the traditionalist picture of the Western economy likewise combined change and continuity. Mikoian, a member of the party Presidium, stated that the capitalist economy was approaching a new crisis, but then went on to repudiate the way that Stalin had treated this issue in *Economic Problems of Socialism in the USSR*. Stalin's analysis, Mikoian asserted, failed to account for the "complex and contradictory phenomena of modern capitalism," particularly the growth of production and technological progress in the West since World War II. The economic changes occurring in the capitalist world had considerable current importance. But partly because the Institute of World Economy and World Politics had been liquidated under Stalin, Soviet specialists were giving these tendencies only superficial attention. It was necessary, concluded Mikoian, to investigate Western economic trends more fully and objectively than in the past.[39] Khrushchev struck a similar note.[40]

Some action had already been taken along these intellectual lines, and more was forthcoming. In the fall of 1955 a party-government decree had set up a new Gosplan institute to analyze the economic results of the competition between socialism and capitalism, as well as the economic ties between the two camps, and to make recommendations on how to treat these matters in Soviet economic plans.[41]

About the same time, the Academy had moved to create an Institute of the Economy of Contemporary Capitalism, but this effort had evidently been blocked.[42] The statements of Mikoian and Khrushchev at the 20th Party Congress helped overcome such obstacles. In the spring of 1956 the Academy Presidium authorized the creation of an Institute of the World Economy and International Relations (IMEMO), which resumed research on many aspects of Western economic and political development that had been monitored at Varga's institute before its abolition.[43] The creation of the new unit indicated that some leaders wanted more objective information about Western economic trends than they had been receiving. In view of the role that Varga's institution had played in past policy debates, together with the policy preferences which Mikoian had long exhibited, it appears that an added motive for establishing the new institute was to generate scholarly justifications for a more moderate Soviet foreign policy.[44]

Although a noteworthy departure from the Stalinist stereotype of Western economic stagnation, these steps did not signify a loss of faith in the USSR's ability to surpass the West technologically. A combination of declining Soviet growth rates and more objective reporting did produce strong doubts on this score in the late 1960s and 1970s, but in the mid-1950s such fears were still unknown. The 20th Party Congress reiterated the regime's long-standing commitment to outstrip the most advanced capitalist countries in per-capita production "in the shortest historical period," and writers on this theme cited with evident satisfaction the worried comments of many Western observers about the rapid pace of Soviet economic development.[45] Soviet commentators scoffed at the view of some Western economists that Soviet growth rates would gradually decline, predicting instead that capitalist growth rates would soon drop. Khrushchev told the 20th Congress that many of the temporary factors which had stimulated Western postwar economic expansion were nearly exhausted, and other spokesmen vigorously backed this assertion.[46] Even Varga, although he doubted the prospect of another Western economic collapse like that of 1929–33, predicted that the next decade in the West would resemble the slump-ridden 1930s more than the boom decade following World War II.[47] Soviet ruling circles still believed firmly that their system's superior technological dynamism would have a decisive impact on trends in world politics.[48]

Thanks partly to disagreements about the international situation, however, elite members disagreed about Khrushchev's emerging tactics for outdistancing the Western industrial economies. The resulting debate, insofar as it centered on domestic technological priorities, was basically a replay of the struggle between Malenkov and his opponents a few years earlier. But added to the question of domestic priorities was another matter that had not figured in the earlier dispute: the value of Western technology transfers as a means of outstripping Western economic achievements.

The International Situation and Domestic Technological Priorities

Technological priorities were part of a dramatic confrontation in mid-1957, when Molotov, Kaganovich, Bulganin, Malenkov, and others joined in an effort to depose Khrushchev. The confrontation was sparked by conflicts over bureaucratic prerogatives, economic policy, and foreign relations. Early in 1957 Khrushchev launched a drive for a major reorganization of industry that promised to boost the influence of the party apparatus and undercut the political base of several of his Presidium rivals in the state bureaucracy. The measure was a powerful stimulus to the formation of the diverse coalition of leaders Khrushchev later branded as the "antiparty group."[49] Having launched his reform drive, Khrushchev not only began to think boldly about overtaking the United States in per-capita output, but proposed extraordinarily demanding timetables for overtaking U.S. per-capita meat production—a yardstick not previously used in the party's comparisons of Soviet and Western economic progress. Khrushchev's Presidium rivals, reportedly even Malenkov, opposed this gambit on the grounds that it would siphon off resources from heavy industry, which they claimed was the key element of the technological race with the West.[50] They probably accused Khrushchev of underestimating the ease with which the USSR could outstrip the United States economically, and on this point they must have had many sympathizers in the party. An editorial that attacked the antiparty group immediately after its defeat in June 1957 remarked that the moment would come when the Soviet Union would produce more total output per capita than the most developed capitalist country. It would come, said the editorial, "sooner than some bourgeois politicians and economists, and also some people in our country, suppose."[51] Once again, although with some of the principals in new roles, the leadership was debating the domestic priorities by which it could surpass Western technological achievements.

The clash within the Presidium also highlighted the differences between Khrushchev and his opponents over foreign policy. Molotov reportedly denounced Khrushchev's agricultural plans because they would weaken heavy industry, thereby undermining the USSR's military might and international position.[52] After being expelled from the Presidium, the leading members of the antiparty group were collectively condemned for resisting the policy of peaceful coexistence—an uncharacteristic accusation, since in all previous succession struggles the losers had been charged with being soft on imperialism rather than with obstructing efforts to coexist with it.[53] Molotov was said to have been especially adamant in opposing the principle of coexistence and disputing the idea that war was no longer inevitable. He had opposed diplomatic measures aimed at reducing East-West tension, including high-level meetings between Soviet and foreign leaders.[54] Khrushchev and his allies were surely distorting the record when they extended such charges to Malenkov. Malenkov had previously manifested a strong predilection for nontraditionalist ideas about the West. Oblique evidence indicates that he still did not agree with other members of the antiparty group on some foreign policy questions, and his participation in the conspiracy probably helped persuade the military to throw its support behind Khrushchev as a more reliable defender of the miltary's interests and outlook.[55] But the charge of opposing a more moderate policy toward the West matches the long-standing political inclinations of Molotov and Kaganovich, and to a lesser extent the inclinations of Bulganin.

Khrushchev's victory over the antiparty group strengthened his tendency to think in broad terms about the Soviet race for economic and technological primacy vis-à-vis the West. So did other developments. In October 1957 the USSR launched Sputnik, the first of a series of earth satellites. The accomplishment bolstered native pride in Soviet technology and caused a wave of consternation in the West. The next month Khrushchev delivered a very upbeat report on Soviet accomplishments. The USSR had long since surpassed the capitalist states of Europe in national economic output. Within 15 years it would match current per-capita American output; and even though U.S. production might grow in the interval, the USSR would outstrip the American economy "in a very short time." Khrushchev also asserted that the Sputnik launching was "of great importance" because it showed that the USSR was winning the competition with the capitalist world.[56] At the same time, he tried to shape Western reactions to the event to suit his own policy preferences. Appealing for a shift of Soviet-Western relations to the path of peaceful coexistence and disarmament, he sought to prevent the American fears aroused by Sputnik from driving up U.S. military spending.[57] By touting the economic and nuclear implications of Soviet space achievements, Khrushchev aimed to exact concessions from the West, particularly diplomatic recognition of the German Democratic Republic. But he also hoped to curb the traditionalist priority of heavy industry and military spending at home, and in order to do so he had to avoid provoking a rapid Western military buildup.

These hopes were reflected in an article that L. F. Ilichev, one of Khrushchev's protégés, published early in 1958, a few months after the first Sputnik launch. Soviet satellites, argued Ilichev, were not merely scientific accomplishments. Rather, they had "great social and political significance." The USSR's demonstration of its ICBMs was likely to cause "a radical reappraisal of many ossified conceptions" in both diplomacy and military strategy, said Ilichev. Although he concentrated on anticipated changes in Western military strategy, he implicitly left open the possibility that Soviet strategy might be affected as well.[58] Like other publications at the time, the article contained just a hint that the advent of Soviet ICBMs might remove the need to develop the other branches of the Soviet military establishment as rapidly as in the past.[59] Recent Soviet scientific feats had strengthened the forces of peace in the capitalist world, making their position "more stable than ever before," and had compelled the U.S. government to begin rethinking its hard-line policies toward the USSR. The West would reappraise its stance toward the USSR because it now knew that any attack on the country would provoke a devastating retaliatory strike by Soviet rockets. Thus, concluded Ilichev, *"The world has entered a new stage of coexistence."*[60]

This analysis resembled the belief, advanced unsuccessfully by Malenkov a few years earlier, that Soviet progress in heavy industry and nuclear weaponry had imparted greater stability to East-West military relations. No doubt members of the elite wondered privately whether it foreshadowed the same policy of decelerated heavy industrial and military development that Malenkov had tried to establish. The question seemed germane because Khrushchev, once he had obtained military backing in his showdown with the antiparty group, cashiered the Minister of Defense and clamped down on the military hierarchy.[61] About the same time, many commentators, citing Sputnik, began to use Soviet achievements in science and technology, rather than the current operational stock of military rockets and other weapons, to argue that the world correlation of forces was changing in favor of the socialist camp.[62] It was also in early 1958 that Khrushchev first publicly toyed with the idea of reducing Soviet conventional forces by gradually converting the army into a territorial militia.[63] Although he did not follow up on this notion immediately, it was not the sort of idea an experienced politician would express without forethought, and it may have been part of the "radical reappraisal" of ossified military conceptions that his protégé almost simultaneously suggested might result from recent Soviet space feats. At any rate, two months after broaching the idea of reducing the army, Khrushchev mounted a major campaign to upgrade chemical production vis-à-vis the heavy industrial sectors, principally metallurgy and machine building, that had customarily received top priority. This campaign constituted an effort to give the appearance of maintaining the primacy of heavy industry while actually shifting economic effort to a sector that would favor agriculture and consumer goods.[64]

Some elite members refused to accept Khrushchev's viewpoint, apparently doubting that recent Soviet space successes had the sort of strategic significance Khrushchev claimed. In a revealing statement, one Khrushchev supporter repeated that Soviet space advances had caused a qualitative change in international relations and the problem of Western military security, thereby forcing the more farsighted Westerners to accept the necessity of peaceful coexistence. But other people, he said, had overlooked this change: "Not all may have yet grasped the influence which man-made earth satellites will exert on world relations. Western commentators—and not they alone—seem to ignore the fact that the importance of Soviet scientific and technological advances is not confined to the purely scientific sphere. . . . *They are also of major social significance.*"[65] Since the views censured did not match the positions of the USSR's main Communist critics abroad, the non-Western skeptics the writer was referring to probably belonged to Soviet ruling circles.[66] We do not know who they were, but it is a fair guess that they were among the persons who remained committed to the priority development of heavy industry and military power. After the defeat of the antiparty group, Presidium members such as Frol Kozlov continued to oppose any softening of the country's foreign policy and the emphasis on heavy industry, and Khrushchev still found it necessary to attack persons who denied that priorities should be shifted from steel and machine building to the chemical sector.[67]

Debating the Role of Western Technology

The leaders' latent differences over the international political environment had implications not only for domestic technological priorities
but for foreign economic policy. Early in 1958 Khrushchev and Mikoian
explicitly renounced the doctrine of capitalist encirclement. Mikoian
urged "all to understand that the Soviet Union long ago emerged from
'capitalist encirclement,' in both the geographic and political senses of
the concept."[68] Agreeing that new circumstances had made the doctrine
misleading, Khrushchev remarked that it was no longer clear "who
encircles whom, the capitalist countries the socialist states, or vice
versa."[69] This was the first time since World War II that any leader
except Stalin had directly questioned the reality of capitalist encirclement, and the two politicians manifestly intended the change in doctrine
to facilitate an expansion of economic relations with the West.[70]

Having downgraded the threat from imperialism, Khrushchev and
Mikoian argued that the regime should more actively seek Western
technology. In March 1958, timing their statements to coincide with a
major Western review of the embargo on strategic exports to Communist
countries, Mikoian and Khrushchev called on the United States to end
its trade restrictions and advocated a broad expansion of Soviet-American commerce. Seeking to justify this initiative to both foreign and
domestic audiences, Mikoian contended that though there were two
world markets, the socialist and the capitalist, there was a single world
economy. The USSR should take a larger part in this economy, and
Western economic interests could be counted on to overcome the resistance Western governments might raise to expanded East-West ties.[71]
Khrushchev made a similar appeal. The Western embargo, he added,
had failed to slow Soviet technological advance, and an expansion of
trade would soften the current American economic recession.[72]

This initiative dovetailed with the chemicalization drive Khrushchev
kicked off two months later. As part of the drive, he called for a drastic
upgrading of chemical R & D. But Khrushchev and Mikoian were unwilling to wait until indigenous R & D could provide the technology
needed to overtake the West in a sector they deemed critical. Because
the USSR required a large quantity of new chemical equipment in short
order, Khrushchev told the Central Committee, it should seek to purchase complete industrial plants from the United States, Great Britain,
and West Germany, and it should engage Western specialists as consultants. This would save a great deal of time and money on design.[73]

What Khrushchev did not say was that purchasing such equipment on credit would help circumvent the resistance to his chemicalization plan by reducing the need to divert capital investments from the domestic metallurgical and machine-building industries. A month later he wrote President Eisenhower a letter that proposed expanded Soviet-American trade, along with U.S. government-backed credits, and that focused on obtaining technical assistance for the chemical industry.[74]

Khrushchev's proposal received enthusiastic support from some other elite members. Mikoian, predictably, gave it a vigorous endorsement.[75] Less prominent figures explained that expanding technological ties with the West would reduce international duplication of R & D and speed the new scientific-technological revolution without seriously alleviating the internal weaknesses of capitalism. Moreover, they said, the capitalist world was now incapable of mounting an effective economic blockade against the socialist countries.[76] The latter assertion was probably intended not only to whet the appetites of Western corporations fearful of losing out to each other in new international markets, but to reassure persons inside the USSR who were asking anxiously what kind of diplomatic leverage expanded trade and assistance would give Western governments.

This anxiety must have been one reason why other elite members objected to Khrushchev's proposal. Two months earlier, Mikoian had condemned certain Soviet "economists" for the "crude error" of regarding the socialist and capitalist world markets as absolutely separate. By rejecting the development of economic ties between the two blocs, he said, such dogmatists contradicted the Leninist idea of peaceful coexistence.[77] At the May unveiling of the chemicalization program, Khrushchev similarly assailed "some comrades" who had challenged his notion of buying Western equipment and technical aid on the grounds that such transactions would bolster capitalist regimes.[78] The main targets of these rebukes were not rank-and-file officials, but members of the party Presidium.

No doubt this is why less powerful figures dared to challenge the line Khrushchev had just laid down. A scant five days after his speech, a foreign-affairs journal published an indirect attack on the proposal to expand technological ties with the West. In a wide-ranging exposition of the Marxist theory of economic base and political superstructure, the writer warned that a misunderstanding of the Marxist position on the relation between the economy and foreign policy was "one of the sources of various revisionist and reformist conceptions in the area of international relations." This criticism obviously reflected disagreement about the internal drives shaping imperialist behavior, since the writer cautioned against underestimating "the expansionist, aggressive motives of the foreign policy of monopoly capital." The criticism was also aimed at the overestimation of domestic economic considerations in the making of Soviet foreign policy. Unlike most Soviet writing on the theme of economics and foreign relations, the discussion blurred the distinction between capitalism and socialism. The article's criticism was cast in very general terms, giving the strong impression that it was directed at misinterpretations of the link between the economy and foreign policy in socialist societies as well. Foreign policy, the author asserted broadly, was not simply a product of internal policy; rather, each influenced the other. He then quoted at length Marx's attacks on the reprehensible acts of various 19th-century Western powers, some of them the same countries with which Khrushchev had just recommended that the USSR expand trade relations.[79] The point was hard to miss. Foreign-policy revisionism, the author was saying, could be found very close to home.

These disagreements carried over into the 21st Party Congress early in 1959. Describing the USSR's political position vis-à-vis the West, Khrushchev struck a very confident note. The Soviet Union was stronger internally and externally than ever before, he declared. Capitalist encirclement was a thing of the past, the USSR had the means to repel any enemy, and the doctrine of the noninevitability of war was even more relevant than when it had first been promulgated in 1956.[80] Moreover, future Soviet economic progress would bring a "decisive preponderance" in the international correlation of forces in favor of the USSR, further deterring any aggressive intentions some imperialists might still harbor. Effusively praising Soviet space achievements, Khrushchev remarked that Sputnik had demonstrated the USSR's leading role in world technological progress, and he stated that the country had now entered the decisive phase of its economic race with capitalism. Soviet industry would continue to grow more than four times as rapidly as American industry, he predicted, and the USSR would surpass the United States in per-capita output within twelve years, "or perhaps sooner." This was a much more demanding timetable than Khrushchev had proposed a year and a half earlier.[81]

At the Congress Khrushchev again advocated the fuller use of Western technology as one means of achieving this rapid economic triumph. Highlighting the possibilities for a large expansion of trade and cultural ties, particularly with the United States, he observed that the Seven-Year Plan (1959–65) offered the opportunity to double Soviet foreign trade. He also voiced approval of the trip Mikoian had recently made to the United States to drum up American support for more trade, adding that domestic American opposition to the trip had failed completely.[82] Mikoian used his own time on the rostrum to treat the issue of Soviet-American trade in greater detail. Reminding the audience of Khrushchev's 1958 letter to Eisenhower, he expressed regret that the U.S. State Department had done nothing to follow up on Eisenhower's response to the letter. It was impossible, said Mikoian, to barter American goods for Soviet concessions over Berlin or the Far East. But the USSR was eager to expand American trade without diplomatic strings.[83] Obviously Khrushchev and Mikoian were still seeking to obtain greater access to American technology, particularly chemical technology, by recalling Khrushchev's proposal to Eisenhower some eight months before.

Khrushchev's timetable for outdistancing the U.S. economy encountered reservations at the Congress. Presidium member O. V. Kuusinen, observing that "someone" might harbor doubts about the superior technological dynamism of the USSR, answered by itemizing the economic weaknesses of American capitalism.[84] But this did not allay the misgivings of other Presidium members. One member, A. B. Aristov, endorsed the idea that the Soviet Union would soon overtake the United States economically. But he also cautioned that America had enjoyed "major successes" in science and technology, and he warned that the USSR could not resemble a frantic jockey who miscalculated the strength of his horse and, after overtaking his oppponent at a turn, failed to last to the end of the race. Aristov feared that Khrushchev's eagerness to surpass the United States in total output per capita would reduce the prospects for surpassing America in heavy industrial output—a change Aristov implicitly opposed.[85]

Aristov saved his strongest objections, however, for the idea of increasing Soviet reliance on Western technology. He addressed the Congress after Khrushchev had advocated a major expansion of Soviet-American trade, and he probably knew that Mikoian, whose speech was scheduled to occur a few hours after his own, intended to spotlight Khrushchev's 1958 bid for American technical assistance. Aristov remarked pointedly that "some of our comrades" thought it necessary to proceed exclusively by borrowing from foreign technical thought. He paid lip service to the idea that such opportunities should not be passed up in industrial branches where the USSR still lagged. But he emphasized that this "is not the main path in the struggle for new technology, for the progress of our native science. If we follow this path, then we will inevitably only . . . doom ourselves to lag behind capitalist technology." To make its great breakthroughs in space and aircraft technology the USSR had relied primarily on its own scientific creativity and had followed its own independent path, Aristov argued. The example of Soviet metallurgy showed that the country had passed beyond the stage of wholesale reliance on foreign technology. Although the USSR had copied its first blast furnaces from American designs and built them with American technical assistance, it had now surpassed the United States in this kind of technology. Moreover, he said, Soviet designers in the machine-building industries, who knew foreign technology well, rejected the idea of simply copying it.[86]

There can be little doubt that this message was addressed to Khrushchev and Mikoian and that other members of the Presidium shared Aristov's sentiment. Aristov's public defiance of the First Secretary was transparent, and political prudence would dictate that so overt a challenge be mounted only when other powerful figures would help the critic avoid retribution.[87]

The Persisting Expectation of Surpassing the West

Although sharp, these debates centered on the means of surpassing the West, not on the feasibility of doing so. On a general level, Khrushchev and other elite members still agreed that the traditionalist economic goal of outstripping the West was entirely realistic, and the new Party Program adopted in 1961 reflected their underlying optimism. The program proclaimed that a new scientific-technological revolution was under way and boldly reaffirmed the belief in the superior technological dynamism of the Soviet system. Working from this premise, it forecast that the USSR would surpass the United States in per-capita production within a decade. Moreover, it said, the rapid development of Soviet productive forces would allow the country to build the foundations of full communism by 1980. Internationally, the Soviet economic triumph over the United States would be a central element in a decisive worldwide political shift in favor of the socialist camp. The economic achievement would strengthen Soviet influence in the noncommunist world and neutralize the aggressive impulses of imperialist circles.[88]

Sustained by this confident vision, the political leadership sponsored a further expansion of comparative East-West economic studies in the early 1960s. In 1962, shortly after adoption of the Party Program, the government established an Academy-affiliated research council to co-ordinate such studies on a nationwide basis.[89] The authorities regarded East-West comparisons as a useful way to substantiate the official claim that the socialist bloc was rapidly overtaking the industrial West in productive capacity. They also viewed economic comparisons as an aid in adjusting Soviet internal technological priorities. The writings of the comparative economists highlighted the structural differences between the Soviet and most Western economies—particularly the much greater salience of the chemical sector in the West—and Khrushchev seized on these writings to support his argument that the USSR could speed its growth if it shifted investment toward newer, more dynamic industrial branches.[90] The studies, however, had no visible impact on the regime's general expectations of overtaking the West. By the mid-1960s, when Soviet growth rates were clearly down from the very fast tempo of the 1950s, a few economists had begun to raise the possibility that Western growth rates were speeding up.[91] But not all economists agreed with the new projections of Western trends, and the ideas of the more apprehensive economists do not seem to have affected the thinking of a substantial segment of the political elite until after Khrushchev's removal. Most members of the elite continued to expect that the USSR would outstrip the West technologically, even if somewhat later than Khrushchev had forecast. It was only after his fall, as the Soviet advantage in growth rates continued to diminish and as non-traditionalist ideas about Western economic performance gained a wider hearing, that this general assumption became the object of widespread controversy within the elite.

Further Debates over Technological Policy

Meanwhile, however, disagreements over the proper means of sur-passing the West continued. Visiting the United States in the fall of 1959, Khrushchev made a dramatic appeal for an East-West accord on arms limitations. No doubt he hoped to reap diplomatic benefits from the proposal whether or not it proved acceptable to the West.[92] But he obviously intended it as a prelude to a real curtailment of some Soviet military capabilities. In January 1960, without waiting for signs of West-ern willingness to match Soviet moves, he announced a drastic change in the USSR's military establishment. Consistent with his recent assessments of imperialism, he proposed that the country sharply cut its military manpower and rely primarily on ICBMs, rather than on the existing combination of conventional and strategic weapons, for defense. The USSR, he argued, was more fully safeguarded against foreign threats than ever before, and the American people now understood that Soviet ICBMs had made them as vulnerable militarily as any other country. The opponents of "tranquility and peace" in the United States were rapidly losing influence. Thus, he asserted, a reduction in Soviet military forces, or even an increase in Western military strength, would not encourage imperialist aggression against the USSR. Alluding to resistance within the Soviet elite to his views, he remarked that while "some Soviet citizens" might entertain such fears, they were groundless.[93] The current situation, said Khrushchev—probably in response to persons who were recalling his own past attacks on Malenkov's "New Course"—was fundamentally different from the situation in 1952–53.[94]

Khrushchev's demarche resulted from his optimism about the receding imperialist threat rather than from doubts about Soviet economic dynamism; but this merely served to make the economic benefits look more appealing. Having adopted a sanguine view of Western intentions, Khrushchev felt the country could safely put more effort into surpassing the West in nonmilitary output. As he presented it, the new policy was made possible by Soviet achievements in developing the economy and rocket technology. The USSR already possessed greater technological dynamism than the Western powers, and if they refused to cut their own unproductive military spending, this would simply allow the Soviet Union to outstrip them more rapidly in civilian production and consumption.[95] Khrushchev clearly expected the new policy to permit a shift of investment and R & D priorities toward the needs of civilian industry, since he referred to the prospect of large numbers of people returning from the military to civilian enterprises and scientific institutes.[96] Presumably at his initiative, the speech was quickly followed by an unusual press campaign that highlighted the discrepancies between the technological levels of Soviet consumer goods and the products of heavy industry, including military goods.[97]

Although Khrushchev's Presidium associates paid lip service to his military program, a few plainly entertained the doubts he had vaguely attributed to "some Soviet citizens." Suslov remarked that while the cutback had been introduced *"through a unilateral procedure . . . we are* by no means advocates of 'unilateralism.' " By warning against complacency and calling on "any" Western power to make reciprocal cuts, he signaled his disagreement with Khrushchev's argument that Soviet force reductions could be carried through no matter how the imperialists changed their military posture. Suslov also remarked that Western monopolists had not ceased their efforts to promote the arms race, and both he and Kozlov avoided endorsing Khrushchev's recent Camp David meeting with Eisenhower as a contribution to the lessening of international tensions.[98] Soon afterward Mikoian, a foreign-policy ally of Khrushchev who in all likelihood backed the cutback, came under sustained political and bureaucratic attack. Meanwhile two top military officials were emboldened to withhold their endorsement of the proposal—an act for which they were finally sacked.[99] It was also at this moment that Molotov reportedly tried to publish a rebuttal of the views on international politics expressed by Khrushchev at the 20th Party Congress.[100] Molotov no longer held a position of power, and the article never appeared. But he probably acted because he knew that highly placed party members were questioning Khrushchev's desire to base choices about military technology on a sanguine view of the international environment.

While Khrushchev managed to fend off such opposition in the short term, he did not succeed in defeating it. He beat back efforts to rescind the military cutback after the downing of an American U-2 spy plane in May 1960 worsened U.S.–Soviet relations and weakened his political position.[101] But in mid-1961 a new confrontation with the West over Berlin, where the USSR was again pressing the West to accept the Soviet interpretation of the Central European status quo, led to suspension of the manpower reduction and an increase in the military budget. Khrushchev implied that he wished to resume the cuts at some later point by declaring the suspension "temporary"; but it was only late in 1963, after the Cuban Missile Crisis and after a disastrous agricultural shortfall, that he announced new reductions in military spending and manpower.[102] The announcement coincided with a resumption of his political offensive to upgrade the chemical industry, agriculture, and consumer goods at the expense of heavy industrial branches such as metallurgy and machine building. It was clear that domestic opponents had delayed the full implementation of this policy, and that they were still attempting to do so.[103]

As Khrushchev resumed his drive to alter technological priorities, technology transfers from the West again sparked disagreement within the leadership. This time, however, the dispute evidently centered not on the usefulness of Western technology per se, but on whether the need for it warranted diplomatic concessions to the West. Judging by Soviet actions, Khrushchev had won the earlier argument about the economic utility of Western technology. After Soviet wooing had helped generate a significant relaxation of the Western trade embargo in 1958, the USSR had purchased a sizable number of complete industrial plants, particularly chemical plants, along with technical assistance for their construction. It had also begun to receive some Western credits, including government-backed credits from Great Britain, for this purpose.[104] These were important steps. But in Khrushchev's eyes, they fell short of Soviet needs. The amount of credit available for technology transfers to the USSR depended heavily on the policies of the Western powers; and the American and West German governments, still sharply at odds with the USSR over Berlin, the German question, and other issues, opposed extending long-term financing.[105] To make matters worse, a disastrous harvest compelled the Soviet regime to spend a great deal of its hard currency to import Western grain in 1963 and 1964, thereby diverting funds that had apparently been earmarked for the purchase of an unusually large quantity of Western capital goods.[106] At the December 1963 Central Committee plenum where he announced his new push to upgrade the chemical industry, Khrushchev indicated that he had faced resistance within the party over the decision to make the grain purchases.[107] He had won the argument over those purchases, but the price of his victory was a reduced Soviet capacity to import industrial technology without Western credits.

From Khrushchev's standpoint, however, recent events only heightened the need for Western chemical technology. He viewed Western know-how as an essential part of his effort to increase the economy's dynamism by enlarging the relative weight of newer sectors like chemicals that offered a higher economic return. No less important, he regarded the expansion of Soviet chemical output as a way to protect Soviet agriculture from the vagaries of nature and ensure reliable agricultural yields in the future. At the December 1963 plenum he signaled his strong interest in Western technology to upgrade the chemical industry. Aware of possible diplomatic pitfalls, he firmly vowed that the USSR, able to fulfill its economic plans without foreign help, would never accede to demands from West Germany and other countries that it make political concessions in exchange for wheat and chemical equipment.[108] Nevetheless, Khrushchev also stated that the USSR advocated a broadening of business ties with the capitalist countries. It would "gladly give orders to the firms of these countries for a complete set of chemical plants and for a series of other enterprises," so long as Western credits were available. He added in his concluding speech to the plenum that the USSR would find "quite a number" of Western businessmen willing to deal with it on acceptable terms.[109]

Despite Khrushchev's adamant rejection of possible diplomatic concessions, his increasing desire for Western technology began to influence Soviet foreign policy. At the December plenum the Central Committee had given a weaker commitment to the chemicalization program than Khrushchev wanted, and by February 1964 it was evident that the military and heavy industrial lobbies were resisting implementation of the targets the Central Committee had authorized.[110] This pressure on his economic program was one of the principal factors that prompted Khrushchev to explore a rapprochement with West Germany. In mid-March the Soviet ambassador to the Federal Republic met with Ludwig Erhard, who had recently succeeded Adenauer as Chancellor, to discuss Soviet-German relations, and soon afterward the USSR began to inquire privately into the possibility of a meeting between Erhard and Khrushchev.[111] At the end of May, Erhard told an interviewer who asked about the possibility of "a special German economic offer to the Soviet Union" that West Germany was ready to sign a new trade treaty and would accept "sacrifices" if "economic means" could improve conditions in East Germany or facilitate steps toward German reunification.[112] No doubt more generous financing for Soviet–West German trade was one sacrifice he had in mind. Meanwhile signs of friction between the USSR and the German Democratic Republic surfaced in April, about the time of the Soviet soundings concerning a summit meeting with Erhard.[113] In June the USSR and East Germany signed a treaty of friendship which satisfied some of East Germany's diplomatic needs, but which nevertheless softened the support the Soviet Union had previously given East Germany on the German question.[114] Nor did Khrushchev hasten to ratify this treaty. Instead he shelved it while Aléxei Adzhubei, his son-in-law, traveled to West Germany, appealed for better Soviet–West German understanding, and agreed tentatively with Erhard on a summit meeting with Khrushchev later in the year.[115] In early September Khrushchev accepted the West German demand that the meeting be in Bonn rather than Moscow. In accordance with West German wishes, the agenda was to be unrestricted; discussion of all outstanding issues, including the reunification of Germany, would be permitted.[116]

Although signs of internal opposition to the trip now began to appear, Khrushchev pressed on. A mysterious Soviet assault on a West German diplomat temporarily slowed Khrushchev's courtship of Bonn, and when the Soviet–East German treaty was ratified near the end of September, *Pravda*, asserting that the treaty would block West German plans "to swallow up the German Democratic Republic," reported Mikoian's statement that "anyone" who expected to improve Soviet–West German relations at the expense of the German Democratic Republic was "deeply mistaken."[117] But Khrushchev was not deterred. Just before the end of September he addressed a combined meeting of the party Presidium and the Council of Ministers on the basic guidelines of economic planning for the immediate future; many economic officials and technical specialists were also present. As paraphrased in the published account, Khrushchev put heavy stress on reorienting the economy toward consumer needs, and he strongly underscored the importance of introducing the most advanced foreign (as well as domestic) technology into industry. Observing that large investments were being made in the chemical and "other progressive" industries, Khrushchev insisted that technological "mistakes" be avoided and that world-standard technology and equipment be installed in the new plants. Soviet specialists, he said, had an obligation to ensure that this was done. "And here," Khrushchev said, "it is impossible to be reconciled with manifestations of autarky. Autarky is harmful in the economy and especially pernicious in the development of science and technology. The better the study and introduction of the newest achievements of world science and practice are organized, the more successfully scientific-technological progress will proceed. Not one country, not even the most developed, will be able to move forward quickly if it does not skillfully use the world achievements of scientific-technical thought."[118] This was the most vigorous attack on autarky by a Soviet leader since the dawn of the Stalin era, and Khrushchev meant it primarily as a call for closer technological ties with the West rather than with the USSR's CEMA partners. He emphasized world technological achievements, not just socialist ones; and *Izvestiia*, under the chief editorship of the same son-in-law who had recently served as Khrushchev's emissary to West Germany, quickly followed up with an unusual article featuring two cases in which technical ties with CEMA countries had failed to yield the required results. The case discussed at greatest length implied that

East Germany had promised more help to the Soviet chemical industry than it could actually provide. Indeed, the article intimated that East German assistance had been useless.[119]

The evidence suggests that Khrushchev's pursuit of West German technology sparked sharp resistance from his Presidium associates. Whatever they thought about Western technology in principle, they feared that he was prepared to pay West Germany an unacceptable diplomatic price for it. His address to the leadership on economic policy and the press warnings against deals at East Germany's expense evidently occurred almost simultaneously.[120] Moreover, his attack on autarky and his sermon on behalf of new priorities and world technology encountered obvious opposition. The speech was reported after a delay and in a lengthy paraphrase, suggesting disagreement about its contents. Despite Khrushchev's clear statement of his policy preferences, the Presidium and Council of Ministers refused to take any definite action; instead they merely referred the task of developing planning guidelines to Gosplan, which was to report back with recommendations.[121]

That the disagreements concerned foreign technology as well as domestic priorities is indicated by statements from other Presidium members. Only four days after the press reported the meeting, Suslov declared that West Germany could not use a commercial deal to undermine Soviet–East German solidarity, "even if all the gold in the world" were offered.[122] The following day Brezhnev attended the 15th anniversary celebrations of East Germany in Berlin. Brezhnev forcefully repeated that East German interests would never be sacrificed in any diplomatic bargain. He then remarked that the USSR had already ensured the preferential development of the chemical and other progressive industries—in clear contrast to Khrushchev's recent speech—and that these plans were being "successfully introduced into life." Obviously he did not view Soviet neglect of world chemical technology as the serious danger that Khrushchev claimed. Instead he went out of his way to stress the fruitful cooperation that the USSR and East Germany had already achieved in economic affairs and technological innovation. The East Germans, he said, had given the Soviet Union extensive help "in the area of the chemical and shipbuilding industries and agricultural production." Brezhnev chose these examples to make a point; in all three, recent Soviet purchases from the West had been unusually large. The development of East Germany, Brezhnev continued meaningfully, "is becoming an increasingly important factor in ensuring the victory of socialism over capitalism on the field of economic competition."

Later he again emphasized that the ties between the two countries were helping close the gap in the economic race with the West.[123] The latent message was that the need for technology from the West was not so pressing that the USSR should consider sacrificing East German interests.

It is far from clear how much Khrushchev actually proposed to concede to West Germany in order to gain the technology he desired. What is clear is that his Presidium associates feared that he intended to make substantial concessions, and that they opposed his diplomatic gestures toward the Federal Republic for that reason. The dispute over Western technology was thus a factor, albeit only one of many, which helped crystallize the Presidium conspiracy that deposed Khrushchev in mid-October. Now that the Western powers were starting to treat technological assistance as a bargaining instrument rather than something which should always be denied—in short, as a carrot rather than just a stick—the relationship between technology transfers and Soviet foreign policy was becoming more complex.

The Reconsideration of Technological Strategy

The xenophobic policy of Stalin's last years had contributed to continuing Soviet lags in many technological fields. After 1953 the new leaders had to weigh the technological benefits and domestic political costs of lowering the barriers to foreign contacts. No less important, they faced fresh questions about the social function of indigenous R & D.

Stalin's successors plainly felt that Soviet technological strategy should be modified. Bulganin, for example, told the Central Committee in 1955 that officials and specialists were doing "great harm" to the pace of technological advance by underestimating foreign science and technology. Part of the harm came from the needless duplication of foreign achievements. As Bulganin said, "some research institutes and design organizations have spent a considerable amount of time and money in . . . the creation of what has already been published in the foreign press and is already in use."[124] Even when there was no international duplication, closer contacts could improve the effectiveness of Soviet R & D. The party's condemnation of technological "boastfulness and conceit" at institutes and enterprises indicated that many specialists had taken the idea of Soviet technological superiority too literally, and that this belief had weakened their commitment to innovation.[125] Exposure to foreign achievements would spur Soviet specialists to exert themselves more fully.

A broadening of international contacts promised to improve the effectiveness of Soviet R & D specialists in other ways as well. In the late 1940s the regime had put great emphasis on preventing the divulgence of scientific information to foreigners. But this all-inclusive secrecy had exacted a price, not only in reduced access to foreign technology, but in less effective internal operation of the USSR's own R & D establishment. In 1956, Bulganin lamented "the incorrect tendency of certain scientists to monopolize the supervision of individual branches of science," and indicated that this tendency was encouraged by the practice of scientific secrecy. Systematic dissemination of information about ongoing Soviet research, he said, had great importance in scientific work. "However, in many cases this is handicapped by the fact that materials are unnecessarily classified as secret. Unreasonable secrecy leads to parallelism in work and sometimes protects unconscientious workers from scientific criticism."[126] Extreme secrecy thus contributed to a paradoxical situation. Soviet R & D organizations might expend extra resources in duplicating one another's work, yet they could still remain "monopolies," that is, organizations protected from the stimulus of competition with one another through open discussion and publication.

Widening Access to Foreign Science and Technology

Responding to such problems, the authorities mounted a campaign to stimulate Soviet specialists to stay abreast of the latest R & D findings abroad. In a sharp change of course, the press now proclaimed that "a scientist can only enrich his work by drawing on the experience of foreign science."[127] Party spokesmen called for fuller exchanges of publications with the West and excoriated technical journals for heaping uncritical praise on Soviet achievements while neglecting new technological developments abroad. They also endorsed closer relations between Soviet and foreign scientists.[128] Soon after the 20th Party Congress, the Council of Ministers adopted a more liberal statute on secrecy that differentiated military from nonmilitary technical secrets, exempted from secrecy research findings that were not "major," and specified the persons who could authorize publication of major findings from nonmilitary R & D.[129] The party chiefs obviously wished to minimize the absorption of foreign political ideas through international ties, since they proclaimed that peaceful cooperation in science and technology should not be confused with ideological coexistence, which was absolutely impermissible.[130] But they were manifestly striving to improve access to the fruits of foreign R & D.

Scientific spokesmen responded enthusiastically to this new policy, which they themselves had dared to advocate obliquely even before Stalin's death. Backed by the Academy Presidium, Academy President A. N. Nesmeianov urged better exchanges of scientific literature, delegations, and especially better personal contacts.[131] Such advocates became particularly outspoken after the 20th Party Congress heard Khrushchev denounce many of Stalin's policies. B. Kedrov, whose attempt to differentiate Western science from Western ideology had been defeated in the late 1940s, returned to the offensive. Kedrov repeated his earlier argument that authentic Leninism combined the rejection of bourgeois ideology with the adoption of the useful elements of bourgeois culture. "Many serious mistakes," he said, had been committed during 1949–53, when the violation of this Leninist tenet had caused a nihilistic rejection of foreign science and technology, especially in biology and physics. In fact, asserted Kedrov, genuine party-mindedness did not require a wholesale repudiation of Western ideas.[132] A substantial number of scientists agreed. Many statements at a gathering of Academy scholars in 1956 dealt with improving foreign scientific ties and travel. The meeting adopted a resolution favoring these goals, and shortly afterward the Academy Presidium resolved to take further steps in this direction.[133] Many top scholars were eager for fuller communication with their foreign colleagues and seemed unconcerned about the threat of ideological contamination. Indeed, some like Kedrov hoped to use these channels to gain a better appreciation of current trends in Western social thought.[134]

The regime's ideological watchdogs, however, asserted that relaxing the controls over the selection of foreign professional journals had allowed Soviet specialists to subscribe to publications which were conduits of bourgeois influence.[135] They also argued that some Western specialists and businessmen in the USSR were really intelligence agents and particularly stressed the efforts of Western intelligence services to gain technological information and suborn Soviet travelers abroad.[136] The political police's lack of enthusiasm for wider communication was indicated by KGB Chairman I. Serov, writing soon after Sputnik. Serov emphasized the superiority of Soviet science and technology, which he said had heightened the efforts of American espionage to penetrate the Soviet Union. Ignoring the distinction between military and non-military information, he demanded that the "intolerable" phenomenon of "chatterboxes" who openly discussed scientific secrets be eliminated.[137] Other writers criticized the leaking of "much valuable secret information" to foreign intelligence services through articles in open journals. The necessary communication among Soviet specialists could be ensured, it was argued, without publishing such R & D results in the literature available to Westerners.[138] In short, the USSR would lose more from relaxing the barriers of secrecy than it stood to gain, either from improved communication with foreign specialists or from improved communication among Soviet specialists themselves.

No doubt virtually every Soviet official, scientific as well as political, subscribed to the principle that the state should protect the technical secrets vital to its security. But serious differences existed over what this meant in practice, and they cropped up in the disparate ways the regime handled real cases of foreign espionage. In mid-1960 an American U-2 reconnaisance plane was shot down while taking photographs over Soviet territory. This affair did not involve any lapse of the institutional procedures aimed at protecting Soviet scientific secrets. If anything, it signified the USSR's first success in bringing down one of the high-flying U-2s, which had been traversing Soviet territory with impunity for several years.[139] Nonetheless, the incident provoked a series of harsh statements by officials associated with the secret police. Challenging Khrushchev's notion that the role of the police should be cut back, one article questioned the new emphasis on increasing technological contacts with the West. Citing cases in which Soviet specialists had gullibly revealed R & D secrets to outsiders, the author claimed that this sort of naivete was "doubly and triply" dangerous now and demanded that the secrecy in enterprises and institutes be tightened. No doubt the vigorous police response reflected both a genuine fear of freer cultural contacts and a desire to safeguard the KGB's special role as a collector of Western technology through covert channels.[140]

In contrast to the U-2 incident, the Penkovskii affair, which constituted a major breach of the institutional rules of secrecy by Western intelligence services, was treated with relative restraint. Oleg Penkovskii, while serving as Deputy Chief of the Foreign Section of the State Committee for the Coordination of Scientific Research, had made contact with British and American intelligence through a British businessman and had passed them exceptionally sensitive material on Soviet missiles and other subjects.[141] Yet in May 1963, when he was convicted of treason and sentenced to death, the tone of the Soviet press accounts was surprisingly mild. An editorial in *Pravda* remarked that Western intelligence agencies were striving to obtain Soviet secrets, and it called in general terms for greater political vigilance. But it also stated that the USSR continued to advocate "the complete development of international contacts" and "the broad exchange of cultural values" and stressed that Soviet citizens did not equate vigilance with suspicion of every foreign visitor.[142] Reiterating this theme, an interview with the chief military prosecutor went on to quote his regretful question: "Is it necessary to say that such dirty actions . . . cause great harm to the cause of increasing trust among peoples . . . and hinder the development of scientific and cultural cooperation and international trade?"[143] The difference in tone from the coverage of the U-2 incident was striking, and must have derived partly from fluctuations in the relative power of those who favored and opposed technological ties with the West. In 1960, using the U-2 affair as a weapon, Khrushchev's conservative rivals in the Presidium had badly undermined his position, whereas in mid-1963 he was on firmer political ground.[144] Conversely, the secret police, who bore no responsibility for antiaircraft defense, had been free in 1960 to press their case against open international contacts. In 1963, however, they were politically vulnerable, having failed to frustrate an extraordinary Western intelligence coup.[145]

Soviet external technological ties were shaped by such pressures. The acquisition of embodied technology increased. Equipment imports from CEMA climbed substantially. Starting from a much smaller base, equipment imports from the industrial West evidently grew even faster, raising the share of Western equipment in Soviet domestic investment.[146] In most sectors the Western share was still only a very small fraction of total investment, but in a few, such as chemicals, where the USSR had begun to import whole plants, it was substantial.[147] As for foreign scientific literature, the Academy's exchange of books with foreign countries tripled between 1950 and 1954 and continued to grow rapidly thereafter.[148] Most foreign periodical literature was channeled into a central institute of scientific information, where police censors deleted all "subversive" information from each issue before it was reproduced in a large printing for ordinary users.[149]

The government also expanded institutional ties between Soviet and foreign R & D specialists. In 1955, soon after a Central Committee plenum had criticized neglect of foreign science and technology, the USSR joined the International Council of Scientific Unions. It likewise joined more specialized international R & D organizations, until by 1965 it belonged to almost twice as many as in 1955.[150] Ties were strengthened with the other members of the CEMA, which had previously given little attention to exchanging proven technologies and virtually no attention to coordinating R & D on new ones. In 1956 a Joint Institute for Nuclear Research was established in the USSR, and the CEMA set up several commissions to coordinate national R & D programs as part of a wider effort to harmonize the economic plans of CEMA members. The USSR's attempt to force its socialist partners to accept national specialization through supranational planning foundered in the early 1960s because of East European opposition. But it did produce more voluntary R & D cooperation and more institutional contact between bloc specialists.[151]

Personal contacts between Soviet and foreign specialists expanded as well. In the second half of the 1950s the government signed cultural agreements with other CEMA members, the United States and Great Britain, facilitating the exchange of specialists with the outside world.[152] The available data suggest that the number of Soviet scientists traveling abroad under the auspices of the Academy of Sciences rose from a mere 175 in 1954 to 2,287 in 1964. These foreign trips were divided about evenly between Communist states and capitalist countries. The number of foreign scientists and technical specialists visiting the USSR showed an equally dramatic rise during the decade.[153]

All these measures were important steps away from the technological autarky of the late 1940s, but they were not sufficient to provide easy access to foreign science and technology. For example, the regime was disseminating many more foreign scientific and technical journals than in the past. But these journals still had to pass through the time-consuming process of Soviet censorship and reproduction, which helps explain why Western R & D results still filtered into Soviet laboratories and factories with delays greatly exceeding the time required to learn of domestic R & D findings published in Soviet journals.[154] Similarly, by the mid-1960s the proportion of Soviet scientists traveling abroad had grown strikingly but remained very small in comparison with the volume of international travel done by Western scientists.[155] The transfer of science and technology through personal channels therefore remained extremely limited. Nor had the deleterious effects of secrecy regulations been eliminated. The security apparatus, though reduced in scale, was still obsessed with preventing leaks to foreigners, and this preoccupation continued to restrict communication between Soviet and foreign specialists through channels such as correspondence, as well as communication among Soviet specialists themselves.[156]

The Scope and Role of Domestic R & D

Although this measured increase of technological borrowing promised real advantages in the many sectors where the USSR still lagged well behind the West, it could not meet all Soviet technological needs. In fields where the country was approaching Western levels of sophistication, further Soviet advances would increasingly have to be sustained by domestic R & D rather than by foreign developments. In response to this emerging requirement, the regime spent very heavily on R & D, and it reconsidered the technological importance of basic research.

After 1953 total R & D spending grew at an extremely high rate, averaging between 15 and 18 percent annually up to 1960. In the early 1960s the rate declined somewhat but was still at 13 percent in 1962.[157] The slower growth in the 1960s caused some friction between science officials, who naturally thought that more was better, and economic planners, who were beginning to feel that added research spending was yielding too few concrete benefits.[158] But on the whole, the authorities were caught up in a wave of enthusiasm for science and technology which was exceptional even by Soviet standards, and they willingly spent a growing share of the state's resources on R & D.

The new leaders' more open approach to technological policy encouraged a reconsideration of the part that Soviet science, especially Soviet theoretical research, should play in technological progress. During Stalin's last years some writers suggested that fundamental research, because it would ultimately lead to technological benefits, might be considered a direct productive force. But such views were spurned in favor of concentration on "practice" and immediate results, with the consequence that the mission of Soviet science continued to be defined largely in terms of applied research.[159] Only after Stalin's heirs began to scrutinize the process of innovation did this view come under serious criticism, much of it coming from leaders of the Academy.

In 1954 Academy President Nesmeianov asserted that the Soviet research effort, while it should combine basic and applied science, ought to give greater attention to fundamental research.[160] A few months later a party secretary from Sverdlovsk, a major center of heavy industry, disputed this notion with complaints that some Academy institutes had succumbed to the "chronic illness" of withdrawing "into the realm of 'pure science.' "[161] But Nesmeianov stuck to his position, arguing that the USSR had to modify its technological strategy. Noting that the country had achieved major successes in catching up with Western technology during the last five-year plan, he pointed out that in 1955 the party had outlined new goals for further technological progress. He acknowledged unequivocally that technological borrowing was one appropriate means of meeting the party's goals. "We can and must use this method when we are faced with the task of drawing abreast in specific spheres of technology," he said. However, "in order to overtake, we must also have our own wealth of accumulated scientific knowledge. . . . We are therefore faced with the necessity of sharply increasing our own store of Soviet science, and our attention to theoretical science and to scientific research, in the Sixth Five-Year Plan." At present there was a tendency to view the tasks of domestic research in narrowly practical terms, confining it to a "crawling empiricism" that was bad for both science and technological advance.[162]

Some top policymakers still doubted the need to change the orientation of Soviet R & D, but the idea was apparently gaining support. On the eve of the 20th Party Congress, *Pravda* ran an editorial stressing that scientific research should be brought closer "to the concrete needs of the economy."[163] Nesmeianov was not deterred, and at the Congress he repeated his argument that the USSR should now shift domestic R & D toward more theoretical work. Moreover, he took the extraordinary step of stating that *Pravda*'s recent editorial on Soviet science had failed to grasp the need for such a change.[164] This was a daring and even provocative act, yet Nesmeianov suffered no political harm. We can safely infer that he challenged *Pravda*'s view only because he knew that powerful figures in the party hierarchy agreed with his position.

The following year, Chief Academic Secretary Topchiev cast the argument in more philosophical terms. Contending that "a huge role belongs to science" in building communism, Topchiev paraphrased Marx to the effect that "creation of society's real wealth depends on the general state of science and the degree of development of technology or the application of this science in production . . . science becomes more and more a 'direct productive force.' "[165] By treating technology as an extension of science, Topchiev's statement reversed the formula that had governed Soviet discussions of the role of science for almost three decades. The implicit corollary was that science made its greatest social contribution not by catering to industry's presently perceived technological needs, but by pursuing theoretical advances that would later transform existing industrial technology.

Opponents quickly charged that this position violated the view of the relation between scientific and industrial progress sanctioned by the party.[166] Against the interpretation of Marx put forward by Topchiev, they cited Engels to the effect that while "technology depends to a considerable extent on the state of science, science depends to a far greater degree on the state and requirements of technology."[167] Many opponents of a shift to more theoretical research were themselves applied scientists. And they obviously did not agree that further Soviet technological advance required the accumulation of the sort of "scientific wealth" advocated by Nesmeianov and Topchiev.

As we shall see below, this dispute was closely bound up with a concurrent debate about the proper mission of the Academy of Sciences, and after considerable equivocation both disputes were resolved within a few months of each other. Early in 1961 the Academy was given a mandate making its foremost task the promotion of theoretical science, and the Party Program issued later in the year confirmed that "science will become, in the full sense of the word, a direct productive force."[168] The acceptance of this formulation implied a recognition by the authorities that further technological progress would have to be propelled, in considerable measure, by the country's own basic research. But recognizing this need was only the first step in meeting it. In order to encourage wider creation and application of indigenous technology, the leadership had to upgrade the effectiveness of research and production institutions.

The Reorganization of Indigenous Innovation

Reevaluating the Academy's Role

A reconsideration of the Academy's role began soon after the death of Stalin. One of the first consequences was an apparent reordering of applied research to reflect the party's new economic priorities. Late in 1953, after the party put new stress on improving food and light industry, the Academy moved quickly to devote more attention to work bearing on these sectors.[169] It soon became clear, however, that a deeper reevaluation of the function of the Academy was under way. Some individuals were questioning whether the activities of the Academy should be so closely geared to any sector of the economy. These persons were proposing that economic officials and institutions should respond more fully to the novel possibilities opened up by Soviet fundamental research instead of pressing scientists to concentrate on short-term economic needs. According to this view, the expansion of the Academy's theoretical investigations should be accompanied by an enlargement of its power vis-à-vis the state economic apparatus.

The top administrators of the Academy led the effort to redefine its place in the R & D system. Without claiming that the Academy should concentrate exclusively on basic research, they contended that it could best contribute to technological progress by enlarging its portfolio of theoretical work.[170] For a year or two Academy spokesmen cautiously avoided advocating administrative changes that would curb the demands of industrial officials on the Academy for applied research. This tactic would have been risky in a period when most public blame for failed innovations still gravitated, in the usual Stalinist fashion, to the Academy. Academy leaders concentrated instead on identifying R & D problems that they deemed critical for the future and tried to draw ministerial officials into R & D planning on these topics.[171]

In 1955, however, top Academy officials launched a campaign to win bureaucratic power commensurate with their vision of the Academy's proper role. Early in that year the ministerial apparatus had lost its leading patron, Malenkov, and come under sharp criticism for technological sluggishness. In response to this change of political atmosphere, the Academy leaders revived their institution's long-standing claim to be the national center for the administration of R & D. Nesmeianov asserted that the Academy "can only fulfill its role in the country's scientific orchestra if it is the conductor, not merely a single performer." Not only should it work closely with all other Soviet scientific institutions, but "this cooperation should be of a definite nature." The seventy major scientific and technological lines of research planned by the Academy in consultation with the ministries should "form the pivot not only of the Academy's plan but also of the plans of several branch institutes and . . . the plans of the Union Republic Academies of Science." A "steadily increasing number" of researchers from outside the Academy should take their cues from its work on these problems.[172] Topchiev stated that the Academy should be empowered to evaluate the work of other research agencies, assign them tasks, and receive their reports.[173]

This effort to bring more of the country's R & D planning under the aegis of the Academy met resistance from the outside agencies whose work the Academy leaders wished to direct. Nesmeianov later revealed that "external circumstances" had seriously hampered the Academy's initial attempt at R & D coordination, and it is safe to assume that one of these circumstances was the resistance of the industrial ministries.[174] But substantial resistance also came from within the Academy itself. The Academy was in the throes of a renewal aimed at eliminating the suppression of dissenting views on scientific issues, the branding of scientific opponents as political deviants, the establishment of research monopolies and interlocking editorial directorates, and other negative consequences of the authoritarianism that had characterized much of science under Stalin.[175] Prompted by this spirit of renewal, many academicians, who regarded the internal decentralization of Academy planning as more pressing than its extension to outside R & D establishments, wished to reduce detailed supervision of their work by the Academy Presidium.[176] They did not believe that the recent effort at extra-Academy R & D planning had been successful, and several division bureaus and institutes were reluctant to accept the burden of guiding projects undertaken by ministerial research units.[177] Nesmeianov and Topchiev thus faced both external and internal resistance to expansion of the Academy's role. Nevertheless, they persevered for several reasons.

In their campaign to change the relationship between basic research and industrial technology, the two spokesmen were also seeking to guard against a diminution of the Academy's existing powers. The chief danger lay in the possibility that creation of a planning agency specializing in R & D might circumscribe the Academy's domain. A state committee enjoying some authority in this field had been created in 1947 and abolished in 1951. Retrospectively, we can see that the political leadership's widening concern over innovation was causing it to consider reestablishing such an agency, so that in early 1955 the charter of the body must have been under discussion. In view of the continuing problems of cooperation between the Academy and the ministries, the new agency might be given authority over both the scientific and industrial bureaucracies. It was partly to counter this threat that the Academy leaders put forward their own institution's claim to coordinate the national R & D effort.

This tactic evidently achieved less than the Academy administrators had wished for. The creation of the State Committee for New Technology (Gostekhnika) was decreed in May 1955. No doubt the Academy leaders were pleased that the powers of the new agency were focused on the performance of applied research and its utilization by the ministries.[178] But it was still true that Gostekhnika's responsibilities infringed on the Academy's applied research activities, and that the new agency blocked the bid to make the Academy the general overseer of Soviet R & D. Several months after Gostekhnika had been created, President Nesmeianov said the new five-year plan "must be a turning point in the rate of application of science's discoveries to technology" and discussed these problems at length, but still managed not to mention the existence of Gostekhnika.[179] His ostentatious silence suggested reluctance to work in harness with the new body.

This public posture proved difficult to maintain in the face of the party's growing concern about innovation. Khrushchev, ticking off the negative consequences for technological progress, told the 20th Congress that the lack of coordination between the Academy and the industrial ministries was "utterly intolerable." Bulganin noted that the Academy, Gostekhnika, and other agencies must together examine and solve the problems of administering R & D.[180] The injunctions of the party leaders did not eliminate all jurisdictional conflict between the Academy and Gostekhnika,[181] but they did serve notice that the Academy administrators must accept a division of labor with the new planning agency whether they liked it or not. Under this prodding, Nesmeianov hastened to ensure the Congress that the Academy "stands ready to help" Gostekhnika in promoting new technology.[182]

The Debate over Applied Research by the Academy

The demand for such a division of labor in turn prompted the Academy leaders to urge that the Academy's primary mission be shifted to basic research. In the past Nesmeianov had emphasized the need to expand basic research but had implied that this was a task for the whole R & D establishment, not just the Academy. The Academy, he had insisted, should coordinate the whole R & D effort, and its divisions should work to overcome "the mutual exclusion of 'theoretical' science and 'applied' science wherever this exclusion exists."[183] At the 20th Party Congress, however, Nesmeianov sharply altered his notion of the kind of research the Academy should do. Linking its mission more closely to the "intensive development of theoretical science," he enthusiastically underscored the theoretical origins of current technological progress in atomic power generation and other fields, but he dealt far more coolly with the Academy's applied-science projects, "some of which are of great economic importance and some of which are not." He drove this second point home by citing a ludicrous request from the ministry of domestic trade for the Academy to design automatic doors for a restaurant.[184] The Academy, he said a few months later, should refrain completely from "petty assistance to production, from final development of the achievements of foreign science for introduction into industry," and from branch research projects, which could better be performed in ministerial institutes.[185] This was an unorthodox position, and one regional party official directly challenged it at the Congress.[186]

The most obvious debate over the Academy's role, though, occurred within its own ranks. The focus of the debate was multidisciplinary applied research, particularly the expanding program of the Division of Technological Sciences. Between 1951 and the end of 1955 the Technological Division had added more new members to the Academy than had any other division.[187] Between 1951 and 1956 it had also led in the creation of new institutes, and its contingent of specialists had outstripped the rate of personnel growth in all other divisions.[188] Nesmeianov was obviously disturbed by these trends, which he felt were at odds with the need for more basic research. Academy institutes set up on disciplinary principles, he claimed, were preferable to institutes organized on multidisciplinary lines, such as the Technological Division's Oil Institute. The Division's activities had often crossed the boundary that should separate research of far-reaching significance from ministerial topics of limited consequence.[189] Later Nesmeianov said that the Academy, having set up too many institutes in the Division of Technological Sciences, should concentrate on improving their work rather than creating still more institutes. He also voiced a desire that the industrial ministries pay more for the very expensive research projects that were being undertaken by the Division to meet industrial requirements.[190]

These propositions provoked widespread disagreement among Academy members. Some expressed firm support for freeing Academy institutes from tasks that ministerial R & D units could perform, but others disputed this view. The Academician-Secretary of the Technological Division, A. A. Blagonravov, said that the question of doing multidisciplinary research must be decided "concretely in each case"; he did not endorse the principle espoused by Nesmeianov. Noting that they were working on economic problems singled out by the party, the heads of institutes in the Technological Division demanded more support from the Academy Presidium.[191] Another applied scientist criticized the tendency to escape into "pure science" by undervaluing technological research and shifting it to the ministries.[192] The director of an Academy establishment working on jet and rocket engines protested what he felt was an artificial distinction between branch R & D and genuine scientific research.[193] Such opposition greatly complicated the effort to redefine the Academy's research mission.[194]

The effort was rendered more difficult by the persisting discontinuities between the research of the Academy and the production practices of industry. Even after the industrial reorganization of 1957 the links between the two remained weak. According to Academy officials, such contacts were "unfortunately . . . rather sporadic" and required "a fundamental improvement" in the receptivity of industry to new scientific ideas.[195] In the past the party authorities had tried to bridge the gap by building up technological research within the Academy and creating a separate Division of Technological Sciences, but these efforts had enjoyed only modest success. Given the already tenuous relations between the Academy and the industrial apparatus, the party chiefs probably feared that a swing by the Academy toward more theoretical research would worsen rather than improve the pace of technological advance. At any rate, the top leadership had still not shown any visible enthusiasm about the idea of such a change in the Academy's role.[196]

In 1959, however, the Academy leaders moved one step closer to this goal when the General Meeting resolved to "consider the advisability" of transferring some of the Technological Division's laboratories to Gosplan or the Academy's local affiliates.[197] In June Khrushchev gave this idea qualified support. Observing that the time had come to reorganize the Academy, he told the Central Committee that "questions of metallurgy and the coal industry" should be dealt with outside the Academy.[198] Khrushchev took this position not simply to bolster basic research, but to strike a blow for the new economic priorities he favored—building up the chemical industry at the expense of higher-priority industries like iron and steel, and restricting the growth of military expenditures. The Academy reorganization offered him a chance to reshape research priorities to match his broader economic preferences.

Even though Khrushchev had come out in favor of a reorganization, the Academy members were still split over the issue, and disagreement probably also existed at the top of the party. In early August *Izvestiia* ran a lengthy article by the distinguished chemist N. N. Semenov, arguing for the Academy's reorganization and the transfer of most technological institutes to other agencies.[199] The next day a *Pravda* editorial, remarking that "life demands" improvement in the Academy's work, criticized the Technological Division and recalled Khrushchev's statement that the Academy and its divisions should be reorganized.[200] Nonetheless, no change occurred. Instead came a new round of debate between the two scientific factions represented in the Academy Presidium.[201] Led by Semenov, the proponents of theoretical science asserted that the Academy's large program of technological research was an "anachronism."[202] The most outspoken opponent of this idea was I. P. Bardin, an eminent metallurgist who had been singled out by Khrushchev as exemplifying the misdirection of the Academy's work. Bardin claimed that a shift toward theoretical investigations would mark a dangerous reversion to the prerevolutionary divorce of science from social needs, and his scientific allies pointedly cited the backwardness of the chemical industry as the fate that might befall all Soviet industry if theory and practice were not linked inside the Academy.[203] Most strikingly, Bardin stated flatly that the need to conduct metallurgical research within the Academy "does not provoke doubts from anyone." These words flew in the face of Khrushchev's public expressions of doubt, and Bardin manifestly knew it.[204] The evidence strongly suggests that Bardin made this defiant statement because he had the backing of party leaders who opposed Khrushchev's effort to downgrade the metal industries and viewed the Academy's research priorities as part of that larger struggle. A. B. Aristov, the only other member of the party Presidium who had addressed the June plenum where Khrushchev objected to metallurgical research inside the Academy, had ostentatiously concentrated on the need for more rapid development of the metal industries and had studiously avoided discussing the chemicals sector.[205] Unless there was resistance to the Academy reorganization at the highest levels, it is difficult to explain how a step advocated by both Khrushchev and *Pravda* could have been stymied for another year and a half, as in fact occurred.

The dispute over reorganizing the Academy also appears to have had military ramifications.[206] The Academician-Secretary of the Technological Division, who had personally devoted many years to research on the military applications of rockets, vigorously defended the Division's achievements. He pointed out that many of its "enormous" accomplishments had been attained by Academy members working in nonacademy institutes—particularly in aircraft and rocket engineering—and he called for the work of these members to be more closely connected with the work of Academy institutes in the future.[207] His prescription was for the Academy to do more applied research and evidently

more military research. Further, according to Khrushchev's later recollections, Bardin had spent "many years engaged in useful defense
work," some of which must have been done at the Academy institute
which he had headed since 1939 and which Khrushchev was now
trying to remove from the Academy.[208] The differences between the
two men over applied research in the Academy probably took on added
sharpness because of Bardin's leading role in a technical commission
which concluded sometime during the late 1950s that Khrushchev's
proposal to base Soviet ballistic missiles in underground silos was un-
workable. About a year afterward Khrushchev, who recollects being
"flabbergasted" at the commission's report, learned that the United
States had begun to replace its missile launching pads with underground
silos.[209]

The combined resistance from all quarters palpably slowed the campaign to change the Academy's research orientation. The number of
new academicians selected in 1960 for the Division of Technological
Sciences easily outweighed the number added to any other natural-
science division, and the Academy's top administrators became surprisingly equivocal about removing institutes from the Technological
Division.[210]

Academy officers had special cause to tread carefully, not only for
personal reasons, but because the issue affected the Academy's standing
as an agency for the national planning of basic research. In 1957 Gostekhnika was reorganized into the State Scientific and Technical Committee, whose name implied that it might take over responsibility for
basic research as well as technology. Teaming up with V. Kirillin, a
corresponding member of the Academy who also headed the Central
Committee's Science Department, Topchiev vigorously advocated preserving the Academy's authority over the planning of fundamental
research.[211] This was essentially what was done, leaving the State Committee to concentrate on problems more closely connected to technological innovation.[212] But as it gradually became apparent that this
division of responsibility for R & D planning was still not satisfactory,[213]
some officials proposed more decisive steps that again threatened the
authority of the Academy. At the 21st Party Congress in 1959 two
influential regional party secretaries advocated the creation of "a central
state agency which could—through a system of main scientific research
institutes—coordinate and direct all the scientific work in the country,
down to the work of plant laboratories."[214] Shortly afterward Kirillin
came out against the idea.[215] Nonetheless, the proposal must have given

the Academy leaders pause. Since 1956 they had sought to make the Academy's primary mission the cultivation of basic research and had accepted the need for another planning agency to oversee applied science and development projects. Now, however, if it further weakened its claim to be a center of Soviet science, applied as well as theoretical, the Academy might be subordinated to another agency authorized to plan activities in both these realms. This must have been one reason the Academy's top leaders moved so gingerly to force a cutback in the Division of Technological Sciences after 1959. The issues of the Academy's research mission and its institutional standing had become intertwined—and so they stayed.

The 1961 and 1963 Reorganizations of Research Administration

After a considerable delay, the Central Committee and Council of Ministers dealt with both questions in a major resolution adopted in April 1961. Demanding better planning and coordination of research, the resolution stated that the many applied research institutions within the Academy were distracting it from vital theoretical problems. The decree therefore authorized the transfer of "a number of institutes" out of the Academy, instructing it to concentrate primarily on research in the natural and social sciences.[216] This critical provision paved the way for the removal of a large number of institutes from the Academy, especially from the Division of Technological Sciences. Between 1960 and 1961 the total number of researchers working in the Academy dropped by a fifth.[217] From the standpoint of Nesmeianov, who had struggled to win top priority for fundamental research within the Academy, this change must have been gratifying.

The simultaneous creation of a new state agency to overcome "shortcomings in the work of the U.S.S.R. Academy of Sciences and other research institutions," on the other hand, was not. The new State Committee for the Coordination of Scientific Research was given a much broader mandate than the State Scientific and Technical Committee it superseded.[218] Assisted by the central organs for economic planning, the agency was empowered to work out draft plans "for research projects in the country" and for introducing research findings into the economy. Along with measures for financing and supplying major research undertakings, the State Committee was to submit these plans directly to the Council of Ministers. Strikingly, the Academy was given no place in devising these nationwide research plans.[219] As one of many ministries and departments, it now found itself participating in national R & D planning solely as a claimant for resources—hardly the role of "conductor of the scientific orchestra." The April resolution was a victory of sorts for theoretical research, but it was also an institutional rebuff for the Academy. The rebuff was clear not only from the contents of the resolution, but from Nesmeianov's sudden replacement as Academy President by M. V. Keldysh, a mathematician with links to the new State Committee.

One likely motive for Nesmeianov's removal is that the political leaders, familiar with his long-standing opposition to the idea of subordinating the Academy to another science-planning agency, doubted his willingness to help make the new arrangements work. Fragmentary evidence suggests that another reason may have been that his conception of the Academy's mission embroiled him in a conflict over the conduct of military R & D. This notion is consistent with the fact that the debate over the technological sciences heated up in late 1959 and early 1960, as Khrushchev announced his disarmament appeal and unveiled proposals to concentrate reduced defense spending on nuclear forces. Judging by events earlier in the decade, Nesmeianov, as part of his campaign to upgrade theoretical research, may have proposed to cut back the Academy's role in military R & D in ways the political leadership found objectionable. In 1956, when Nesmeianov stepped up the pressure for more theoretical research, a few hints indicated that the Academy was involved in a struggle with other hierarchies, probably including the ministry producing nuclear weapons, for administrative control of atomic research.[220]

At any rate, by 1958 some Academy members were seeking to slow the pace of weapons development. Academician Andrei Sakharov, a top specialist on nuclear weaponry, tried unsuccessfully to persuade Khrushchev in 1958, and again in 1961 and 1962, to cancel a series of nuclear test explosions.[221] The regime did observe a moratorium on atmospheric tests from October 1958 to September 1961 but carried out the tests Sakharov opposed. In 1959 another academician, after describing the Pugwash meetings between Soviet and Western atomic scientists, urged that Soviet scientists be given the same opportunity as Western scientists to take public positions on pressing political issues such as disarmament and the termination of atomic-weapons tests.[222] In the same year, both Nesmeianov and the chemist Semenov made forceful speeches on the need to restrain the arms race. By itself this may mean only that the two were helping put across the official line to foreign audiences; Nesmeianov's speech, for instance, occurred after Khrushchev had made his dramatic disarmament appeal. But the two speeches were phrased very strongly, and they came from the two academicians who were the most outspoken advocates of curtailing applied research in favor of more theoretical science.[223] The defenders of applied research within the Academy apparently failed to make similar statements either before or after Khrushchev's disarmament appeal.

In public, Nesmeianov never directly challenged the official line about the regrettable necessity for Soviet military R & D, but Khrushchev's memoirs imply he was not fully committed to it. In 1961, according to Khrushchev, Nesmeianov was criticized at a meeting of the Council of Ministers. Afterward Nesmeianov, without actually resigning, suggested that perhaps M. V. Keldysh, a prominent academician directly involved in military R & D, should be made Academy president. The remark seems to have been a bureaucratic ploy rather than a real resignation. But the matter was sufficiently serious that the authorities decided a few days later to install Keldysh in Nesmeianov's place. While Khrushchev does not state directly the reasons for dissatisfaction with Nesmeianov, he prefaces his account with the observation that Igor Kurchatov, an academician who played a central role in developing Soviet nuclear weapons, had always understood that defense had to come before the general advancement of Soviet culture and technology: "Like Kurchatov, Keldysh was irrevocably committed to our concept of what needed to be done in the development of nuclear missiles, and consequently he was held in especially high regard." Then, making an implicit connection, Khrushchev describes the decision to replace Nesmeianov with Keldysh. The latent message seems to be that Nesmeianov had failed to implement the political leaders' policies for nuclear weapons research with sufficient vigor.[224]

On the basis of this fragmentary evidence we can speculate that Khrushchev, although he favored upgrading theoretical research and improving civilian output relative to military technology, did not want these two efforts to get out of hand. Moreover, given the attitudes of other leaders toward technological priorities, some of his Presidium associates must have thought that even limited changes along these lines would be a mistake. Khrushchev manifestly wished for the Academy to concentrate on theoretical work, since he later repeated his objection to its tendency to expand its technological research. He may even have backed the decision to transfer some institutes working on atomic, rocket, and electronics projects out of the Academy after 1961.[225] But he also wanted to ensure that Soviet R & D, especially R & D on the atomic warheads and ICBMs he was trying to substitute for conventional weapons in order to cut military spending, would not be disrupted. The appointment of Keldysh may have been Khrushchev's way of guaranteeing that the recent changes in the Academy's role would not harm this part of the military R & D effort. Unlike Nesmeianov, Keldysh had spent a large part of his career as a leading researcher in aerodynamics. He had done all his professional work at institutes outside the Academy system, including fifteen years as the head of the main institute for research on rocketry, and he had developed close personal ties with the members of the new State Committee for the Coordination of Scientific Research whose backgrounds were in military R & D.[226] Finally, in contrast to Nesmeianov, he was firmly committed to nuclear ICBMs and could be counted on to ensure the Academy's cooperation whenever projects in this field needed high-powered scientific assistance. He would not invoke the primacy of basic research to try to restrict such aid.[227]

Whatever provision Khrushchev made for coordinating research on ICBMs, however, the new State Committee quickly encountered the more difficult problem of coordinating programs across the whole range of R & D topics. In 1962 the Chairman of the Committee complained of "many inadequacies in coordination," some of them stemming from the refusal of R & D organizations to cooperate with each other.[228] The Academy was one of these organizations.[229] In 1963 the State Committee complained that the Central Statistical Administration had given it information on the execution of fewer than half the projects contained in the national plan for R & D in 1962. No data were received, it said, on projects planned in the "natural and social sciences"—the Academy's special province.[230] Whether the cause was negligence by the statistical agency or the Academy's refusal to supply the information, in such circumstances the State Committee could not fulfill the broad coordinating role mapped out for it in 1961.

In a resolution adopted in April 1963, the authorities tried again to resolve these problems. The resolution affirmed and extended the principle that the establishments of the Academy system should concentrate primarily on theoretical issues.[231] But it also strengthened the Academy's authority vis-à-vis the State Committee for the Coordination of Scientific Research. This step was probably simplified, and possibly even precipitated, by the discovery of Penkovskii's espionage activities within the State Committee, which must have badly undercut the Committee's political standing.[232] At any rate, the resolution empowered the Academy to coordinate the research being done by the industrial bureaucracy in the natural sciences (as opposed to engineering R & D). Rather than funneling all suggestions for research projects and financing through the State Committee, the Academy was again authorized to submit these proposals directly to the Council of Ministers. And it was entitled to confirm state plans for the USSR's most important research projects in the natural sciences, "in agreement with the State Committee."[233] By giving increased attention to theoretical research and vesting primary authority for its supervision in the Academy, the resolution divided responsibility essentially along the lines advocated by Nesmeianov and Topchiev in the late 1950s.

The decision to make the expansion and guidance of basic research the principal mission of the Academy was a significant departure from the Stalinist approach to R & D. It was, however, only one step in adapting the administrative structure to the demand for more extensive indigenous innovation. Because of its new research orientation, the Academy would now pay less attention to the applied research needed to bridge the gap between theory and practice. This change, in turn, placed a greater burden on the industrial bureaucracy. Upgrading basic research would yield beneficial economic results only if industry could develop the findings of basic science into effective new products and processes.

The Industrial Bureaucracy and Innovation

In late 1954 the ministerial bureaucracy came under growing pressure to step up the pace of technological advance.[234] Linked to Malenkov's declining fortunes, the pressure culminated in two Central Committee resolutions issued in May and July 1955. The ministries, said the May resolution, were using their R & D organizations "badly." Technological stagnation in some ministries had "doomed to backwardness" a series of economic sectors. This situation harmed the state and had arisen even though extensive technological innovation was necessary for the economic victory of socialism over capitalism.[235]

To speed up the pace of innovation the resolution created the State Committee for New Technology (Gostekhnika), already mentioned above. In addition to conducting design competitions for especially important types of technology, Gostekhnika was empowered to oversee the ministries' innovative efforts to upgrade the USSR's international technological position. The May decree also ordered the ministries to increase the resources available to R & D units and improve the supervision of those units. Last, it warned officials from the level of minister to institute and plant director that they bore "personal responsibility" for the rapid introduction of new technology.[236] The July edict sharpened this warning.[237]

Several months later the 20th Party Congress provided evidence that industry's technological performance was still unsatisfactory. Delivering the main Congress report, Khrushchev remarked that past industrial successes had fostered "conceit and complacency" in some officials, leading to "an underestimation of the need for continuous improvement of production." Khrushchev derisively labeled these officials "men in mummy cases" and added that the policies endorsed at the July 1955 Central Committee plenum "were merely a beginning on a great and important job." Coupled with his call for a "ruthless struggle against bureaucracy, that intolerable evil that is doing great harm to our common cause," these words implied that further organizational changes were in the offing.[238]

This prospect was plainly unwelcome to other powerful politicians. Later in the year, when proposals for economic reform were already before the party Presidium, *Kommunist* urged an uncompromising struggle against "nihilistic" criticism of the state apparatus by "insufficiently politically mature people" serving the purposes of imperialism. Khrushchev's ideas were the main target of these strictures, which were backed by members of the future antiparty group.[239] But he continued to press his case, and the next year he forced through a major restructuring of industrial administration.

The 1957 Reorganization of Industry

Khrushchev's blueprint for reform called for an industrial decentralization that was drastic by Soviet standards, but not comprehensive by the standards of comparative economics. He aimed to improve the performance of the central economic bureaucracy by transferring many of its powers and personnel to a large number of regional economic councils (sovnarkhozy), but without relinquishing vital economic decisions to the play of market forces. Khrushchev expected this change to eliminate the inertial policies and ill-informed decisions the ministries had sometimes inflicted on enterprises. The new structure of industrial administration, by bridging the gap that had previously separated Moscow-based R & D experts from the concrete problems of local enterprises, would enlarge the role played by science in the development of production.[240] Politically, Khrushchev saw the reform as a chance to strengthen the regional party apparatus, from which he derived much of his personal support, and to undercut the central state apparatus, where the clients of several of his Presidium rivals were still entrenched.

The political aspect of Khrushchev's proposals hardened the determination of other leaders to resist them. In February 1957 the Central Committee met to consider the reform question and heard a major address from Khrushchev. Lobbying hard for a far-reaching transfer of power to the sovnarkhozy, Khrushchev sweepingly asserted that "there will be no further need to have Union and republic ministries controlling industry and construction." He also argued for decentralizing control over R & D. "The majority of planning, scientific, testing and design organizations at present subordinated to Union and Union-republic ministries," he said, should be transferred to the jurisdiction of the economic councils.[241] But some elite members were unwilling to accept these prescriptions. Khrushchev's speech was published after an unusual six-week delay, and even then in a format emphasizing that it contained only guidelines for further discussion rather than definitive policies.[242] Moreover, in the published version Khrushchev still found it necessary to rebut unnamed persons who advocated forming Moscow-based committees to administer "key branches of heavy industry," and his supporters had to conduct a running battle against the charge that decentralization would destroy the unity of the economy.[243]

Top-level opposition of this kind prevented Khrushchev from obtaining all the decentralization he wanted. The reform discussion was colored by the rising tensions that culminated in his showdown with the antiparty group in June 1957.[244] The resistance of these Presidium rivals, coupled with the need to outflank them by enlisting the military establishment on his side, caused Khrushchev to give ground to the persons who wished to avoid decentralizing the military industries. Although the May 1957 reform abolished most industrial ministries, it spared seven, six of which had an important role in military production.[245] Khrushchev did not accept the status quo for these sectors, but he did temper his earlier proposals. The economic councils, he now said, would manage the enterprises in the defense sectors, but the ministries would administer the R & D programs for these branches.[246]

He also endorsed a form of R & D decentralization in other branches that was more modest than he had earlier sought. While repeating that the majority of industrial R & D units should go to the sovnarkhozy, he agreed that aside from the R & D bodies retained by the surviving ministries, the rest of the "major research institutes" should be transferred to Gosplan.[247] Allowing for institutes kept by the ministries, this means that Gosplan probably acquired about half the central R & D manpower for civilian industry.[248] In these strictly civilian sectors Gosplan was to supervise R & D programs while the sovnarkhozy administered the enterprises. This was roughly the same division of labor between the center and the sovnarkhozy that Khrushchev hoped to impose on the ministries serving military needs.

The ultimate scope of the reorganization, however, still hinged on the vagaries of its implementation, which revealed continuing disagreement. The reform law stipulated that enterprises from the surviving ministries should be transferred to the sovnarkhozy "according to a list approved by the USSR Council of Ministers"—which implied that the enterprises would not be transferred automatically according to their geographical location.[249] The underlying intent seemed to be at odds with Khrushchev's desire to have the sovnarkhozy administer all enterprises; and after several months he took new steps to enforce his conception of the reform. At the end of 1957, having defeated the antiparty group and then ousted Minister of Defense Zhukov from the party Presidium, Khrushchev apparently felt secure enough to abolish most of the surviving industrial ministries directly involved in military production. The four were reorganized into "state committees," and party commentaries stressed that their functions had been "reexamined and narrowed."[250] It is not clear, though, that Khrushchev had succeeded in stripping the state committees of contol over all their enterprises. In 1957 about 13 percent of the share of industry that had been administered directly from Moscow during the previous year remained under Moscow's control. Reportedly the percentage did not decline in 1958 and 1959.[251] While we cannot be certain, it is quite possible that the staffs of the state committees had managed to retain control over some enterprise production, just as they had withstood Khrushchev's earlier efforts to disband them along with other industrial ministries.[252]

Persisting Problems

Such infighting molded industrial administration into a pattern that made more rapid innovation hard to achieve. The difficulty was probably most serious in civilian manufacturing, especially if it is true that military production was more centralized after 1957 than nonmilitary production. Even when the major R & D bodies and the industrial enterprises had been controlled by the same ministry, the linkage between them had been tenuous. Now this problem was compounded by placing the most important industrial R & D units under Moscow's jurisdiction and most of the enterprises for nonmilitary industry under control of the sovnarkhozy. Administratively overburdened, Gosplan gradually shifted control over the R & D units to several additional state committees created after 1958 to set technological policy for particular industrial branches, but this did not solve the problem.[253] Without direct control of the enterprises, these state committees had trouble applying the ideas developed by major R & D establishments. For example, institutes subordinated to the State Committee for Automation and Machine Building developed a new computer for automated industrial processes. Yet the State Committee was unable to have several small batches of these computers produced at a plant controlled by the Severodonetsk sovnarkhoz, because sovnarkhoz officials ignored the State Committee's order and refused to include this task in the plant's production plan.[254]

The criticism the central economic agencies received for poor R & D management made such incidents doubly irritating, and the agencies began to press for restoration of their authority over the enterprises.[255] In 1962 the party responded by giving the state committees more control over local R & D and new plant design, but leaving control over industrial production with the sovnarkhozy.[256] Demands for more centralization continued, however, as shown by Khrushchev's counterarguments. "State Committee officials who want not only to be responsible for science, for design ideas, but also to manage enterprises are holding us back," he asserted. Such officials were striving to return to the "impossible" practices of the industrial ministries.[257]

Representatives of the sovnarkhozy were also dissatisfied with the current management of innovation, but the solution they proposed was further decentralization. In handling major technological projects, the economic councils lacked authority to involve the center's R & D organizations, which frequently refused to cooperate, even when their help was essential.[258] Promising to overcome such impediments, spokesmen for the regional party apparatus and the sovnarkhozy pressed for greater diffusion of control over industrial R & D, and their efforts had a noteworthy institutional effect.[259] After changing little in the mid-1950s, the total number of R & D organizations shot up from about 2,800 in 1957 to around 3,800 in 1961 with most of the additional organizations being set up by the sovnarkhozy.[260] Given the counter-vailing pressures exerted by the new state committees in Moscow, it is uncertain how much R & D manpower at the sovnarkhoz level actually increased. But it probably grew significantly as the sovnarkhozy tried to build up R & D agencies under their own control.

In any event, the performance of the sovnarkhozy did not provide a persuasive argument for the fuller decentralization regional officials desired. A further relaxation of central control over technological de-cisions might have produced benefits if the sovnarkhozy had had a strong interest in fostering innovation, but this was not the case. The councils continued to operate in a planning system which placed a premium on the volume of output without subjecting them to com-petitive, market pressures for technological improvements. Any dim-inution of the center's ability to force innovation consequently invited the sovnarkhozy to neglect this kind of activity. The councils were eager to have more R & D organizations of their own, but this did not mean they were actively committed to innovation.[261] For instance, of the 800-plus targets for new technology in the 1959 plan of the Ukrainian sovnarkhozy, scarcely more than 50 percent were met.[262] Not only did the economic councils fail to fulfill many of these targets, they also tried to avoid embarrassment by having the lagging projects removed from their plans in the course of the planned year. In 1960 a top science planner reported that the Ukrainian sovnarkhozy were pushing to have one-quarter of the current targets for new technology rescinded; the situation, he lamented, was the same in the Russian republic.[263] These pressures were obviously weakening the central planning of innovation, since at the end of 1959 the Ukrainian Gosplan excluded more than 300 unfulfilled projects from the plan for the introduction of new technology.[264]

A further reduction of central control over R & D would have been effective only if the incentives at lower levels of the economic hierarchy had been changed to stimulate spontaneous innovation. In the late 1950s the political authorities started to recognize the importance of such incentives, but the actions they took were too timid to have a substantial effect.[265] At mid-decade the authorities began to set so-called temporary prices, not included in published price catalogues, to spur better innovation. By the early 1960s, however, these prices had gotten out of control and become so widespread that they made eco-nomic calculation extremely difficult. Their most visible effect on innovation was to encourage enterprises to relabel old products as new ones in order to obtain a higher, temporary price.[266] A system of special enterprise premiums for fulfilling plans for new technology was set up in 1957, but the premiums were much smaller than those for fulfilling the regular production plan, and innovation therefore continued to be subordinated to the exigencies of meeting short-term output quotas. In 1960 the government made production bonuses contingent on the fulfillment of the plan for new technology as well as the plan for regular output. But this policy caused managers to write as few innovations as possible into their plans in order to protect their production bonuses; and since new technology was the realm in which the central planners were least equipped to detect managerial evasion, the measure actually reduced the pressure for innovation at the plant level. It was rescinded in 1964.[267] On balance, the incentives for officials at lower levels of the system continued to be stacked against the spontaneous promotion of technological innovation. The authorities had recognized this problem in principle but had not solved it.[268]

Several years after enacting a major industrial reorganization, the Khrushchev regime had made little headway in accelerating the rate of technological advance. In most sectors the hallmarks of organic administration—widespread initiative by subordinate specialists, lateral cooperation across bureaucratic lines, slack resources for innovation, and the easy circulation of information—were still missing. Complaints about the technological inertia of the sovnarkhozy closely resembled earlier criticisms of the ministerial apparatus, and plans for new technology continued to be badly underfulfilled.[269] New equipment was frequently described as not meeting "contemporary requirements," and was sometimes reported to be less economical than the equipment it was intended to replace.[270] Design and testing facilities in the machine-building sectors remained inadequate, while relations among R & D organizations were hampered by autarkic technological behavior on the part of the sovnarkhozy.[271] Finally, the poor circulation of information continued to be a major bottleneck to innovation.[272]

Conclusion

Stalin's death allowed new questions to be raised about the relentlessly traditionalist doctrines codified during his last years. One noteworthy result was a reinvigorated discussion of the probable political and military conduct of the Western powers. A number of elite members gradually concluded that the threat from imperialism had diminished, due largely to Soviet progress in military and industrial technology, and were therefore willing to consider major modifications in the regime's technological policies. Others, however, doubted that the Western danger had diminished as much as leaders like Khrushchev believed, and they consequently resisted efforts to alter the technological policies carried over from Stalin's time. Khrushchev's period in office did not resolve the internal discussion of Western intentions. Rather, it placed the question squarely on the political agenda for the men who deposed him.

Under Khrushchev the official outlook on the Western economies, though it became more sophisticated, changed less than official thought about Western politics. The party retained the traditionalist expectation of ultimate victory in the technological race with the West. The emergence of more objective scholarly analyses of Western economic achievements was, if anything, a symptom of increased confidence rather than growing doubt. At the same time, this development raised the possibility that if Soviet economic performance worsened in the future, such studies, by threatening the elite's economic faith, might become a subject of intense political controversy.

Events during the Khrushchev era focused new attention on the relationship between Soviet technological strategy and diplomacy. The regime eased the country's hermetic isolation in the hope of acquiring more Western know-how and stimulating more effective domestic R & D. But it soon discovered that some Western governments, as they relaxed their embargo on technology transfers to the USSR, were trying to use such transfers as lures to gain Soviet diplomatic concessions. Khrushchev's colleagues prevented him from striking such a bargain over the German question. But the broader issue of exchanging political concessions for technology was bound to reappear as more Western governments resorted to this tactic, especially if the USSR could not upgrade the technological capacity of its own science and industry. Khrushchev presided over several efforts to do so, but with few positive results.

The reorganization of the Academy and the restructuring of the industrial bureaucracy were vigorously contested, both at the top of the party and at lower bureaucratic levels, and some of the resulting administrative compromises made innovation harder rather than easier. Still confident that the USSR was closing the technological gap with the West, the party leadership was not prepared to consider a major market-oriented reform. Whether mounting anxiety over Soviet technological performance would produce a different attitude toward reform remained to be seen.

5

The New Leadership, 1964–1968

The men who deposed Khrushchev in October 1964 ushered in a new style of leadership. Determined to avoid unsettling political shake-ups, they strove to study problems systematically and make decisions collectively.[1] Rational deliberation, they felt, would enable them to arrive at policies more effective than Khrushchev's, and this judgment lent an aura of calm and unanimity to their leadership.

The party oligarchs must have felt that recent political and technological trends gave them good reason to ponder the views and policies laid down by Khrushchev. Had Khrushchev's nontraditionalist evaluation of Western intentions toward the USSR been correct, or did events such as America's post-1960 military buildup and intervention in Vietnam mean the West was more dangerous than he had assumed? Economically, was the traditionalist optimism upheld by Khrushchev still justified? Although still very respectable, the Soviet growth rate had declined, and the industrial West, engaged in governmental promotion of technological change on a new scale, was showing impressive signs of economic vitality. In particular, U.S. growth was speeding up.

Capitalism, Socialism, and Technological Advance

Although the new leaders reaffirmed their allegiance to many of Khrushchev's theses on East-West political relations, they also became more apprehensive about Western intentions toward the socialist world. In this sense, there was a limited movement back toward the traditionalist interpretation of international politics. At the same time, some elite members experienced their first serious doubts about the traditionalist premise that Soviet socialism is technologically more dynamic than Western capitalism.

The elite's views on these two general issues had far-reaching policy implications. The new leaders exhibited a much more cohesive outlook than had existed under Khrushchev, but the fit between their views was not complete, and the succession jockeying inside the Politburo helped inject their differences into policymaking.[2] Attaching special significance to the imperialist threat, several figures stressed the need for more military power and a large degree of economic self-sufficiency in the socialist bloc. In their view, the Soviet system remained economically far more dynamic than its Western rivals. Others, assigning less weight to the Western military danger, worried more about the USSR's ability to close the productivity gap with the West and laid greater emphasis on expanding East-West economic ties. In other words, distinguishable traditionalist and nontraditionalist viewpoints persisted within the elite, although the divisions between the advocates were less dramatic than in Khrushchev's time.

Weighing the Western Military Challenge

The new Soviet leaders continued to endorse the doctrines of peaceful coexistence and the avoidability of world war. After 1964, however, as the widening U.S. involvement in Vietnam heightened international tensions, the oligarchs coupled this endorsement with the assertion that imperialism, particularly its American and West German variants, was becoming more aggressive, and they muted the theme of East-West accommodation.[3] They also allocated more resources to developing and producing new military technology. After a decline of 3.7 percent between 1964 and 1965, the overt defense budget increased by 4.7 percent in 1966 and 8.2 percent in 1967.[4]

Despite a commonly expressed conviction that imperialism was becoming more dangerous, the leaders evaluated the Western military challenge differently and disagreed over increased outlays on Soviet military technology. First Secretary Brezhnev was the most prominent spokesman for the traditionalist view. Politburo members A. N. Shelepin, P. E. Shelest, and M. A. Suslov may have believed even more dogmatically in the traditionalist outlook, but it was Brezhnev who most fully expressed the demand for a rapid military buildup. After the United States began to bomb North Vietnam in 1965, he showed a marked apprehension that U.S. claims of strategic superiority might undermine the USSR's international standing and weaken its influence over other countries.[5] In view of imperialism's dangerous schemes, he stated, "concern about the further strengthening of the defense of our Motherland . . . assumes paramount importance." Military technology was undergoing an unprecedented round of "the deepest qualitative changes," and it would be a mistake not to take into account that the imperialists were reequipping their military establishments with new weapons. The USSR had recently made "important strides" in anti-missile defense. But the regime, while devoting special attention to nuclear rocketry, was also arming its conventional forces with the latest types of tanks, aircraft, and artillery.[6] Suslov likewise argued that the USSR must keep its defense capacities "on the very highest level," improving them continuously, and accept sacrifices in popular welfare to do so.[7] Shelepin drew parallels between Nazi Germany and the United States, which he claimed was bent on world conquest, and promised that during the forthcoming five-year plan the regime would devote "untiring attention" to strengthening the armed forces and defense industry.[8] Military officials echoed this call for rapid advancement of weapons technology.[9]

Other leaders took a softer line. Shortly before the United States began to bomb North Vietnam, Prime Minister Kosygin remarked that the principles of peaceful coexistence were exerting an increasing influence on international relations, and he hailed the prospect of mutual American and Soviet reductions in defense spending as "a definite positive step in the direction of reducing international tension."[10] After the American attacks on North Vietnam early in 1965, Kosygin criticized the United States but seemed reluctant to abandon the possibility of improved relations, commenting that "comparatively recently" Washington had affirmed the need to broaden international contacts in the cause of peace.[11] Viewing international politics from a nontraditionalist angle, Kosygin was unenthusiastic about a military buildup on the scale advocated by Brezhnev and others. In March he observed that the Soviet armed forces "are equipped with the most contemporary types of weapons, unsurpassed in their power." He said nothing about increasing the defense effort.[12] At mid-year he promised that the USSR would continue to strengthen its armed forces, but that this policy might change if disarmament agreements could be reached.[13] More important, Kosygin obliquely suggested that Soviet military technology presently required only limited improvement. He agreed that R & D since World War II had led to unprecedented progress in armaments; but by repeating that the USSR currently possessed the most up-to-date weapons, he subtly suggested that the fruits of this progress had already been incorporated into Soviet weapons programs, rather than requiring a greater military R & D effort, as Brezhnev implied. The regime had recently reorganized the defense industries, Kosygin remarked. "Today our military industry is on the level dictated by the international situation and the contemporary development of military affairs."[14]

Kosygin made the first of these 1965 statements four days after delivering a diatribe against Gosplan's alleged errors in preliminary planning of the new five-year plan for 1966–70—errors that may have been partly connected with the targets for higher military spending.[15] Kosygin's general political outlook and lack of enthusiasm for enlarged expenditures on military technology were shared by Politburo member N. V. Podgornyi. Like Kosygin, Podgornyi criticized American actions in Vietnam and pledged Soviet aid to North Vietnam. But during 1965 and early 1966 he also argued that current trends were leading to the political isolation of aggressive circles in the West, thereby contributing to a Western reappraisal of foreign policy even "in the countries belonging to military blocs." Podgornyi rejected the idea that the defense effort still required sacrifices in popular welfare, and he conspicuously avoided advocating a larger military program to match the increase in international tensions.[16]

The leaders' differences over military technology may have diminished in 1966, but they did not disappear. At the 23rd Party Congress Brezhnev, highlighting the "growing aggressiveness of imperialism," called for "further development of defense industry, improvement of nuclear weaponry and all other types of [military] technology." A few months after the Congress approved this goal, Brezhnev asserted that the regime must "devote still more effort . . . to strengthening our defense might" in order to keep the armed forces "on the very highest level of contemporary military technology."[17] Painting the international scene in even darker colors, Shelepin warned that American imperialism was behaving "more and more irrationally." Because the United States was bent on world domination and was continuing "an unrestrained arms race," the USSR must strengthen the armed forces further.[18] Shelest likewise warned that the threat from American and West German imperialism demanded an unflagging military buildup. The USSR, he said, must constantly be prepared to destroy any aggressor attacking the Soviet Union or other countries of the socialist commonwealth— a definition which, we should note, included any country attacking North Vietnam.[19]

Kosygin, in contrast, took a less adamant position. At the 23rd Party Congress he agreed that the Vietnam war had served as a pretext for the United States to develop a new stage in the arms race, and he noted approvingly that the Soviet defense industry was being continuously upgraded. However, he remarked that if the international situation were not so bad, "we would surely undertake a substantial reduction of military expenditures, a corresponding expansion of capital investments in the peaceful branches of the economy, and a further expansion of the share of [popular] consumption in national income." The arms race, he asserted, was "harmful and dangerous," and the USSR was doing everything in its power to end "this senseless waste of human labor, energy, and resources."[20] Other leaders had not sounded this theme. Moreover, Kosygin was noticeably less eager than some other Politburo members to risk a direct confrontation with the United States over Vietnam. Whereas Shelest had defined the mission of the Soviet armed forces as the defense of the whole socialist commonwealth, Kosygin defined that mission only as the defense "of the Soviet people."[21] At a time when the United States was bombing North Vietnam, this nuance was of no small consequence. Kosygin, in short, was eager to prevent a further escalation of Soviet-American tensions and to preserve the option of curbing the Soviet-American arms race should those tensions decline.

Views of the Economic Race with the West

Disagreement over the proper rate of the military buildup stemmed not only from the leaders' varying perceptions of Western intentions, but also from their views of Soviet economic performance vis-à-vis the West. Led by Brezhnev, the advocates of rapid military development retained a fundamental optimism about the outlook for surpassing the West economically. Kosygin and a few other elite members, on the other hand, were beginning to have serious doubts about the USSR's prospects in the economic race with the capitalist powers.

The fullest brief for the traditionalist view of Soviet economic prospects came from Brezhnev, who asserted that the USSR was outstripping the West. Since 1917 Soviet industrial growth had averaged 10 percent a year, compared with a mere 3.4 percent in the United States. Thanks to the socialist marriage of science and production, the USSR was successfully building the material-technical base of communism and moving into "leading positions in the key areas of world scientific-technological progress."[22] Brezhnev noted that the competition with socialism had prompted Western governments to place "special hopes on the application of methods of state regulation of the economy, on scientific-technological progress, on the increase of military production," but that "this has not led and cannot lead to capitalism's cure from its basic diseases." Although the main capitalist countries had grown faster since World War II than before, "it is impossible not to see that the economy of capitalism continues to remain unstable. Periods of a certain upsurge are replaced by recessions." Together with inflation and other ills, these trends showed that "the latent destructive forces of the capitalist economy are continuing to operate, and it will not escape new shocks." In recent years the USSR, despite certain economic troubles of its own, had continued to gain ground on the West. Its high, stable rates of growth were beyond the West's ability and constituted "an indisputable advantage of socialism over capitalism."[23] Politburo members Suslov, Shelest, and Kirilenko supported this optimistic view.[24]

Led within the Politburo by Kosygin, another group of officials questioned the traditionalists' economic optimism. During 1965 and 1966, Kosygin emphasized that the regime faced serious problems in administering a larger and more complex economy. The contemporary scientific-technological revolution, he said, was becoming the main focus of the economic competition between socialism and capitalism, and the USSR faced "essential inadequacies" in technological innovation that were exerting a "serious influence" on economic growth.[25] Without rapid innovation "it is impossible successfully to solve the task of creating the material-technical base of communism." Moreover, the "course of the economic competition of the two world systems depends in enormous measure on the degree of the development of science and on the scale of the use of research results in production. . . . It is necessary in the near future to create a well-arranged system for . . . the most rapid . . . introduction of the results of scientific research into production."[26] Clearly Kosygin believed that domestic and international pressures demanded an acceleration of Soviet technological innovation and that the optimistic East-West economic prognoses inherited from Khrushchev must be reassessed.[27] Kosygin's statement about the impossibility of building the material-technical base of communism without more rapid innovation was in marked contrast to Brezhnev's contention that the material-technical base was already being successfully built.

Top R & D officials echoed Kosygin's anxiety. Warning that a tremendous worldwide acceleration in the application of research was occurring, Academy President Keldysh observed that the West was "mobilizing all forces and means" to stave off its collapse and was achieving "stormy technological progress." Keldysh stressed that "all countries, not just the socialist but also the highly developed capitalist ones," had undertaken the state organization of science. This meant that the better organization and utilization of Soviet research ranked among "the basic questions of the further development of our country."[28] D. Gvishiani, a deputy chairman of the State Committee for Science and Technology and Kosygin's son-in-law, cautioned that the need to improve innovation should not be construed as a narrow issue. Rather it concerned "the prospects for the further development of the socialist countries . . . [and] the very fate of peace and socialism." In the competition between socialism and capitalism, the system that attained vanguard positions in technology would triumph. The USSR, he said, must improve the administration of innovation. By itself, further quantitative expansion of Soviet R & D resources "cannot give the necessary effect."[29]

Attitudes toward Acquiring Western Technology

These differences within the leadership were linked to latent disagreements over the need for Western technology. Broadly speaking, all the top policymakers believed that the USSR should tighten its economic links with other socialist countries while building up technological ties with the more accommodating capitalist nations such as France, Italy, and Great Britain (but not the United States or West Germany). Leaders with a traditionalist cast of mind, however, put more emphasis on economic relations with other socialist partners, whereas nontraditionalists showed more desire to obtain Western knowhow.

Brezhnev exemplified the traditionalist outlook. At the 23rd Party Congress he briefly mentioned the desirability of widening economic and scientific ties with the West,[30] but argued more forcefully for enlarged technological cooperation with other socialist countries. Economic relations within the socialist bloc had entered a "new stage," and CEMA's significance was "increasingly growing. . . . Only in this way can the national economies of the socialist countries keep pace with the stormy scientific-technological revolution of our day, assuring the conditions for further successes in the economic competition with capitalism."[31] Brezhnev identified the imperatives of the scientific-technological revolution with closer intrabloc relations rather than with expanded East-West ties.[32] No doubt one reason for his attitude was the belief that the USSR and its socialist allies, being technologically more dynamic than the West, need not press for larger technology transfers from the capitalist world.

Kosygin, worried about Soviet technological performance, showed a much stronger interest in economic relations with the West. He told the 23rd Party Congress that "it is becoming more and more obvious that the scientific-technological revolution . . . demands freer international economic intercourse and creates the prerequisites for broad economic exchange between the socialist countries and the countries of the capitalist system." During the last five-year plan, he continued, "foreign trade helped us to solve a series of important economic tasks." However, due partly to bad decisions by trade and industrial officials, "we are still not using sufficiently the possibilities which the development of foreign economic ties is opening before us." The USSR should buy more licenses for Western technology, in order to save "hundreds of millions of rubles on scientific research."[33] While Kosygin endorsed the idea of strengthening economic relations with other CEMA members, he identified the imperatives of the scientific-technological revolution with a major expansion of East-West economic links. During a visit to France he remarked that neglect of foreign scientific achievements and an attempt at self-isolation "on the part of any state" would inevitably produce economic failures, as well as "a loss of a sense of reality in politics." The USSR, he said, favored broad international cooperation in science and production, "and not only on a regional scale in particular areas of the world"—clearly an allusion to CEMA. France and the USSR, he suggested, might develop special production facilities "and even whole [economic] branches" to supply each other over the long term.[34] Such overtures served the useful diplomatic purpose of encouraging the tensions between France and the United States, but they had a serious economic motive as well. While Kosygin concentrated his appeals for wider economic ties on Western Europe, he also held out some hope for fuller technological relations with the United States. Once the United States ceased its aggression in Vietnam, he said, "a large group of questions" could be addressed, including expansion of Soviet-American scientific and economic exchanges.[35]

Soviet Economic Needs versus Socialist Bloc Cohesion

The conflict over how to interpret Soviet economic performance and Western technological trends sharpened in 1967 and burst into the open in 1968. One area of controversy was whether to give higher priority to acquiring Western industrial technology or to concentrate on shoring up the economic and ideological cohesion of the socialist bloc. This dispute was intensified by Western efforts at diplomatic "bridge-building" to Eastern Europe and further exacerbated by the Czechoslovak reform movement, which threatened the foundations of political legitimacy throughout the bloc. A second source of disagreement was how fast to push the buildup of Soviet military technology and whether to enter into strategic arms limitation talks (SALT) with the United States. Both issues pitted traditionalists against nontraditionalists, and the debates over the two questions were intertwined. For the sake of clarity, however, I will examine the issues separately, beginning with the dispute over the acquisition of Western technology.

In February 1968 Kosygin provoked a clash over the pursuit of Western technology by delivering a lengthy nontraditionalist evaluation of the economic race with the West. The leadership had just begun assessing current scientific and technological trends as a basis for formulating the 1971–5 five-year plan, and Kosygin was obviously trying to influence his colleagues' ideas about the plan.[36] Underscoring "many unsolved tasks," Kosygin pointed out that the USSR still trailed the United States in national income, industrial output, and especially labor productivity. Low Soviet labor productivity, he asserted, stemmed not from the inferiority of the work force, but from the shortcomings of the persons who administered production. More efficiency and faster growth were necessary because the current trade-offs among defense, investment, and consumption were restricting the improvement of living standards. Kosygin clearly thought better technology was an answer to this problem, but he was not optimistic about improving the rate of indigenous innovation. Voicing deep dissatisfaction with the yield from Soviet R & D, he singled out the weaknesses of the computer industry.[37] Scientists and engineers must expend resources "wisely, in order to receive more rapidly the very greatest effect from science." "In this respect," he added, "it is necessary to subject to criticism the activities of many of our . . . institutes, design bureaus, and enterprise engineering services. Often scientific and engineering cadres occupy themselves with reproducing innovations which were created abroad long ago and which, incidentally, are far from the best."[38]

Western dynamism made the sluggish Soviet technological performance an urgent problem, Kosygin argued: "We must not be content with the fact that from the beginning of our scientific-technological research until its assimilation into current production, many years go by. In this way they can outstrip us." In the West "the monopolies are forced to wage a sharp fight for profit, to react quickly to the demands of the consumer, and to put out modern types of production, to seek the most rational forms of productive organization."[39] The contrast between these words and the traditionalist assessments offered by other leaders was sharp indeed. Others contended that the Soviet Union was steadily closing the technological gap with Western states beset by increasing economic instability. Kosygin was warning that technologically the Western powers might outstrip the USSR.

As a counter to this danger, Kosygin mentioned recent Soviet experience in developing economic and political ties with several West European countries. In addition to making diplomatic inroads into the American-led NATO bloc, he noted, such relations offered valuable access to advanced technology. It would be "shortsighted" not to take advantage of all the latest foreign R & D achievements, and industrial leaders and researchers were obligated to avoid this pitfall. "We should use all the best new technology and take every opportunity to buy licenses . . . so as to accelerate technological progress in our economy."[40] Kosygin was not writing off Soviet R & D, but he was calling for a major shift from domestic to Western sources of new technology.

The speech provoked a vigorous riposte from Politburo traditionalists. Addressing an audience that included Kosygin and other Politburo members, Brezhnev conceded that certain branches of Soviet science had shortcomings, and many technological novelties were being applied slowly. However, in discussing such problems "particular officials clearly underestimate the achievements of scientific-technological thought in our country and the other countries of socialism. At the same time these people are inclined to exaggerate the achievements of science and technology in the capitalist world." In reality, Brezhnev contended, socialist planning ensures "unprecedented possibilities" for technological progress.[41] Soviet researchers had made "very great" scientific discoveries; along with other CEMA specialists, they had done "a great deal" to advance technology in many fields, such as electronics and telemechanics.[42]

As part of this effort to tone down the image of Western dynamism, Brezhnev argued that the surge of Western economic growth since 1945 had come to an end. It was being replaced "by a series of crisis slumps whose scope surpasses everything that was known over the last quarter-century," making a "deep economic crisis of the capitalist system" a distinct possibility. He urged that "we must talk about this at the top of our voices, comrades, because this is the truth, this is a persuasive weapon in the ideological struggle." Imperialism, he explained, seizes tenaciously on any ideologically immature elements of the Soviet intelligentsia.[43] Undoubtedly Kosygin was among the anonymous "officials" Brezhnev was censuring. Like most elite members, Kosygin felt no sympathy for the political revisionism gaining currency among Soviet dissidents, as well as in wider circles of Czechoslovakian society.[44] Brezhnev, however, evidently felt that the ideas advocated by Kosygin were encouraging such revisionism. From his standpoint, revisionism was both an internal problem for each socialist country and a potential threat to the socialist bloc's solidarity against Western diplomatic blandishments.

In particular, Brezhnev feared that a vigorous Soviet bid for Western technology might encourage some East European regimes to strike unacceptable diplomatic bargains with the West in order to obtain similar benefits. He seemed especially concerned about the stepped-up efforts of a new West German government, which still refused to recognize the rival German Democratic Republic, to use economic inducements to gain diplomatic recognition from the German Democratic Republic's socialist allies.[45] Apart from assailing this tactic and touting East Germany's ability "to develop successfully in keeping with the requirements of the scientific-technological revolution," Brezhnev had warned wavering East European officials that imperialism was using increasingly sophisticated means to undermine the political and economic unity of the socialist bloc.[46] During 1968 the Politburo's economic deliberations, the mounting East European appetite for Western technology,[47] and the turbulent situation in Czechoslovakia made Brezhnev still more apprehensive. Two days after Kosygin's February speech, Brezhnev stated that the difficulty of reconciling national peculiarities within the socialist bloc meant that intrabloc economic cooperation had exceptional significance. Such cooperation, he cautioned, "has become an objective requirement of our countries, but it is not established automatically but under the influence of the conscious will of communists. This matter manifests the responsibility of the socialist countries . . . and their ability to apply Marxist political categories in the practice of economic ties."[48] The next month, representatives of the Warsaw Pact countries met to discuss the foreign policy implications of the Czechoslovak reform movement. The representatives, expressing concern about the role of West Germany and the "aggressive intentions" of imperialism against the socialist bloc, reportedly criticized Czech-oslovakia for its willingness to accept West German loans and offered bloc loans instead.[49] A few days later Brezhnev, in his attack on persons who exaggerated the level of Western technology, again singled out intrabloc ties as a valuable source of advanced foreign technology.[50]

The leaders were plainly divided. Should they adopt an unyielding stance toward the West in order to strengthen the socialist bloc against ideological erosion and Western diplomatic inroads? Or should they put more emphasis on obtaining the Western know-how needed by the Soviet economy, hoping in the process to strengthen the forces in the West that favored a more conciliatory policy toward the USSR? Traditionalists and nontraditionalists gave conflicting answers to these questions.[51]

The Question of Strategic Arms Limitations

The debate over Western intentions and comparative economic per-formance was also linked to Soviet policy toward strategic arms lim-itations. In January 1967 President Lyndon Johnson proposed that the United States and the Soviet Union begin negotiations to slow the race in strategic weapons. An exchange of private messages between Johnson and Kosygin indicated a Soviet desire to explore this possibility, but not a readiness to set a time and place for talks.[52] By early 1968, however, it became public knowledge that the United States planned to deploy an antiballistic missile (ABM) system that might conceivably become the basis of a large defensive network intended to neutralize recent Soviet missile gains.[53] Moreover, U.S. officials, announcing the devel-opment of multiple offensive warheads (MIRVs) that would purportedly counterbalance or even outweigh the recent Soviet missile buildup, moved to finish testing and begin producing these weapons.[54] It thus appeared that the United States, which had just completed the de-ployment of a large number of ICBMs and shown signs of wanting to moderate the competition in such weapons, might embark in another round of strategic arms expansion. This made the definition of American intentions and technological capacities an essential starting point in Soviet decisions on new weapons programs and SALT negotiations.

The American initiatives posed difficult choices for the Soviet leaders, and they had trouble agreeing how to respond. Kosygin was cautious but showed some receptivity to the idea of strategic talks. He endorsed increased Soviet defenses, saying that the USSR could not ignore the continuing U.S. aggression in Vietnam and deteriorating Soviet relations with Communist China.[55] But he also said disarmament would enable the "enormous means" currently spent on the arms race to be diverted to meet the needs of "millions of people in many countries of the world."[56] In June, following the Glassboro meeting at which Johnson made a strong personal appeal to Kosygin for an early beginning of SALT talks, Kosygin commented that limits on ABM and offensive systems must be linked—which seemed to imply that under the right conditions he would favor negotiations.[57] In October, when he sharply criticized American conduct in Vietnam, the conclusions he drew for Soviet military policy were surprisingly mild. Observing that the complicated international situation had to be reflected in "all the activity" of the regime, he cited as examples foreign policy, economic development, and only in third place, strengthening the military. The party was doing "everything possible," he remarked, to improve the technology supplied to the defense establishment.[58] In an address on the anniversary of the Revolution, Kosygin praised the country's "glorious Armed Forces" but conspicuously failed to call for their further strengthening.[59]

Several other Politburo members disagreed, however, and their opposition evidently kept Kosygin from taking up the renewed American offer at Glassboro.[60] Four days after Kosygin returned to Moscow with a report on his meetings with Johnson, Brezhnev, adducing recent U.S. actions in Vietnam, argued that "the treachery of imperialist reaction" required "still greater attention to the strengthening and improvement of the armed forces." Although the armed forces currently possessed awesome weaponry, Brezhnev warned, "this must not cause complacency. In military affairs, more than in any other [field of activity], we must not mark time. The technological equipping of the army and navy must constantly be on the level of the newest achievements of scientific-technological thought."[61] The timing of this warning strongly suggests that it was addressed to Kosygin. Shelepin backed Brezhnev by describing the international situation in especially stark terms. Its key features were "an unceasing arms race," "the growing aggressiveness of imperialism, headed by the ruling circles of the U.S.A." in Vietnam and in many other countries, and a repressive assault by imperialist monopolies on the rights of Western workers.[62] Suslov similarly asserted that the American bloc was seeking to escape deepening internal contradictions through an aggressive foreign policy, intensified expansion, and new military adventures.[63] These traditionalist appraisals of the international scene implied that the USSR should force the development and production of military technology.

None of these leaders mentioned the possibility of strategic arms limitations, partly because they remained confident the Soviet Union had the wherewithal to match the United States in an arms race. Brezhnev and Shelepin reiterated their faith in the economic superiority of the socialist system,[64] and Suslov argued the case in detail. He conceded to the nontraditionalists that the capitalist camp must be evaluated "soberly and realistically." But he found few economic trends to worry about. Over the last half-century the socialist system had passed "an unprecedented historical test" of its defense capacity and economic effectiveness. Enemies had tried it "by fire and sword," "by 'cold war' and a harsh competition in the field of development of the economy and science." The socialist system, Suslov asserted, "has passed through all these trials with honor, and has demonstrated indisputable advantages over the capitalist system in political, military, socioeconomic and scientific-technological respects." Recent changes in imperialism only reinforced this conclusion. The shift from monopoly capitalism to state-monopoly capitalism "further sharpens the internal and external contradictions of imperialism. Neither the strengthening of bourgeois state intervention in the economy, the creation of international state-monopolistic associations of the type of the 'Common Market,' nor the arms race in the U.S.A. and other capitalist states can stop the growth of the economic and political instability of contemporary imperialism," he claimed.[65] In other words, Suslov dismissed the possibility of a basic shift in the West's politics and its capacity to marshal technological resources for economic and military ends.

During 1967 no other leader directly disputed these traditionalist conclusions, but the confident opposition of a few academic specialists indicates that someone at the top felt the conclusions were wrong. In October, about the time Kosygin was implying his reluctance to expand military programs, the Academy's Institute of the World Economy and International Relations (IMEMO) convoked a large conference to offer the kind of policy-related analysis the party had recently urged social scientists to provide.[66] There the Institute director, N. N. Inozemtsev, expounded a nontraditionalist view of Western politics. He began with a bow to traditionalist orthodoxy, saying that Lenin's assertion that imperialism manifests a tendency "toward increasing reaction on all fronts" was still valid. But, he hastened to add, Lenin's proposition that this tendency "inevitably gives rise to an opposing tendency" was also still true. The opposing tendency consisted of an intensifying struggle through which the democratic masses in the West rebuff the monopolists' political offensive. This countertendency, Inozemtsev suggested, had a direct bearing on the struggle for a reduction of international tension and the affirmation of the principles of peaceful coexistence in international affairs. Obviously convinced that these principles were gaining ground, he offered the overall judgment that the USSR faced "a much more favorable international situation" than in the past.[67]

A. G. Mileikovskii, like Inozemtsev a corresponding member of the Academy, elaborated. In the industrial West "bourgeois reformism," the willingness to make concessions to the workers in order to preserve the system, had become the dominant political trend. The most reactionary imperialist circles, he granted, still embraced the ideas of "small wars" (that is, limited wars) and an atomic blitzkrieg. But this danger was being neutralized by the successes of socialism and of world democratic forces. The result was a new capitalist phase centering on economic competition with socialism, "without war, in a situation of relative peace."[68] Mileikovskii's ostentatious failure to mention the Vietnam conflict made his dismissal of "small wars" and his accent on "relative peace" especially striking. Moreover, by stating that the Western proponents of atomic war had been politically neutralized, he implicitly suggested the feasibility of agreeing with the United States to limit strategic arms. His portrayal of the West's moderate internal politics and temperate foreign conduct contrasted dramatically with the traditionalist image of mounting repression and imperialist aggressiveness presented by Brezhnev, Suslov, and especially Shelepin. Lest the relevance to Soviet policy go unnoticed, Inozemtsev stressed that the points made by himself and his colleagues were not simply theoretical but had "paramount practical significance."[69]

Inozemtsev's incisive treatment of economic issues further buttressed the case for curbing the East-West race in military technology. The USSR, he said, faced "a series of new questions" concerning economic administration and the promotion of science and technology. As for Western economic trends, any attempt to examine these trends abstractly, in isolation from the adaptive steps outside pressures had forced the West to take, would be wrong. Competition with the socialist camp had compelled Western regimes to adopt measures they would otherwise have spurned. In particular, government regulation of the capitalist economy and promotion of R & D had become "a question of life and death" for the West. Inozemtsev then subtly indicated that even party leaders could misjudge such matters by reminding his audience of Stalin's predictions in *Economic Problems of Socialism in the U.S.S.R.* Stalin had predicted that the economic output of the Western powers would shrink. In actuality, Western postwar industrial growth had averaged almost 6 percent, significantly higher than during the interwar period. "It is completely obvious," said Inozemtsev pointedly, "that if we based our strategic estimates on the assumption that capitalism's productive forces are bottled up and that capitalism entails an inevitable regression in the scientific-technological field, then this would have the most negative consequences both for our country and for the world communist movement."[70] Although Inozemtsev may have meant the term "strategic estimates" to encompass more than strategic weapons, he must have regarded the strategic military balance as a key part of these assessments. His reference to such estimates, and especially to possible errors in them, was extremely rare for a Soviet commentator. Together with his depiction of the West's political moderation, the passage implied that the USSR would be wise to enter into SALT negotiations, and the allusion to the "world communist movement" was probably meant to suggest that this step would serve the interest even of embattled North Vietnam.

Inozemtsev clearly felt that some Soviet policymakers had not yet taken his lesson to heart, since he connected it explicitly with the "special significance" of IMEMO's forecasts of future economic and political trends in international relations.[71] Despite Suslov's claims to the contrary, Inozemtsev was arguing that the traditionalist assessment of Western technological capabilities was neither "realistic" nor "sober." Many participants in the IMEMO conference probably recalled that Suslov had been a major exponent of Stalin's *Economic Problems* when it appeared in 1952, and they cannot have failed to grasp that Inozemtsev's comments directly contradicted the traditionalist interpretation of faltering Western economic and technological development being advocated by Suslov, Brezhnev, and others.[72]

The IMEMO symposium was almost certainly part of a coordinated effort to present nontraditionalist views both to Politburo members and foreign policy specialists. Held late in October 1967, the symposium was published in abbreviated form in January 1968. The account of the proceedings was probably meant to pave the way for Kosygin's controversial speech the following month by airing the scholarly justifications for nontraditionalist policies. Many points made at the symposium were reflected, albeit less dramatically, in Kosygin's address.

Kosygin's February speech showed his concern about the strategic military implications of the East-West race for growth and innovation. He granted that the West was experiencing internal strains, including the financial problems traditionalist leaders cited as evidence for their views.[73] He cautioned, however, that Soviet observers "cannot oversimplify the situation. Imperialism still possesses a large economic and military potential." Significantly, Kosygin chose to cite Western military potential rather than forces in being, thereby playing down the present threat and indicating that it could become much greater in the future. The distinction took on added meaning in view of his telling remark, discussed above, that the West could conceivably outstrip the USSR technologically. In short, he was worried about the Soviet ability to match future Western strides in military (as well as civilian) technology. Kosygin's attitude toward the prospect of a full-blown arms race differed sharply from the traditionalists' belief in the USSR's unquestionable superiority in developing science, the economy, and defense.

Kosygin was plainly concerned about the economic burden of Soviet defense spending—especially given the massive 15.2 percent increase in the 1968 defense budget that the traditionalists had just pushed through.[74] Current budgetary trade-offs, he said, were very difficult. Investment could not be cut, since this would slow economic growth. Defense spending could not be cut in view of "a certain complication" of the world scene. And yet consumer welfare must be increased more rapidly. As shown above, part of Kosygin's solution to this dilemma was to acquire more Western industrial technology. But he also hoped to ease the economic pressures by restricting the arms race, particularly by starting Soviet-American SALT negotiations.

To this end, Kosygin suggested that Western intentions toward the socialist world were not implacably hostile. He paid lip service to the traditionalist thesis that imperialism was becoming increasingly aggressive, but he also made a number of statements that ran against this thesis. The USSR had recently accumulated interesting experiences of cooperation with "a series of major capitalist states on the basis of the principles of peaceful coexistence." No matter how much the United States tried to disrupt these cooperative ties in order to maintain its system of military alliances, the ties would continue.[75] Kosygin, however, did more than justify peaceful coexistence as a tactic to frustrate U.S. foreign policy; he showed an interest in improving relations with Amer-

ica itself. Discussing the Soviet-American draft of a nuclear nonproliferation treaty, he said: "As a result of long and, one must say, rather
difficult negotiations, we succeeded in obtaining the agreement of the
U.S.A. on a draft agreement . . . which removes the question of creating
NATO multilateral atomic forces, forecloses access by the West German
revanchists to atomic weaponry and stipulates reliable supervision over
states' fulfillment of [the treaty] obligations." The obvious implication
was that though Soviet-American negotiations might be arduous, the
USSR could persuade the United States to make concessions that could
be reliably policed and were of real strategic benefit. Moreover, Kosygin
tried to get around the sticking point of Vietnam by intimating that
changes in American policy toward that conflict might be in the offing.
It was still difficult to determine, he said, whether the United States
would elect to seek a political settlement of the war. But the understanding that military victory was impossible was growing "in the
United States of America itself. Increasingly wider and more influential
circles . . . are coming out in favor of seeking to solve the Vietnamese
question by means of political negotiations."[76] This was a far more
optimistic prognosis than that of Shelepin, who had recently highlighted
the "unceasing arms race" and condemned American ruling circles for
leading an aggressive crusade against the communist world. It also
contradicted Brezhnev's statement that the American pretense of interest
in a Vietnamese political settlement had been unmasked "once and
for all."[77]

In a March address, Brezhnev countered Kosygin's worries about
technological lags and conflicts between defense and economic needs
by accenting the USSR's superior technological dynamism. He also
remarked critically that "our scholars and propagandist-internationalists
could pay more attention to . . . those socioeconomic and political processes which are currently unfolding in the capitalist world and making
it feverish." Soviet commentators should proclaim this "truth" at the
top of their voices.[78] This admonition was clearly aimed at IMEMO's
analysts, who had made forecasts of Western trends contradicting
Brezhnev's forecasts and had energetically pointed out the policy implications of their predictions. Suslov seconded Brezhnev with an attack
on those "apologists of capitalism" who were falsely trying to "defend
its right to existence" by citing stepped-up Western economic development since World War II,[79] and *Kommunist* pointed the moral. The
USSR's technological level gave it a "realistic prospect" of complete
victory over capitalism and "presently permits the Soviet Union

simultaneously to develop rapidly the production of producer goods, the production of consumer goods, and to strengthen the defensive might of our state."[80]

Sustained by this view, traditionalists saw little reason for strategic arms limitations. In his speech, Brezhnev made no reference to the draft nuclear nonproliferation treaty that Kosygin had implied might set a precedent for a future SALT agreement. Three months later Brezhnev voiced his skepticism more clearly. The Soviet military absorbed large expenditures, he said, but these were necessary, and as long as imperialism resisted social progress, they would continue. He added emphatically: "Again and again we are reminded of the necessity for this by the contemporary international situation, especially by . . . the aggression of American imperialism in Vietnam, of Israel in the Middle East, and the activation of the forces of reaction . . . in West Germany."[81] Suslov also underscored the need for an untiring defense effort.[82] In answer to Kosygin, Brezhnev acknowledged that American opponents of the Vietnam war and the arms race were raising their voices, and he granted that progress on the nonproliferation treaty was a positive sign. However, "while giving these phenomena in U.S. political life their due, we cannot close our eyes to the fact that the proponents of aggression, or as they are called, the 'hawks,' retain their position."[83] In other words, the political prospects for strategic arms control, as well as the economic need for it, were being exaggerated by nontraditionalists.

The debate over whether to enter into SALT negotiations peaked in May and June of 1968. In May, First Deputy Foreign Minister Kuznetsov told a U.N. audience that the USSR was "ready to reach an agreement on practical steps for the limitation and consequent reduction of the strategic means for delivering nuclear weaponry."[84] This was not a firm Soviet commitment to begin SALT talks, however, since private bargaining over the possibility of the talks was still going on between Johnson and Kosygin in June.[85] Kuznetsov's statement apparently represented an effort by some Politburo members, acting with the aid of the Ministry of Foreign Affairs, to force the issue,[86] and it provoked a militant rejoinder from the defense establishment.

Military writers had already published articles showing resistance to the notion of strategic arms curbs, but these warnings paled beside a remarkably belligerent article signed for printing two days after Kuznetsov's U.N. speech.[87] The author, Colonel I. Grudinin, argued that the heightened military danger created by imperialism demanded a further expansion of the armed forces to attain a "stable supremacy" of Soviet military power. Emphasizing that the introduction of nuclear rockets and missile guidance systems had triggered drastic, ongoing changes in weaponry, he asserted that the armed services must be equipped "with the newest military technology. So long as aggressive forces exist, military technology that is becoming obsolescent must be replaced in a timely fashion with new technology, and new technology with technology that is newer still." Since this was "one of the most important laws" of military preparedness, simple expansion of the armed forces could not overcome serious inadequacies in the quality of military technology. Nor, for that matter, could quality be substituted for quantity. Driven by a "military psychosis," the imperialists were expanding their armed forces, particularly their rocket forces, at an accelerated rate, and the USSR had to have a number of weapons adequate to meet this threat. In case his point was not yet clear, Grudinin concluded by assailing unnamed Soviet writers who misunderstood the Leninist precept that the USSR must have "an *overwhelming* preponderance of force," particularly in nuclear rockets. The USSR and its allies had a "real potential" to destroy any aggressor. But to make this potential a reality, the country must constantly increase the fighting capacity of every military unit.[88] Permeated by traditionalist values, the article was a clarion call for an all-out military buildup and the avoidance of SALT negotiations.

Less than a week later, the IMEMO journal published an article by Inozemtsev countering this view. Inozemtsev contended that during World War II fascism had suffered a "crushing defeat" that had made a lasting impact on the political features of the imperialist states. Forced to abandon direct repression, the regimes of the industrial West had offered "real concessions," making the situation of Western workers "immeasurably more favorable" than during the 1930s.[89] The implication was that no belligerent, fascist regime was likely to come to power in the developed West in the foreseeable future. Presumably Inozemtsev also meant to intimate that the Western proponents of a nuclear blitzkrieg remained politically neutralized. Along with this benign depiction of Western politics, Inozemtsev painted a picture of Western economic and technological prowess based on state guidance of the economy and stimulation of technological change. He still professed that the "historical doom" of capitalism was assured. "But it would be incorrect and dangerous to draw from this a conclusion about any sort of 'automatic' collapse of imperialism, not to see its enormous vitality, its striving to adapt to new conditions," he warned.[90] Plainly he was unpersuaded by Brezhnev's argument two months earlier that there had been "no revival of the 'vitality' of capitalism."[91] Inozemtsev's latent message was that SALT negotiations with the foremost imperialist power were politically feasible and technologically necessary. Refusal to undertake them would derive from a basic error in formulating Soviet strategic estimates and would entail a dangerous exaggeration of the Soviet ability to compete successfully against the United States in an all-out missile race.

Finally, after more Soviet-American negotiations and internal Soviet bargaining, at the end of June Foreign Minister Gromyko announced the USSR's clear commitment to undertake SALT talks. This step, which coincided with completion of the nonproliferation treaty, was hailed by nontraditionalists. Kosygin again lauded the nonproliferation treaty as "a major success for the cause of peace," expressing the view that it would smooth the way for strategic arms negotiations,[92] and other Politburo members shifted publicly toward his position. Kirill Mazurov, for example, voiced concern over America's intensified efforts to widen its lead over the socialist camp in science and technology, and he hinted that the constellation of political forces in the West had recently swung against maintaining the arms race. Like Kosygin, he expressed a wish to limit the arms competition and praised the nonproliferation treaty. Keeping a noteworthy silence about strengthening the Soviet military, he observed that the Soviet initiative to limit strategic arms had been "positively received" by the outside world, presumably including the United States.[93]

Resistance persisted, however, within the Politburo and at lower levels. In his announcement of Soviet willingness to engage in talks, Gromyko, noting that some observers had argued that the arms race was inevitable, gave a vigorous riposte to the "good-for-nothing theorists who try to reproach us on the grounds . . . that disarmament is an illusion."[94] But some opponents were unconvinced. About a week after Gromyko's speech Shelest proclaimed that the forces of imperialist aggression "are resorting on a constantly increasing scale to actions which are extremely dangerous for the cause of peace." Shelest cited Vietnam, the Middle East, and West Germany as cases in point. Unlike Kosygin and other nontraditionalists, Shelest maintained a stony silence about the recently concluded nonproliferation treaty and the announcement that SALT talks would begin, and he argued that the international situation required a further strengthening of Soviet defenses.[95] On the basis of earlier statements and comments made after the talks began, we may reasonably suppose that Shelepin and Suslov likewise opposed the decision to begin negotiations.[96] With backing from the military, these top-level opponents were powerful enough to delete several references to the forthcoming talks from the published speeches of SALT supporters.[97]

The decision to enter the talks, in short, did not signify consensus about the desirability of an agreement or a deceleration of the race in military technology.[98] Perhaps the explanation of why the nontraditionalists managed to prevail under these conditions is that Brezhnev now adopted a less categorical stance on the question. He vigorously asserted that the hawks were still in charge in America and urged a rapid buildup of Soviet military technology. But in July he also briefly alluded to the nonproliferation treaty and the agreement to begin SALT talks, thus implying qualified support for the decision to start negotiations. Brezhnev apparently backed the talks for tactical reasons— perhaps to encourage antimilitary sentiment in the West or as a "peace" gesture to minimize the damage to Soviet diplomacy should the USSR decide to invade Czechoslovakia.[99] His attitude toward East-West relations was just beginning to evolve toward a less traditionalist position. Whether the USSR should actually conclude a SALT agreement soon became the subject of another cycle of internal debate, in which the Soviet economy's comparative technological performance was again the focus of sharp contention and in which Brezhnev played a very different role. Before examining that debate, however, we must bring the other parts of our analysis up to 1969.

Technological Strategy

The controversy in leading party circles over Soviet R & D capacities and external economic relations encouraged secondary elite members, who had previously hesitated to speak out, to discuss technological strategy more openly. Top science planners signaled that they desired more contact with foreign scientists and engineers. Petr Kapitsa, a member of the Academy Presidium, expressed this sentiment in an article commemorating the 200th anniversary of the death of the famous Russian scientist Lomonosov. In a tour de force of Aesopian communication, Kapitsa asked why Lomonosov's scientific discoveries had not won recognition in his own time. A major cause, he said, was Lomonosov's inability to travel abroad and maintain personal contacts with foreign scientists. This impediment also kept Lomonosov from achieving his full scientific potential. By contrast, suggested Kapitsa, another eighteenth-century scientist, Ben Franklin, won recognition by traveling to London and speaking to the Royal Society about his discoveries. No doubt some readers knew that Kapitsa had been prevented from traveling to England to continue his research there in 1934, and many more must have grasped the implied contrast between the conditions of scientific work in Russia and America. While Kapitsa did not openly voice a demand for freer scientific travel, he implied it by arguing forcefully that personal contacts were indispensable for scientific progress and by asserting that the task of making Soviet science truly advanced was still unsolved.[100] Kapitsa had a special stake in this matter; he had not been allowed to go abroad since 1934, and he wished to go to Denmark the following month to receive a prestigious scientific award (which he was finally permitted to do).[101] But the article could not have been printed in a large-circulation newspaper just to communicate his personal wishes. It fit neatly into the long history of pressures from the Academy for expanded international contacts, and we may fairly conclude that it signified broad support for this goal within the scientific establishment.

This support, however, did not entail a blanket endorsement of increased reliance on Western technology. Soviet scientists wanted more contact with the West in order to build up Soviet R & D across the board, rather than to facilitate R & D cutbacks in selected domestic sectors. Academy President Keldysh, for example, said that more rapid innovation required "maximum use of all world science. At the same time, we must strive to achieve the very highest level in the world in the decisive fields of science and technology."[102] A year later, perhaps reacting to Kosygin's effort to curb the R & D budget, he shifted his principal emphasis to the primacy of domestic research: "Although dependence on the accomplishments of world science . . . can allow us to fulfill individual complex programs, a general and high level of technological progress in our country has become possible only on the basis of . . . the planned development of fundamental research along an extended front."[103] Keldysh made no mention of savings on R & D that might accrue from wider contacts with foreign science, either East European or Western. Kosygin and other political leaders could count on scientists to back closer contacts with the West, but not to support efforts at specialization that would restrict the scope of their own research.

Top science officials also indicated their reservations about national R & D specialization by implying that under the right conditions, Soviet research could meet the economy's needs without foreign products and licenses. M. D. Millionshchikov, an Academy Vice-President, remarked that there were still cases in which major Soviet scientific discoveries did not produce high-quality Soviet technology or were first applied abroad. Chemistry, he said, was an example. Asking what set such lagging fields apart from those in which the USSR was technologically advanced, he answered that the key was close coordination among research units and the generous provision of resources, especially for development. Maintenance of R & D organizations, he said, entailed "a certain burden for our economy. However, as a result equipment of high quality is created . . . . Economic calculations will show, in my opinion, that expenditures on maintaining project-design institutes and test facilities . . . are fully compensated by the . . . advantages just enumerated."[104] The moral seemed to be that policymakers should avoid the kind of impatience that had led Khrushchev to discount Soviet chemical R & D and to turn to the West for chemical technology.

V. Trapeznikov, First Chairman of the State Committee for Science and Technology, provided figures that conveniently supported Millionshchikov's budgetary claims. The economic return from science expenditures, he asserted, was almost four times larger than that from other forms of capital investment; therefore R & D spending should be nearly doubled.[105] About the same time the State Committee prepared a study that slated "expected" R & D expenditures to grow between 1966 and 1967 at the highest annual rate since 1960.[106] Top R & D administrators wanted more contact with the West, but they wanted the contact in a form that would strengthen, not undercut, their own programs and budgets.

Trade officials also favored closer ties with the West, though their particular institutional interests caused them to envision such ties somewhat differently than did R & D planners. Most trade officials desired a dramatic expansion of imports from the industrial West as well as from other CEMA countries. The journal of the Ministry of Foreign Trade printed a forceful repudiation of the notion of autarky— for the USSR alone or for the CEMA countries as a group—and Ministry officials regularly dwelt on the economic advantages to be obtained from the international division of labor.[107] In view of this emphasis, the dark colors in which these same officials depicted the West's desire to dominate socialist resources and markets seems surprising.[108] The explanation is that Ministry officials wished to expand the flow of technology from the West but wanted to keep it under their own control. Recently Kosygin had chastised them for managing foreign trade badly, and economists had begun to discuss how the "gap" between industry and the trade apparatus might be closed by decentralizing authority over foreign trade.[109] Moreover, the State Committee for Science and Technology had started to edge out the Ministry in managing some transfers of technology from abroad.[110] The stress on the West's manipulative aims was an essential plank in the Ministry's effort to protect its administrative powers, which had purportedly earned "the hatred of the international bourgeoisie," by lumping foreign opponents and domestic critics in the same group of "many enemies" who opposed its monopoly of foreign trade.[111]

Although there evidently were no outspoken advocates of technology transfers in the industrial ministries during the 1960s, it seems plausible that industrialists in backward sectors favored obtaining more Western technology. However, there was also resistance, or at least indifference, in some ministries. The Deputy Chairman of the Committee on Inventions and Discoveries, citing the benefits of buying foreign licenses, noted that these benefits had generally not been appreciated by Soviet industrialists, who sometimes tried to produce their own designs when the necessary technology was already available from abroad. He found the explanation for this phenomenon in "the conservative ideas of many of our designers . . . and economic officials, who have not wanted to understand just how important and mutually profitable international scientific and technological exchange is." Ministerial attitudes toward the foreign sale of Soviet licenses, he contended, were no better.[112] There is thus some evidence that the insularity which had earlier characterized Soviet industrialists persisted in the late 1960s, and that a substantial number of them were not eager to expand technology transfers from the West.

Party ideologists viewed closer Western scientific and technological ties primarily as a threat, particularly after the Prague Spring raised the specter of counterrevolution in the minds of the party elite. In April 1968 the Central Committee called for special efforts to prevent infiltration of bourgeois ideology,[113] and ideologists took this injunction to heart. A key article by one official expressed satisfaction that Soviet specialists were world leaders in many R & D fields, but then cautioned that this was not enough. The political commitment of the scientific-technical intelligentsia must be strengthened against the West's multiplying attempts at subversion. Western propagandists used "any means" for this purpose, including "any international meeting of scholars and specialists, any stay by Soviet persons abroad." Soviet scientific institutions were remiss, suggested the writer, because they sometimes "send abroad persons who do not stand out by their especially principled character" and who accept Western propaganda as truth. The official clinched his point by referring to several scientists from Obninsk who had traveled abroad and then published articles at home which were marked by "the absence of a class approach to the evaluation of capitalist actuality."[114] Rather than call for the restriction of foreign contacts, he urged better Soviet propaganda and more attention to the problem from regional party committees. But it would be surprising if assertions of this kind were not used in private to buttress the argument for limiting contacts with the West.

The political police also showed a suspicion of closer Western ties. An article commemorating the fiftieth anniversary of the founding of the Cheka, predecessor of the KGB, claimed that "tourism, cultural ties, private correspondence and so forth" were being used to promote anticommunist propaganda; foreign spies might be disguised as visiting students or businessmen. Foreign intelligence services also attempted to suborn Soviet citizens traveling abroad, it warned.[115] The article indicated that the traditional hostility of the political police to contacts which might facilitate more effective technology transfer was still strong.

Changes in Technological Strategy

These internal crosscurrents contributed to significant changes in technological strategy. One change, though far from dramatic, affected R & D expenditures. After 1965 research spending increased rapidly, but not as rapidly as in the preceding years. Between 1965 and 1968 the average annual rate of increase declined to 9.3 percent from the 10.9 percent it had averaged between 1962 and 1965, even though the economy as a whole grew faster than in the earlier period.[116] This pace was far below the nearly 17 percent that the science establishment had endorsed for 1966–67. It suggests that Kosygin's dual stress on more effective management of research and greater dependence on foreign technology was damping the growth of R & D. The regime's commitment to a large research program remained strong, and it still hoped to receive much of the technology it wanted from this source. But the generosity of the authorities toward the research establishment was more restrained than it had previously been.

At the same time the authorities showed a greater willingness to rely on negotiable technology transfers from the West. Between 1965 and 1970 total trade turnover between the USSR and the developed capitalist countries increased by 67 percent, whereas turnover with other socialist countries grew by only 43.4 percent.[117] It is true that in 1970 the proportion of Soviet trade with other socialist states was still some three times greater than the fraction with developed capitalist countries,[118] but the balance was changing, and considerably more finished industrial technology was coming into the USSR from the West than had before Khrushchev's fall.

In contrast to their past preference for one-shot foreign trade transactions, Soviet planners also manifested a new desire to maintain closer, long-term ties with Western suppliers of technology.[119] Just as the USSR was seeking to extend cooperation within CEMA from trade to joint R & D programs,[120] so it was seeking sustained access to new technologies being developed in the West. In the second half of the 1960s it began to sign a considerable number of industrial cooperation agreements, as well as general agreements for R & D cooperation, with Western firms. The most dramatic agreements were concluded with Fiat, the Italian automaker, in 1965–66, calling for cooperation between Fiat and the USSR in automotive R & D, and for Fiat assistance in constructing a vast passenger car plant on the Volga. As general consultant for the plant project, Fiat was to design the manufacturing process, assist in designing the model to be produced, oversee construction, and train Soviet personnel to run the plant.[121] Such close cooperation between the auto industry and a Western company indicated the regime's intention to pursue negotiable technology transfers from the West on a wider front than practiced under Khrushchev.

Soviet patent policy mirrored the same tendency. In July 1965 the USSR joined the Paris Convention on Industrial Property, which entitled Soviet inventions to greater protection abroad but also required the USSR to accord foreign patents greater recognition.[122] This step was probably motivated by a desire to elicit more Western cooperation in Soviet technological borrowing and reduce redundant Soviet R & D.[123] Without assurance that their rights would be protected, Western patent holders would be reluctant to furnish the equipment and training often required to make effective use of patented ideas; and the inclination for Soviet R & D agencies to duplicate work done abroad would be greater. From 1966 through 1970 the USSR bought five times as many foreign licenses as it had in the period from 1946 to 1965.[124] Although the base figure was very small, this jump signified an important change.

The number of Soviet specialists traveling to the West also increased. Although exact measurement is impossible, there was clearly a rise in the number of persons coming to the West in connection with specific industrial projects. For the Fiat deal, more than 2,500 Soviet technicians traveled to Italy for training, and about the same number of Western specialists visited the plant site in the USSR.[125] The number of Soviet scientists visiting the West also increased. Whereas 1,180 scientists traveled to the West in 1964, some 1,800 went in 1966.[126] This figure probably grew larger in 1967–68. The total number of scientists traveling abroad, to socialist as well as capitalist countries, doubled between 1964 and 1968, and since past foreign scientific travel had generally been divided about equally between the two groups of countries, it is likely that travel to the West continued to climb until 1968.[127] Nonetheless, the amount of foreign travel remained strikingly small; only about a quarter as many Soviet as American scientists traveled outside their native land in 1966.[128] It was also noteworthy that the volume of travel by Soviet scientists to socialist countries was not substantially different from the amount to the West. These statistics demonstrate that the regime was only grudgingly relaxing the restrictions on person-based technology transfer, and they also shed light on the travel impediments to technology transfer within CEMA about which East European spokesmen had complained.[129]

Although not spectacular, these changes in technological strategy were significant because they indicated a new willingness to enter into negotiated arrangements for acquiring technology, together with a reluctant decision to allow a limited increase in personal contacts with foreigners for this purpose. These initiatives foreshadowed more dramatic policy shifts. Those shifts would come, however, only after the new leaders had tried their hand at improving the capacity of the bureaucratic apparatus to produce sophisticated indigenous technologies.

Bureaucratic Reform and Technological Innovation

The new leaders were determined to find methods of economic administration different from Khrushchev's. Although Khrushchev had renounced wholesale physical coercion, his handling of administrative matters owed much to the Stalinist legacy. His penchant for overly ambitious economic plans and his reliance on exhortation and organizational shake-ups in place of carefully designed reforms resembled the Stalinist mobilizational style. Nearly all elite groups had been hurt by one or another of Khrushchev's shake-ups, and his successors' denunciation of his "hare-brained schemes" elicited a warm response from many officials.

The new rulers aimed to make a transition from the prevailing bureaucratic pattern, marred by the economic defects of mobilizational organization, to institutions that could operate at a higher level of economic efficiency. Viewed from the perspective outlined in Chapter 1, these new institutions were to combine the advantages of the mechanistic and organic patterns of bureaucracy—that is, both to improve the yield from current technology and to speed the introduction of new technology.

An enthusiasm for mechanistic administration led Khrushchev's successors to abandon the compilation of unattainable economic plans, which inevitably caused administrative confusion, in favor of plans roughly matching goals with available resources.[130] They also evinced a strong new commitment to "scientific decision-making" based on firm empirical knowledge of the problem to be solved, including the attitudes of various bureaucratic and social groups, and on a "systems approach" that would ensure consistency among the elements of any reform package.[131] Not least important, the new leaders showed more willingness to single out the structure of institutions rather than the malfeasance of individuals as a source of economic and technological failures.[132] This was part of their policy of "trust in cadres," or avoidance of purges. By concentrating on improving institutions rather than on punishing errant officials, they hoped simultaneously to upgrade the country's economic performance and to strengthen its political stability.[133]

In addition to these features of mechanistic administration, the new leaders sought to incorporate some virtues of organic administration into their reforms. One of their early themes was the need to dispense with "petty tutelage" from above. This slogan implied a diffusion of authority to subordinate organizations, and some diffusion did occur.[134] The leaders also highlighted the importance of technical expertise and the need for more cooperation and information exchange among agencies working on the same administrative level. All these elements are characteristic of the organic pattern of bureaucracy.

But what institutional steps could translate these general prescriptions into real improvements in economic performance? Could the advantages of mechanistic and organic administrative patterns be obtained within a centrally planned system, or must a large measure of market competition be introduced to achieve this end? If Western experience was any guide, making the transition without market competition would be difficult, since market pressures have historically played a large role in reshaping Western business units into forms more suitable to the pace of technological change in their particular sectors. Moreover, how would the effort to make major alterations in administrative behavior be affected by the political bargaining among various bureaucracies? Could the new leaders' calculated, step-by-step reforms succeed where Khrushchev's shake-ups had failed? Or would "trust in cadres" invite protracted bureaucratic infighting that would destroy the coherence of gradualist reforms? These issues were sharply posed by efforts to improve the organization of industry and the management of R & D.

The 1965 Industrial Reform

Despite their common dislike of Khrushchev's administrative style, the members of the political elite could not readily agree on the form a reorganization of industry should take. At the risk of oversimplification, we can say that the elite divided into three groups on this issue. One favored a recentralization of industrial authority in which Moscow ministries would wield as much power as they had before 1957, although the particular levers at their disposal might be modified. Another group favored concentrating power at the intermediate bureaucratic level dominated by the regional party organs and sovnarkhozy. A third group, which favored a radical decentralization of power to the level of individual enterprises, wanted to give a much larger role to competitive markets and restrict central planners to the use of indirect financial inducements; the issuance of administrative orders to the enterprises would have no place. The infighting among these three coalitions made formulating and implementing any reform more difficult.

Within a week of Khrushchev's fall, the advocates of a major decentralization gained some ground. The press announced that the system of enterprise administration begun earlier in the year at a few plants would be extended to numerous light industrial factories. This organizational experiment, first applied at the *Bolshevichka* and *Mayak* clothing firms, freed enterprises from the obligation to obtain approval of their physical output plans and gave them more control over the prices at which their goods were sold. The goods were now to be manufactured and distributed on the basis of orders received from a stipulated group of outlets, rather than on the basis of a production plan handed down from above. Moreover, the broad target that still required higher-level approval was the planned level of profit for the firm. The concept of profit had long been anathematized in Soviet economic thought, and its introduction, even in attenuated form, was a notable departure. This new arrangement fell considerably short of the latitude enjoyed by Western firms, since it did not increase enterprise control over capital investment or the selection of retail outlets.[135] But it was major change from previous Soviet practice, and its extension to other plants was a good omen for the proponents of radical decentralization.

A number of economists welcomed this move. In the discussion of economic theory and administration that had blossomed since 1960, many economists had advocated a greater role for profit calculations and economic incentives.[136] As part of this liberalizing trend within the economics profession, the study of Western economic administration was expanded, and economists studying capitalist systems began to show a lively interest in the domestic application of their findings. As noted above, the Academy of Sciences had set up a special council in 1962 to study the international economic competition between socialism and capitalism.[137] By 1965, according to a Soviet account, the council no longer limited itself to criticizing bourgeois ideologists and comparing levels of development in the USSR and the West. Instead, the council's main interest in capitalism had become "the discovery of that which is rational in the development of its productive forces, and also the new things in the technique and organization of production which can be critically used in our economy." Specialists on capitalism now had the "widest possibilities," which should be "used to the fullest in pre-paring recommendations on the most important national economic questions," including improvement of Soviet economic efficiency.[138] Among the topics being considered by the council were the management of capitalist corporations and the place of material incentives in capitalist industry.[139] Some economists clearly favored a basic change in the balance between markets and central planning, and they apparently drew some inspiration from the study of Western experience.

The many opponents of decentralization, however, quickly responded to the extension of the *Bolshevichka/Mayak* experiment. On 1 December 1964 *Pravda* printed an article that advocated recentralizing control over industry. An editorial note that the article was "for discussion" indicated its controversial nature. The authors maintained that at present jurisdiction over supplies, investment, and technological policy was confusingly divided among various regional and central agencies. Heavily criticizing the failure of the sovnarkhozy to direct the enterprises' technological policies, the writers advocated restricting the sovnarkhozy to the supervision of local industries. Control over all major industrial plants would revert to the Moscow-based state committees, which would effectively become ministries of the pre-1957 variety.[140] In contrast to the outlook underlying the *Bolshevichka/Mayak* experiment, this article argued that the technological discipline and guidance required by industry could be obtained by recentralizing administrative power and clarifying lines of authority. The authors were plainly upset by the conflicts and uncertainties which the present jumbled administrative arrangements inflicted on enterprises; and rather than suggest that a new type of uncertainty be introduced in the form of more competitive market relations, they favored the more familiar solution. Significantly, both authors were enterprise officials.[141] Whether or not they spoke for factory managers as a group, these particular managers saw regularization of the administrative apparatus, rather than a sizable increase in their own freedom of maneuver, as the top priority.

Advocates of centralization could be found in the ranks of professional economists as well. The economists were split over the reform issue, and while some favored a major decentralization, many opposed it. Some of the opponents were convinced that cybernetic concepts and the application of computers could improve economic performance without decentralization.[142] Others resisted decentralization because their professional reputation was invested in the defense of orthodox economic doctrines, because they yearned for the "order" and discipline of the Stalin years, or because they felt the existing economic system was basically adequate.[143] The economists who opposed decentralization were associated, more often than not, with central governmental institutions, especially planning and budgetary agencies.[144] In the eyes of these persons, measures like those introduced at *Bolshevichka* and *Mayak* were steps in the wrong direction.

Although it is impossible to determine how most top leaders lined up on the issue of industrial reform during this period,[145] Kosygin's position was clear. Prompted by his concern over the technological inertia of industry, he endorsed an approach that closely paralleled the *Bolshevichka/Mayak* experiment. Scarcely a week after *Pravda* had printed a plea for recentralization, the Prime Minister, underscoring the need for more rapid innovation, advocated giving greater sway to market forces as the proper solution: "So that enterprises better feel the condition of the market and changes in demand from buyers, it is necessary broadly to practice the establishment of direct ties between enterprises (or associations) and the stores selling their products to the population. Exactly the same direct ties are also necessary . . . in the branches producing the means of production." It was essential, Kosygin maintained, "to evaluate the activities of enterprises and directing organs by the degree to which their ouput meets the requirements of consumers, finds a buyer and corresponds to the world technological level."[146] This statement implied a major change in the relation between market forces and central planning. Coming from a top leader the statement was rather daring, especially since it encompassed not only light but also heavy industry.[147] Soon afterward *Kommunist* struck a similar note by endorsing a larger role for prices and profits and warning against excessive reliance on "administrationism" and "all sorts of organizational restructurings."[148]

Kosygin's resistance to restoring the ministries evoked support from regional party officials, who were eager to preserve and expand the powers of the sovnarkhozy. Shortly after his speech, two important regional officials criticized the argument for recentralization which *Pravda* had published. Citing the article by name, the first secretary of the Moscow city party committee said it contained unjustified slurs on all Soviet industry.[149] The first secretary of the Leningrad city party committee put the case more forcefully. He acknowledged that the current system was failing to accelerate technological change and was inhibiting the country's economic development. But he laid the blame on "excessive centralization." Condemning sentiment for the reestablishment of ministries as "unfounded," he asserted that "the system of sovnarkhozy is progressive and has fully vindicated itself. . . . We advocate sovnarkhozy, but with substantially wider rights," including more control over R & D organizations and technological policy.[150] The regional party organs were weighing in as a powerful ally against centralizing pressures in the reform debate.

Regional party officials, however, were unlikely to back any effort to promote a further decentralization to the level of the individual enterprise. When it came to changes of this kind, regional officials, like the proponents of reorganized ministries, had much to lose. The party officials desired a limited redistribution of power that would increase

their prerogatives, not the more far-reaching decentralization that Kosygin was pursuing. At the same time that the leader of the Moscow
party committee was backing Kosygin's case against recentralization,
the sovnarkhoz which his committee supervised imposed physical output plans on *Bolshevichka* in contravention of the original terms of that
economic experiment. The sovarkhoz overseeing *Mayak* did the same.[151]
The functionaries of the regional party and governmental agencies were
hardly economic liberals, and most would oppose any steps toward
the introduction of market freedom for enterprises. Anyone who sought
that end would have to take on the party as well as the state apparatus.

It is possible that Kosygin attempted to do so. Unconfirmed samizdat
reports indicate that during this period Kosygin proposed that the branch
economic departments at every level of the party hierarchy, from Central
Committee to regional committees, be abolished.[152] This step, by greatly
reducing the capacity of the regional party organs to intervene in industrial affairs, would have created an institutional environment more
compatible with a radical decentralization of economic authority to the
enterprise. If Kosygin did assay such a cut in party power, it would
help to explain the contention that broke out in the spring of 1965
over the relative prerogatives of state and party officials. The proponents
of the party apparatus emphasized that not only what the economy
produced, but how it produced it, was important; production had to
be managed in a "noncapitalist manner."[153] But such criticism may
have been prompted by proposals other than the one credited to Kosygin
by the samizdat reports,[154] and it must remain an open question whether
he actually tried to curb the power of the regional party apparatus.

At any rate, the persistent advocates of recentralization soon won a
significant victory. On 4 March *Pravda* announced that several ministries
would be reestablished in the defense industries; it mentioned no
changes for the nondefense sectors.[155] Just as in 1957, the backers of
centralization had evidently concentrated their greatest efforts on the
defense industries, with substantial success.

This move provoked new resistance from Kosygin. In mid-March he
publicly defended the existing division of authority between the sovnarkhozy and the state committees against demands for recentralization.
He noted that the defense industries had recently been reorganized.
Avoiding mention of the word "ministry" in this connection, he stated
that the Politburo still planned to consider measures to improve the
administration of industry, and that this work "will be conducted gradually, after careful and thoughtful preparation." The Central Committee,

he emphasized, was undertaking this work "very correctly."[156] In other words, industrial reform was still in the future; the creation of ministries of defense industry was not a reform, and people pushing this idea were in error. Kosygin granted that the present state committees suffered from "not a few difficulties," but he stressed that they also "possess many advantages in realizing technological policy and solving problems of technological progress."[157] His statement was an appeal to other members of the Politburo, which regularly shaped the proposals that came before the Central Committee, not to lose the strengths of the current administrative system by acting hastily.

Kosygin was on the defensive. Rather than concentrate on what should be done, he now spoke mostly about what should not be done. One thing which he wanted to prevent was abolition of the sovnarkhozy in favor of a more centralized economic bureaucracy. "Many leading workers," he said, "are expressing the opinion that there will not be the necessary order in planning so long as . . . the sovnarkhozy exist." It was indisputable that the present system needed improvement. "However, to equate inadequacies in planning and economic activity only with the imperfectness of the system of administering industry would be incorrect. If we stand on this view, we will make many more errors. . . . There are many questions it is impossible to decide locally, but one must not disdain initiative coming from below."[158] Kosygin clearly wished to avoid recreating a system of strong central ministries, and he presumably still wanted more reliance on financial controls and market forces. But he was being compelled to fight a rearguard action against officials who distrusted financial controls and wanted to reconstitute the ministries.

These tensions produced a six-month stalemate. Although at several points in mid-1965 individual leaders indicated that the Central Committee would soon convene to deal with the topic of industrial administration, it did not meet until late September.[159] The resolution which the Central Committee finally approved bore the marks of compromise on all the contentious issues, and it provided a test for the leadership's assumption that a coherent reform could be formulated and implemented on an incremental basis.

In some respects the resolution entailed recentralization. It abolished the sovnarkhozy and the state committees on the grounds that neither had been able to coordinate technological policy on a national scale.[160] Hence many powers in the field of technological policy that had been divided between the sovnarkhozy and the state committees were again to be concentrated in Moscow-based industrial ministries. The ministries received the right to approve enterprise targets in a number of vital areas, including capital investment, the mix of major types of products, and projects for innovation.[161] Similarly, the inter-enterprise exchange of goods was to be centrally planned. On the other hand, the reform also decreed the elimination of numerous controls previously exercised by the ministries, and it put new emphasis on using financial levers and incentives to stimulate enterprise performance. A charge on capital was introduced to discourage plants from hoarding resources, and the main targets of the enterprise plans confirmed by the ministries were defined in terms of sales, profit, and profitability.[162] The regime expected this stress on sales and profits to motivate the enterprises to strive for greater short-term economies and for regular technological improvement of their products and processes.[163] On balance, the reform introduced some economic inducements and enterprise initiative into a structure of administration that was still quite centralized.

The ambiguities of the decree allowed the leadership to act, but they ensured that infighting over the operational features of the reform would continue. In commenting on the resolution, various leaders stressed contradictory aspects. Consistent with his traditionalist faith in the strength of the economy, Brezhnev strongly favored centralization. Preparing its suggestions to the Central Committee, Brezhnev said, the Politburo had followed Leninist guidelines. Lenin had "energetically emphasized that to take the right from a single national center to subordinate to itself all enterprises in a given branch of large industry in the whole country 'would be regional anarcho-syndicalism, and not communism.' "[164] Rather than accentuate the importance of strengthening the powers of economic units below the ministries—a subject he mentioned only glancingly—Brezhnev dwelt on the need to make the ministries "the supreme authority in their branches" and "genuinely governmental organs."[165] Instead of arguing for enterprise freedom from excessive supervision, he argued for ministerial freedom from such supervision. Underscoring the obligation of the ministries to "strengthen state discipline, to ensure model order and high standards in the work of enterprises," he expressed no desire to enlarge the number of subministerial associations of enterprises or to expand their powers.[166] Nor did he refer at all to the desirability of direct ties among enterprises. Brezhnev's accent, in short, was on recentralization.

Kosygin, in contrast, underlined the ways the reform might lead to further decentralization. He did not single out the danger of "regional anarcho-syndicalism." Instead he prefaced the Politburo's recommendations with a plea that ministerial staffs be kept small, since "enterprises and economic associations are being presented with large rights and there is no need to create an apparatus which would exercise petty tutelage over enterprises." The view that central organs had only rights and enterprises only obligations must be renounced, he said. The introduction of economic methods of administration made this view outdated.[167] Moreover, he showed his preference for subministerial associations by citing them as examples of the revivified "branch principle," which most commentators equated with the new ministries, and by saying the ministries would transfer "many operational functions" to them.[168] Finally, he repeated his call for direct supply links among enterprises, although he put the case more cautiously than he had the year before.[169]

Just as the top leaders disagreed over which aspects of the reform to stress, so did other commentators. One liberal economist presented a scenario according to which the number of administrative controls would gradually shrink and enterprises would receive financial targets "as indicative figures rather than as precise directives." Direct ties and profit incentives would play an enlarged role in spurring innovation, as would competition. "It is completely rational to assume that competition will develop between enterprises in distributing orders [for goods], based on comparison of the quality guarantees offered, periods for fulfillment [of orders] and prices of output."[170] Another economist asserted that the "rapid tempos of technological progress change the requirements of producers and consumers so violently that not even the most perfect plan can take this into account. . . . After the basic proportions and directions of technological progress have been defined in the national economic plan, enterprises must have the opportunity to put out production in keeping with the requirements of the market."[171] Fluctuations of market prices were not a case of wasteful "spontaneity," he contended; production of the wrong goods was the really dangerous form of spontaneity.[172]

Such attitudes met staunch opposition from those who wished to restore the ministries' old powers. A few months after the Central Committee approved the reform package, A. Rumiantsev, the editor of *Pravda*, assailed unnamed "opportunists of various colors" who were saying the reform contradicted Leninism and signified a rebirth of capitalism in the USSR.[173] He followed with a critique of "some people" who were still denying the need for a switch to financial levers, such as a charge on capital goods and the use of profit as an indicator of enterprise performance. Their outlook, he said, fostered "that nihilistic attitude toward problems of price-formation which existed in our country for a long time and whose consequences we are painfully feeling up to the present time."[174] The statement suggests that even after passage of the reform, conservatives were still using the specter of a capitalist restoration to discredit its provisions.[175] The testimony of a party secretary indicates that after the announcement of the reform, conservative economists continued to lobby party authorities and argue against any reliance on profits and prices.[176]

This view enjoyed powerful support in the new ministries. On the whole, the ministries were led by officials whose past experience had taught them to distrust generalized economic incentives as a means of administration. Of the thirty-three industrial and construction ministers appointed in September 1965, twenty-two had served as ministers or deputy ministers in the ministries existing before 1957, and five more had served at levels just below these.[177] Under such conditions the Politburo's policy of "trust in cadres" was an invitation to reassert old administrative practices, and as industrial plants gradually were shifted to the new system, complaints about arbitrary ministerial behavior began to surface in large numbers. The ministries compelled enterprises to report and fulfill economic targets that the reform had legally abolished, and they sometimes confiscated profits to which the enterprises were legally entitled. Moreover, they frequently introduced sudden changes into current enterprise plans without warning, and if enterprises failed to meet the new plans, the ministries laid the blame on their shoulders.[178] This arbitrary behavior, a carry-over from the earlier mobilizational pattern of administration, seriously hampered efforts to alter the character of the industrial bureaucracy.

What were the attitudes of enterprise managers toward the reform and its future? Although sparse, the evidence is worth considering. Most enterprise personnel who participated in the prereform discussion of profits and sales as enterprise indicators appeared to favor the notion.[179] During implementation of the reform, a survey of about 250 directors of Siberian plants operating under the new system revealed that fewer than one in six objected to having plan fulfillment measured in terms of sales.[180] Judging by this sample, it seems that a key aspect of the reform had solid support among plant managers.

There was little indication, however, that most managers wanted the reform to evolve toward the market competition advocated by the liberal economists. When asked to identify the sources of their problems with the reform, only a third of the Siberian sample mentioned a lack of enterprise independence as an impediment.[181] Moreover, when asked in which areas they wanted further authority, almost 80 percent said they wanted more power to make decisions on labor and wages, but fewer than 50 percent wanted more power over investments and fewer than 20 percent wanted more control over prices; receipt of the right to select suppliers and buyers on their own initiative ranked sixth among the directors' concerns, well behind the need for an improvement in the existing system of material-technical supply and an increase in their power over wages.[182]

What most of these managers seemed to want, in short, was improved central planning, not the abolition of central planning per se. They desired the stability and security of a mechanistic bureaucracy, rather than the political unpredictability of mobilizational administration or the economic uncertainties of market competition. Indeed, it was probably because such competition had not been introduced that so many managers were willing to accept the introduction of sales targets, which would have been far more painful in a competitive setting. Because the great majority of managers had been trained as engineers rather than economists, they lacked the skills to adjust to a system in which competition and marketing were central elements; even the reform's limited swing toward economic criteria caused some enterprise managers considerable difficulty.[183] In view of these facts, it seems fairly safe to conclude that there was little support within managerial ranks for a major infusion of market forces, and that further steps in this direction might arouse the opposition of many plant directors.

Changes in the Administration of R & D Institutions

Occurring about the same time, the struggle over the administration of R & D organizations resembled the infighting over the industrial reform. The links between basic and applied research, and between applied research and production, were still weak, and some officials and social scientists continued to stress that they must be improved to upgrade the pace of technological advance.[184] As shown below, the attempt to change the controls over research establishments revealed that serious obstacles to effective reform existed in the realm of research as well as in the realm of production.

Inspired by the greater operational authority granted to enterprises under the industrial reform, top R & D planners began pushing in 1965 to enlarge the powers of research organizations. Endorsing the new dispensation for industry, Academy President Keldysh suggested that scientific institutions should also be freed from "petty tutelage."[185] Not long afterward the General Meeting backed this idea, and the Academy Presidium set up a special commission to develop specific suggestions.[186] Presidium member Kapitsa was especially outspoken about the need for more institutional autonomy. The object of planning, he said, was to help science "successfully and freely to develop. If one recognizes this to be the major task of the plan, then it is necessary for us to renounce all the petty 'supervisory' elements of the plan and preserve in it only its broad organizing principle."[187] One top official of the State Committee for Science and Technology likewise embraced the idea of broadening the rights of institute directors, and another called for the creation of "new, less conservative, more flexible forms" of scientific organization.[188]

Would granting greater administrative autonomy to R & D establishments suffice to strengthen the links among fundamental and applied research and production? Many advocates of autonomy, especially those based at the institutes, seemed to think so, but this notion was questionable. In order to upgrade the yield from research, the extension of greater administrative freedom to R & D organizations required that they have a strong desire to promote the application of their scientific ideas. It was not clear that the purely internal motivations of researchers would provide the impetus, particularly in view of the suggestion by some Soviet observers that the proportion of researchers preoccupied with material rewards rather than with science as a "calling" was gradually rising.[189] One possible answer was to tie the financial standing of R & D units to the actual economic results of their findings, and in the 1960s this idea gained growing support among economists and economic planners.[190] Partly in response to such thinking, the regime increased the proportion of research spending channeled to researchers via contracts with enterprises. Despite the skepticism voiced by some scientists, the percentage climbed from under 30 percent in 1965 to more than 35 percent in 1967, and was slated to reach almost 40 percent in 1968.[191]

Another means of stimulating R & D units to perform well was to broaden scientific competition among them. This idea had some influential advocates among central R & D planners. V. Kirillin, Chairman of the State Committee for Science and Technology, said that competition would benefit major projects,[192] and his deputy, V. A. Trapeznikov, argued the case with far more force. Trapeznikov noted that one "often" heard that a planned economy could eliminate the duplication of effort and required that a single organization be responsible for each subfield of science and technology. Vigorously disputing this view, he observed that the current system allowed mediocre scientific institutions to succeed bureaucratically without achieving substantive scientific results. The lesson was that plan fulfillment was "far from sufficient" as a basis for evaluating R & D organizations: "In science and technology, just as in art, it is in principle impossible to establish absolute criteria of the perfection of achieved results. . . . Therefore, it is possible to evalute results . . . only on the basis of a comparison of several solutions."[193] For this reason, two or three organizations should be set up to work in each R & D subfield, and design competitions should be staged among them. The "competition of ideas," said Trapeznikov, ought to become the "basic method of organizing scientific and design projects," and it should be applied in other social spheres as well. Otherwise technological monopolies would lead inevitably to stagnation, permitting each design body to force customers to choose either its current results or no results at all.[194] Coming from a high official, this was a remarkably bold attack on the conventional Soviet notion of how R & D should be managed.

The idea of more R & D competition also received support among social scientists. A conference heavily attended by economists suggested applying competition on a trial basis to develop the most important types of new technology,[195] and a few social scientists took a stronger position. One prominent management theorist told a CEMA symposium on R & D that the creative, nonrepetitive character of research necessitated parallel R & D projects. Giving a research monopoly to an institute, he explained, "causes a loss of a sense of the new," something "immeasurably more dangerous" for science than duplication. "Cybernetics has established that systems with 'duplication' possess the greatest number of possibilities—and double or triple duplication . . . increases the system's potential not by two or three, but by many more times." The speaker then spelled out what this meant for R & D establishments. While positive incentives were important, "sometimes people forget that socialism also needs compulsion." Institutes that did poor work should be disbanded. "The collapse of a particular institute is the failure of a group of scholars, and it does not cause the economy any greater loss than the dissolution of a soccer team that has been unsuccessfully organized by a bad coach." Having created the conditions for successful R & D, society was "entitled to bolster the demand for quality . . . with the threat of closing the research institute."[196] Seldom had a proponent of R & D competition expressed its harsh personal implications for researchers so bluntly.

Although ministerial officials voiced concern about innovation, they opposed most of these proposals for institutional change. The ministers of electronics and machine tools praised the idea of paying design bodies a share of the profit their designs brought to enterprises but remained silent about granting more autonomy to R & D organizations or fostering scientific competition among them.[197] A third minister who dealt with the computer industry confidently asserted that the ministries had liquidated the gap between research and production that existed under the sovnarkhozy.[198] His deputy, while far less sanguine about the current pace of innovation in computers, explained the problems of R & D as the result of too little ministerial power rather than too much. The opinion of some scientists that research could not be planned was "incorrect." Even more erroneous, said the deputy, was the view of some industrial designers that design work could not be rigorously planned from above.[199] There was no room in this picture for more institute autonomy or for widespread R & D competition.

On the issue of greater autonomy, spokesmen of the R & D organizations sided resolutely with the reform-minded science planners. One Academy member condemned the recent planning of fundamental science as the compilation of "endless summary thematic plans . . . full of minor details and major confusions." Central planning had basically amounted to "efforts to connect the activity of independent organizations which might have worked excellently without either coordination or coordinating organizations." It would be better for institute councils to confirm their own research plans; coordination with other institutions would be realized through scientific conferences.[200] Heads of other Academy institutes argued along similar lines,[201] as did officials of ministerial R & D bodies. The director of a major ministerial institute contended that permitting the institutes to define the entire list of their projects would be more effective than central planning.[202] Referring to his own establishment, he claimed there had not been one case in which research undertaken at the institute's initiative had not offered "a large economic effect," whereas projects executed according to an assignment from above "sometimes turned out to be useless."[203] These were harsh words, and it is tempting to conclude that most researchers wanted the centralized structure of R & D administration dismantled completely.

Rank-and-file scientists, however, did not advocate complete administrative devolution, as evidenced by their resistence to competition among scientific organizations. One economist asserted that "many scholars" favored parallel R & D projects.[204] Obviously this assertion had an element of truth, since some social scientists favored the idea and since some dissident natural scientists were gravitating toward the idea of a market economy.[205] But the endorsement of competitive R & D was not echoed by nondissident natural scientists working in research institutes. In their many public discussions of R & D improvements, none of these researchers and directors ever advocated the idea. Evidently many scientists were unenthusiastic about interorganizational competition. Perhaps they thought competition was unnecessary. But they were probably also motivated by a desire to garner a maximum share of the resources available in a given scientific sector and to avoid stiff challenges from other researchers.

This suggests a significant similarity between scientists and junior industrial officials. Like their managerial counterparts, many researchers at middle and lower levels seemed to derive a certain sense of security from the bureaucratic hierarchy. The structure provided stable careers with a minimum of competitive pressure,[206] and Soviet surveys of scientists' attitudes in the late 1960s indicated that most researchers felt a fairly high measure of work satisfaction.[207] Although the bureaucracy was a source of recurring irritation, it also protected researchers from effective pressures for better technological performance, and many scientists apparently did not want it abolished. They favored a loosening of central control, but they disliked the thought of facing vigorous R & D competition that would be demanding and, in some cases, professionally disastrous. The threat of having unsuccessful institutes disbanded like the losing teams in a soccer league did not appeal to them. In short, Soviet scientific and managerial specialists were not simply at odds with the bureaucratic structure handed down from the Khrushchev era; they also had a strong vested interest in its protective features.

The tensions among various bureaucratic groups affected the changes made in R & D administration. In November 1966 the Council of Ministers, complaining that the ministries had neglected scientific and technological information, ordered them to set up units to improve its collection and dissemination.[208] The following March, the Council decreed a reduction of the administrative controls over many research bodies. The decree did not alter the principle that central agencies must approve the main research topics of subordinate institutes, but it did streamline the regimen under which the institutes were to work. Institute directors received more power to shape their institute's staff and administrative structure, as well as greater freedom to shift funds among research projects and acquire scientific instruments.[209] Taken together, the two measures suggest that the authorities were trying to foster attitudes toward technological information and expertise which were closer to the organic style of R & D administration.

Like the 1965 reform of the controls over enterprises, however, the March decree was a compromise. It was an attempt to reconcile the views of the State Committee for Science and Technology and the Academy of Sciences, on the one hand, and the ministries, on the other. The initiative for the decree had come from the Academy and the State Committee,[210] and it provided some of the autonomy favored by their science planners. But the decree did not go as far as they wished. It did not endorse competition as an essential ingredient of effective R & D, and it did not encompass design bureaus. This was a victory for the ministries. Ministerial officials had been especially adamant about keeping a tight rein on design bureaus, and after adoption of the decree one State Committee official voiced regret that controls over them had not been relaxed.[211] Three more years had to pass before the Council of Ministers extended the provisions of the 1967 decree to ministerial design units.[212]

In the meantime the ministries worked to restrict the exercise of the new powers offered to their research institutes. As implemented, the March decree governed only part of the research of the institutes. Projects confirmed by the ministries continued to be planned under the old system of detailed indicators, whereas top-priority projects confirmed directly by the State Committee were handled according to the new system.[213] This was a willful misinterpretation of the March decree,[214] and it invited administrative confusion. Apparently the proponents of decentralization had only enough power to override ministerial opposition on R & D projects which the State Committee planned directly. At the end of 1967 *Pravda* rebuked the ministries for dragging their feet. "The ministries," it said, "are still slowly including scientific research organizations in . . . the new system of planning and economic stimulation. The . . . broadening of the rights and responsibilities of the leaders of these organizations will undoubtedly make it possible to organize their work more successfully."[215] Research institutes probably enjoyed some expansion of their prerogatives after 1967, but it was less than they had been granted on paper.

The Technological Impact of Administrative Reform

The existence of such bureaucratic conflicts proves little, however, about the efficacy of the measures that emerged from the reform process. Did the changes in industrial and R & D management during 1965–67 generate more rapid technological innovation and more intensive economic growth? Was there a shift toward the organic pattern of administration that could facilitate such an improvement?

The initial results of the reforms cannot have been very encouraging. The GNP grew slightly more rapidly in 1966–70, at a rate of 5.3 to 5.5 percent, than in 1961–5, when it grew at 4.9 to 5.0 percent.[216] But the improvement was in agriculture, which had benefited since 1964 from better weather and from substantially increased inputs. The growth rate for industry fell in the second half of the decade by about 0.5 percent, from a level Khrushchev's successors had already deemed too low.[217] Moreover, the growth in factor productivity for the economy as a whole was disappointing. While the rate was better than in 1961–5, it was far inferior to the figures reached in the 1950s.[218] The economy, in short, was still growing primarily through the extensive mobilization of new resources, rather than through more intensive resource utilization.

The technological performance of industry was a crucial part of this persistent problem. Despite the leadership's intention to speed up innovation, the reforms slowed the rate at which industry put new products into production. An investigation of more than two hundred enterprises under the new system showed that at the majority of plants the volume of new products fell between 1965 and 1966, both absolutely and as a share of total output; in the machine-building enterprises sampled, the volume of new output fell to 85 percent of its 1965 level.[219] Because the machine-building branches constituted the driving force of technological advance in industry, these results were especially ominous. By 1968 the "coefficient of renewal of production" for machine-building enterprises in Kazakhstan had dropped to less than half its 1966 level.[220] For the country as a whole, between 1965 and 1968 the fraction of the nine machine-building branches' output constituted by new goods dropped from 13.8 percent to 8.2 percent.[221] These figures indicate that the reforms had not created a bureaucratic structure conducive to accelerated innovation. In fact, they had done just the opposite.

One cause was the persisting shortage of slack resources. The regime had attempted to make these resources more available by formulating less demanding national economic plans. But ingrained habits and the structure of the industrial bureaucracy prevented this accommodation from exerting much effect at the enterprise level. Because enterprise targets continued to be planned in terms of quarterly intervals, they posed a threat to any manager who closed down production to introduce new products. No less important, the ministries were unwilling or unable to adjust the targets to encourage innovation. Inherently suspicious of the enterprises and frequently uncertain how much adjustment a particular innovation really required in the enterprise plan, the ministries generally avoided such adjustments. Instead they tended to "ratchet" enterprise sales targets, just as they had jacked up output targets from one year to the next under the old system, thereby prompting managers to avoid innovation whenever possible.[222]

The ministries also diverted slack resources to other uses. Since the early 1960s several industrial branches had possessed a centralized Fund for the Assimilation of New Technology, designed to cushion enterprises against the hardships of innovation by covering some innovation-related costs. In 1968, however, the ministries channeled 15 percent of these funds into expenditures unconnected with innovation, even though many enterprises were suffering serious short-term financial losses connected with recent changes of technology.[223] Like the avoidance of plan adjustments, this action stemmed partly from an instinctive suspicion that enterprises were misusing payments from the Fund, rather than applying them to new technology.[224] Although it often had deleterious effects, the suspicion reflected a real problem. Past experience had shown that enterprises often claimed and stockpiled resources they did not really need. How could the ministries be sure that enterprises would not use the complexities of new technology as a cover for hoarding resources? In the absence of competitive pressures from other enterprises, what guarantee did they have that slack resources granted to an enterprise would actually be employed for innovation?

One notion behind the 1965 reform was that planners would give enterprises a fresh incentive to innovate by setting favorable prices on new products. The defects of the price system, however, contributed powerfully to the deceleration of product change. Prices could stimulate technological advance only if there was a positive correlation between the technological levels and prices of products, and in actuality the correlation was negative. Rather than offer more profit on new products, the price system favored older products that had long been in production and were technologically outdated.[225] The differences in profit margins between old and new products were large, giving enterprises a strong incentive to avoid new products whenever possible. By making enterprises more sensitive to financial inducements without revising the operation of the price system, the reform ironically strengthened some of the impediments to innovation that it aimed to overcome.

These negative effects of the price system were rooted in the conduct of both the central economic organs and the enterprises themselves. The planning agencies and ministries were responsible for matching price with technological level for several hundred thousand products, a task beyond their capabilities. In a large number of cases, the ministries' formulation of quality standards for products was not coordinated with their reviews of the prices of those products.[226] Often this was simply the result of administrative overload, but sometimes it was not. In 1968 the Council of Ministers noted that the ministries were inadequately supervising pricing at their enterprises, and it remarked that "in a series of instances" the ministries had knowingly confirmed prices which provided excessive profitability.[227] Nor was there any indication the ministries were approving such profits for the goods that were technologically most advanced. The government's chief price-planner stated that the majority of the machine-building ministries were seeking high profitability "no matter how it is obtained."[228]

The temptation to push up prices was even stronger for enterprises. By making profits a measure of plan fulfillment, the reform had given enterprises a stong interest in raising profits, whether by technological improvements or by bureaucratic manipulation. Since there was no market competition to match prices with technological levels, the easiest way to increase profits was to hide reductions in the manufacturing costs of old products, in order to prevent the center from reducing prices as products became outdated. A study of one industrial sector showed that during a general price revision the enterprises had manipulated their cost data so as to obtain an actual level of profitability more than twice the rate officially prescribed.[229] Since the costs of old products were easier to control and to hide, new products became unattractive to enterprises as a source of profit—unless the planning authorities were willing to allow very high inflation, which they were not. Enterprise behavior of this kind was impossible for the authorities to prevent, because they depended heavily for information on the very organizations they were attempting to regulate. But a further relaxation of central control without the introduction of market competition threatened to increase technologically regressive choices by the enterprise, rather than to curb them.

Data on the creation of prototypes indicate that the obstacles which slowed the manufacture of new products also impeded the preproduction phases of the R & D cycle. In the years before the reforms, the number of new types of machines and equipment created in the USSR had risen fairly steadily.[230] In 1967, when the first substantial transfer of enterprises to the new system took place, this figure began to decline.[231] By 1969, when more enterprises had been transferred, the number of new machines devised dropped to slightly more than 80 percent of the number created in 1966.[232] The same trend was visible in the creation of new instruments, which had also tended to rise before 1965. By 1969 the number of new models had dropped to less than 65 percent of the number devised in 1966.[233] Some of this trend may have been due to ministerial efforts to phase out duplicate R & D programs inherited from the sovnarkhozy. But much of the decline stemmed from the reluctance of industrial officials to support a vigorous R & D effort. In 1965 and 1966 the ministries parsimoniously spent substantially less on R & D than the economic plan called for.[234] Despite a large increase in total Soviet spending on science and a great deal of talk about increasing the resources available for R & D, the bias of the industrial bureaucracy against providing slack resources for innovation continued to make itself felt.

The only encouraging sign of bureaucratic change came in the handling of scientific and technological information. By the end of the 1960s, the declining rejection rate for patent applications revealed that fewer researchers and inventors were repeating work already done elsewhere in the USSR.[235] But the rejection rate remained very high by Western standards,[236] which suggested that a large measure of avoidable duplication was still occurring. Moreover, the barrier between the research establishments of the Academy and those of the industrial ministries remained.[237] Thus, even in the exchange of information significant problems persisted.

Conclusion

Although the oligarchs' dispute about Western political dynamics revolved around questions clearly posed in the preceding decade, uncertainty about the USSR's comparative economic performance gave the post-1964 debate about the West a dimension lacking under Khrushchev. As much as Western political trends, it was the fear that the Soviet Union might lose the economic race with the West that prompted some leaders to advocate slower expansion of defense spending, a more forthcoming attitude toward strategic arms limitations, and more venturesome diplomacy to obtain Western industrial know-how. One noteworthy development was the intermingling of the Politburo and academic debates on these issues. While many factors obviously shaped the views of the party chiefs, the analyses and ideas supplied by specialists in economics and foreign affairs had enough political significance to provoke public rejoinders from leaders who disagreed with them.

The differences between Politburo traditionalists and nontraditionalists affected the regime's technological strategy and its attitude toward internal reform. The country began to draw more heavily on Western technology through long-term agreements, but resistance from Politburo members and lower officials tempered the rate of this change. As for domestic R & D, the oligarchs tried to enlarge their system's capacity to create sophisticated indigenous technologies. They had taken the political helm with the expectation that the bureaucratic apparatus could be improved through step-by-step reforms introduced with a minimum of conflict and disruption. Instead, however, the reform process provoked sharp bureaucratic conflicts, and the effort to establish more organic institutions proved very difficult. Indeed, in most respects the reforms appeared to worsen indigenous innovation rather than improve it. These unexpected results generated new pressures on Soviet foreign policy and internal politics.

6   The Brezhnev Administration, 1969–1975

Toward the end of the 1960s party leaders started to recognize that their reforms had yielded unsatisfactory results. The long-term decline in the rate of economic growth abated only slightly after 1965, while the Western industrial states continued to grow impressively. Moreover, in 1969 the United States showed extraordinary technological prowess by landing astronauts on the moon. The moon landing overshadowed the space achievements so widely touted by Soviet propagandists, and it vividly demonstrated that the United States had the capacity to create a new generation of sophisticated strategic weapons which would be difficult for the USSR to match. Scarcely less important, the Prague Spring shook the Soviet hold on Eastern Europe. Although the USSR finally imposed political conformity on Czechoslovakia militarily, the crisis made the Soviet elite more aware that the USSR lacked some economic instrumentalities with which Western states, especially West Germany and the United States,[1] were trying to shape developments in the communist and third worlds.

These events intensified the controversy over Soviet economic achievements and needs. Intertwined with conflicts over foreign policy as well as domestic reform, the dispute played a key role in the decision to improve relations with Western Europe and the United States after 1968. It goes without saying that external political developments exerted a powerful influence on this decision. Deteriorating relations with China sharpened the Soviet desire to prevent a Sino-Western rapprochement, while the gradual American withdrawal from Vietnam and the advent of a West German government willing to accept East Germany made the nontraditionalist view of Western political intentions more plausible. In substantial measure, however, the quest for arms control agreements and wider economic relations with the West was precipitated by the expanding debate over the USSR's technological performance.

Socialism and Capitalism: Official Thought in Flux

The new cycle of debate increased the influence of nontraditionalist ideas within the elite. Acknowledging the significance of declining Soviet growth rates, more elite members accepted a somber appraisal of Soviet economic dynamism vis-à-vis the West. Many, though not all, of these observers also concluded that politically the West was becoming more moderate, and that a relaxation of East-West tensions therefore offered a way out of Soviet economic difficulties. This shift of outlook was not universal, however, and substantial resistance to it persisted in governing circles.

Shifting Views within the Leadership

As before, Kosygin was a major exponent of the nontraditionalist position. In April 1969 he made a detailed statement on the economic achievements of capitalism. "Preserving its aggressive nature and not refraining from further . . . military preparations, and above all from the race in atomic missiles," he said, "imperialism is putting great emphasis on the aspects of the struggle of the two world systems connected with the development of the economy, science, technology and education." Thanks to the recent expansion of government economic regulation, the Western industrial states were now attaining higher rates of growth. Although economic contradictions such as inflation persisted, "the intensification of social production and the scientific-technological revolution in certain measure today permit the leading imperialist states to have additional possibilities for maneuver in . . . the economy and social policy." Imperialism, he concluded, was still a strong and serious opponent. "The contemporary bourgeois state is increasingly becoming a kind of coordinating center" that "uses the most diverse means of ensuring adequately high and stable rates of economic growth and the further development of scientific-technological progress."[2] By Soviet standards, this statement was remarkably direct and pessimistic.

Several features of Kosygin's statement stand out. First, he presented a much more positive picture of Western economic achievements than was customary. Omitting the conventional Soviet accent on the disruptive "spontaneity" of the capitalist market, he referred to stable rates of capitalist growth and alluded almost wistfully to the "diverse means" of economic administration employed by capitalist regimes.

Moreover, his remark about the West's new room for maneuver suggested that technological progress had reduced the budgetary pressures on Western governments and that new domestic programs could widen the popular support for these governments.

Most important, Kosygin stressed that the capitalist states had been able to upgrade the economic instruments of foreign policy without cutting back military programs, including strategic weapons programs. By underscoring Western economic capabilities, he intimated that the USSR would face a forbidding task if it pursued all-out competition with the West in both economic and military terms. This judgment bore directly on the question of whether to engage in an unrestricted arms race with the imperialist powers or to join them in a common effort to reach SALT and other arms control agreements. In Kosygin's eyes, the latter course was clearly preferable.

Kosygin's economic assessment also prompted him to advocate wider Soviet technological ties with the West. Although socialist states could not slip into a position of economic dependence on capitalist countries, he said, a policy of autarky toward these countries was "economically unbeneficial" and "could lead to the most negative results, predetermining an inevitable lag of states that entered on this path." Therefore economic ties should be expanded with all willing countries. Kosygin coupled this message with a reminder to other socialist states that more bloc economic integration was necessary to restrain Western influence among CEMA members.[3] He badly wanted the benefits of more trade with the West. But he also wanted to avoid taking a position that might diminish Soviet leverage in Eastern Europe or allow his political rivals to claim he was risking such a loss.

Several top leaders rejected Kosygin's somber appraisal of the USSR's comparative economic performance. But by 1969 he had persuaded some of his Politburo critics that the problem was indeed serious. The most vivid evidence appeared in Brezhnev's speeches. Brezhnev first voiced heightened concern about the economy in a December 1968 Central Committee speech that was published years afterward. Calling the acceleration of innovation "the most important of the tasks of our economic strategy," he underscored its "timely political significance."[4] It was important, in the first place, for domestic reasons. But it also had crucial international implications. Eschewing any contrast between Soviet and Western achievements, Brezhnev said the scientific-technological revolution was occurring not only in socialist countries, but in capitalist countries as well. The competition for technological primacy "has now become one of the main bridgeheads of the historical struggle of the two systems." Recently the main capitalist countries, especially the United States, had "strenuously built up capacities in the most progressive branches of industry" and increased their R & D spending manyfold. Their motive was not simply to solve domestic problems but to strengthen their economic and political expansion "in many areas of the world."[5] These were striking words, coming from a figure who only nine months before had demanded that commentators put more weight on "the incurable diseases of the capitalist economy." Like Kosygin, Brezhnev was beginning to believe that the West's economic dynamism might give it more domestic stability and a new source of international power.

Brezhnev gave the first public signal of his new attitude at a June 1969 international Communist meeting in Moscow. His speech, which contained contradictory elements, drew partly on the traditionalist outlook he had previously espoused. The mounting influence of the military-industrial complex, he said, was making the imperialist states more aggressive and leading them to design new weapons for use "whole decades" in the future. Under these conditions the socialist countries must budget large resources for defense and beware of Western economic pressure intended to undermine bloc solidarity.[6] Intertwined with this line of argument, however, was a nontraditionalist strain absent from Brezhnev's earlier public speeches. Marxists, he said, could not ignore that the West had powerful production capacities which it was developing further through government regulation. This regulation comprised government programming of production, public financing of research and innovation, and steps "toward a certain limitation of the anarchy of the market." In several countries "this is leading to a certain increase in the effectiveness of social production."[7] Brezhnev's tone indicated a new respect for the technological capacities of imperialism, including its ability to devise futuristic military technology.

Brezhnev also indicated that political changes in the West might enable the USSR to avoid the risks of an all-out race for military and economic superiority. Spurred by the contest with socialism, capitalism was being forced "to use new means . . . of struggle which, in many respects . . . even contradict the customary 'classical' features of the capitalist system." Although still employing repression, the capitalists "are moving toward partial satisfaction of the demands of the workers."[8] This more temperate image of capitalist politics tied in with Brezhnev's shifting view of Western foreign policy. Commenting that opposition to the Vietnam war was becoming widespread among the American intellectual and middle classes, he observed that in addition to aggressive circles, there was also a moderate wing in the West that favored solving disputes through negotiation. Appealing to this wing, Brezhnev stressed that the USSR desired peaceful coexistence with all Western countries. Not even the United States was excluded from this offer,[9] he said meaningfully, and gave his firm approval to the SALT negotiations.[10] Finally, he showed a new interest in technological exchanges with the West. He kept the interest low-key to discourage other CEMA members from breaking ranks in an uncoordinated scramble for Western technology, but his change of stance was unmistakable.[11]

Even with Brezhnev's backing, this moderate appraisal of the West remained a matter of controversy in the Politburo. A few months after the Moscow meeting, Suslov issued a ringing denunciation of revisionism. Lenin, he said, had unmasked the essence of contemporary revisionism, which entailed "increasingly subtle forgeries of antimarxist doctrines as Marxism" in questions of political economy and tactics.[12] Although Suslov had many forms of revisionism in mind, one was certainly the tendency to describe Western capitalism too favorably, since he underscored the comparative economic strengths of socialism. Suslov granted that state-monopoly capitalism was striving for greater economic effectiveness but conspicuously refrained from mentioning any success from this effort. Instead he argued that since World War II the USSR had preserved its superiority over U.S. growth rates, just as the socialist bloc as a whole had grown faster than the capitalist world. These trends supported Lenin's prediction that "Soviet power will overtake and surpass the capitalists and that our prize will turn out to be not only purely economic."[13] Suslov's message was that Soviet economic troubles should not be exaggerated. Perhaps he meant to link it to the reminder that Lenin had demanded discipline from all party members, including leaders, and had urged a merciless struggle against "panickers, capitulationists, and opportunists" who violate the party's general line.[14]

Suslov's views had important foreign policy ramifications. The article implicitly denied that there was any compelling economic reason to improve relations with the West, since the socialist world was outstripping the capitalist world. Although Suslov apparently favored a more positive policy toward the West German Social Democrats to enhance their prospects of forming a government, he did not believe that such a policy was necessitated by the Soviet need for Western technology, and he firmly opposed a more conciliatory line toward the United States.[15] Avoiding any mention of SALT, he underscored the "plundering, aggressive policy of imperialism" and emphasized the contemporary relevance of Lenin's censure of "various kinds of apologists of American imperialism" who had tried to obscure the fact that the United States was "the most reactionary, most rabid" variety of the species.[16] This was a striking contrast to Brezhnev's recent endorsement of SALT and his remark that the USSR wished for peaceful coexistence with all capitalist countries, including the United States. Rather than view capitalism as being so technologically dynamic that an East-West relaxation and closer economic ties were advisable, Suslov regarded the capitalist countries as both politically dangerous and economically inferior.

Another Politburo supporter of this view, Petr Shelest, seemed even more uncompromising. The rates of growth in the socialist countries, he remarked, were a "great deal higher" than in capitalist states and had contributed to socialism's "superiority of power" over capitalism.[17] Shelest showed no interest in widening technological ties with the West or in arms control. Rather than support Brezhnev's new emphasis on East-West cooperation and negotiation, he highlighted the dangerous intentions of the capitalist world. For example, soon after the formation of a West German government headed by the more accommodating Social Democrats, he attacked the new coalition's policy in Eastern Europe as an attempt to tear some members out of the socialist camp.[18] In the light of his premises, Shelest's failure to advocate expansion of economic ties with the West was logical. Such ties would simply be an added weapon in the hands of aggressive imperialists bent on destroying the socialist commonwealth.

The Revival of Theoretical Debate among Specialists

As the conflict over the implications of the scientific-technological revolution grew more intense within the Politburo, less powerful elite members also began to clash over this subject. These disputes were part of a wide-ranging debate between the traditionalist and nontraditionalist schools of thought. Just as in the Politburo, the central issue was how to define and respond to the USSR's economic and political situation vis-à-vis the West.[19]

In the spring of 1969 a somber assessment of Western economic performance was published by N. N. Inozemtsev, who in addition to heading IMEMO, helped direct the Academy's Scientific Council on the Economic Competition of the Two Systems.[20] Inozemtsev emphasized that a sharp postwar increase in government regulation had facilitated the "transition to a predominantly intensive type of [economic] reproduction" in most capitalist countries.[21] Widespread technological progress had allowed them to increase popular consumption, thereby refuting the notion of "dogmatists" that the capitalist proletariat was undergoing absolute immiseration.[22] Moreover, the performance of the Western economies would be still better in the future. IMEMO calculations suggested that "after the comparatively low [capitalist growth] rates of the late 1960s will follow higher rates in the first half of the 1970s."[23] It was probably for this reason that Inozemtsev stressed the need to identify the "objectively progressive" tendencies in the growth of capitalist productive forces which could be emulated by the USSR.[24] For the foreseeable future, the country would face powerful technological competition from its imperialist rivals.

Traditionalists disputed this outlook. In the spring of 1969 one commentator claimed that "a certain slowing of the rates of growth of our economy, together with the simultaneous increase in the rates of development of the U.S. economy, which was earlier observed, is now past. In recent years we have witnessed an acceleration of the rates of our economic development together with a simultaneous lowering of the rates in the U.S.A."[25] The prognosis of writers like Inozemtsev, in short, was wrong. This optimistic tone was most likely due to the fact that the author was a colonel specializing in military economics. The military was still resisting entering SALT negotiations with the United States,[26] and his interpretation of the economic race with capitalism was probably colored by a desire to blunt the nontraditionalist economic arguments for arms control espoused by leaders like Kosygin and Brezhnev.

A few months later the exponents of traditionalist views offered a harsher rebuttal to the analysis of capitalism elaborated by Inozemtsev.[27] In a book-length treatise, M. F. Kovaleva belligerently attacked some Soviet authors for exaggerating the changes in state-monopoly capitalism.[28] Under the pretense of combating Stalinism, she maintained, these observers had misrepresented the relationship between the capitalist state and the monopolists and had overestimated the improvement in the economic functioning of advanced capitalist regimes. "In particular works the 'merits' of capitalism in developing the productive forces are stressed, and in certain measure its defects, its destructive force and its social character are smoothed over. This plays into the hands of the revisionists."[29] The central targets of this attack were the late Evgenii Varga, who had been the focus of intense political controversy during Stalin's last years, and S. A. Dalin, an active scholar who was an intellectual descendant of Varga.[30] Given the nature and circumstances of the attack, there can be no doubt that Kovaleva intended it to apply to Inozemtsev as well, although she did not name him.[31] Kovaleva accused the nontraditionalists of a wide range of revisionist sins, including deviation from the party's programmatic position on several points.[32] Fragmentary evidence indicates that this attack may have been sponsored by the Academy of Social Sciences attached to the party Central Committee.[33]

Soon after Kovaleva fired this broadside, the journal of IMEMO, where Varga and Dalin had done much of their writing, countered with strong editorial praise of Varga's ideas. Decrying scholarly dogmatism, it said the 1947 abolition of Varga's institute had done "serious damage to the study of world capitalism in the Soviet Union."[34] A month later, at a major conference convoked by IMEMO and other Academy institutes to commemorate Varga's legacy, the nontraditionalists pressed their attack. In his opening address Inozemtsev said that a "rigorously scientific, objective analysis of the laws of development of productive forces is especially necessary for us today." This was so because economic competition was playing a special role in the East-West struggle, and also because the rapidly developing scientific-technological revolution was "putting its stamp on literally all the processes occurring in the contemporary world—economic, social, and political."[35] Varga's analysis of the role of the state in the capitalist economy had shown that "the most characteristic feature of state-monopoly capitalism is the coalescence of monopolies with the state, the unification of the gigantic power of the monopolies and the state in a single mechanism"; views to the contrary "have not stood the test of life."[36] Lest anyone think that this argument was merely hairsplitting among social scientists, Inozemtsev stressed its bearing on Soviet policies. The disagreement was "not at all only a matter of a theoretical dispute. The question . . . of the relations between monopolies and the state, of the relative independence of the state, has a great deal of significance in defining the possibilities which capitalism commands and in determining . . . the strategy and tactics of the struggle against it."[37] Given the views expressed by Inozemtsev on other occasions, it is clear that he was presenting a brief for the relaxation of East-West tensions and the pursuit of arms control.

Brezhnev Seizes the Initiative

As nontraditionalist scholars stepped up their offensive in academic forums, the party chiefs who held similar views pushed harder in the political arena. By late 1969 it was evident that the year would be the worst for economic growth since Khrushchev's fall, and at the December Central Committee plenum Brezhnev forced the issue. In an unusually blunt speech never published in its entirety, he asserted that economic efficiency was replacing rapid growth at any price as the main criterion of success in the competition with capitalism.[38] This clearly implied that the long-term growth figures cited by optimistic leaders like Suslov and Shelest did not accurately represent the economic race between the USSR and its Western rivals. Apart from external pressures, Brezhnev felt that the domestic shortage of new manpower and the limits on the amount of future investments demanded more efficiency.[39] "This is becoming not only the main but also the only possible means of developing our economy and solving such fundamental socioeconomic tasks as the construction of the material-technical base of communism, the improvement of the welfare of the workers, and victory in the economic competition of the two world systems. . . . We have no other way."[40] According to Soviet conventions, this was a very forceful statement. It frankly acknowledged that without a major improvement in economic performance, the USSR could neither build communism nor triumph over capitalism internationally.

Evidently Brezhnev was more venturesome in dramatizing the problem than in offering solutions. It may be that he voiced a new determination to meet Soviet economic needs through Western assistance,[41] but the published portions of his speech do not touch on this question. As for domestic solutions, he hinted that he favored a centralized approach to economic administration different from the compromise measures of 1965. The earlier reforms, he remarked, "naturally . . . could not fully solve" the problem of increasing the effectiveness of the economy.[42] Expatiating on the new "science of administration" based on rapid information processing and rational decision-making, Brezhnev castigated the central economic agencies and demanded that they assimilate the new methods more quickly.[43] At the same time, he explained that because these were "in the first place political and not technical problems," they required "principled solutions" based on "the Marxist-Leninist line."[44] In short, he was calling for a comprehensive administrative rationalization of central management, without a significant devolution of power. Brezhnev accepted Kosygin's contention that there were economic problems, but he prescribed different remedies. In this way Brezhnev aimed to maintain his political credentials as an exponent of "principled" solutions, that is, solutions in which decentralization and free markets did not have a place.[45]

Brezhnev's emphasis on rational centralization, however, did not exclude attacks on individuals. He remarked that "both many rank-and-file officials and many leaders" needed to have their sense of responsibility raised, and he warned that where officials failed to respond to criticism, demotions would follow.[46] No doubt this warning caught the attention of other Politburo members as well as midlevel bureaucrats. It is rare in Soviet practice for anyone to criticize "leaders," and the threat seemed to apply even to the top of the party hierarchy.

Brezhnev's effort to tackle the country's technological problems quickly ran into stiff resistance. Almost two months elapsed between his speech and the publication of a follow-up declaration calling for a nationwide effort to improve economic efficiency; the long delay suggests that the contents of the declaration were a source of disagreement.[47] Podgornyi echoed Brezhnev's call for more intensive growth, and Kirilenko gave unusually firm backing to his speech.[48] But other leaders rejected his analysis of the Soviet economy as alarmist. According to a contemporaneous samizdat source, shortly after the December plenum three Politburo members (Shelepin, Suslov, and Mazurov) circulated a private letter condemning Brezhnev's speech for producing "only hysteria" without providing any solutions to the difficulties it depicted.[49]

In March the conflict boiled over in a remarkable article that a major party journal published on the theme of the Leninist style of collective leadership. Alluding favorably to the October 1964 plenum that removed Khrushchev, it noted that Lenin had asked permission of the Politburo to criticize the views of other leaders in public.[50] It added that Lenin had stated that only decisions adopted by the Politburo or a Central Committee plenum could be implemented by a party secretary, and it maintained that freedom of discussion should not be construed to justify intraparty factions or criticism which exceeded the bounds of party loyalty.[51] All these points offered some support to Brezhnev's opponents, who wished to use the slogan of collective leadership to curb his power. But the weight of the exegesis was in Brezhnev's favor. The article differentiated the cult of personality from the rightful authority of a party leader, and it stated that Lenin had taken steps entitling party secretaries to choose which matters were to be decided by the Politburo and Central Committee.[52] Asserting twice that truly collective leadership was impossible without criticism and self-criticism, it contended that criticism was the means by which the party regularly overcame inadequate policies.[53] "Communists, and first of all leaders" were obligated to accept criticism.[54] The party condemned cases in which "some excessively zealous leaders" interpreted frank criticism as discrediting, if not the whole party, then the authority of an individual leader.[55] The December 1969 plenum, continued the article, illustrated the party's commitment to businesslike criticism and self-criticism. It then referred to Lenin's censure of Tomsky, who had once occupied the administrative post Shelepin now held, for failing to fulfill directives of the Central Committee.[56]

This justification of self-criticism failed to still Brezhnev's opponents, who continued to reject the nontraditionalist notions that capitalism was undergoing moderating changes and posing a novel economic challenge. Shelepin retorted that the United States exemplified "dangerous and bloody" imperialism. In recent years the United States had taken a clear turn toward the "escalation of armed, aggressive . . . struggle against the world revolutionary process." But the tide of history was running in favor of socialism. "This is attested . . . by processes occurring in the . . . world economy. Although imperialism has also managed to achieve a relative stabilization and even a certain growth of its productive forces, on the whole it is gradually losing to socialism its positions in the sphere of economic development one after another." The share of the socialist camp in world economic production "is growing with every year."[57] P. N. Demichev, a candidate Politburo member and a party secretary with major ideological responsibilities, took a similar tack. The USSR had surpassed U.S. economic performance in terms of "several" indices and held advanced positions in "many" fields of science, such as space exploration. Although it had not yet matched the United States in total economic resources, "we already are confidently accomplishing tasks which capitalism is incapable of solving," including steady growth and the elimination of unemployment. "We do not deny," Demichev said, "that we are still forced to collide with some difficulties. . . . But these are still not the difficulties we experienced in the twenties, thirties, or the early postwar years."[58]

Brezhnev, in short, had not persuaded his opponents of the urgency of Soviet economic problems. Nor had he persuaded them of the need to take the foreign policy steps, such as a SALT agreement and the expansion of trade with the West, which might alleviate those problems. This conflict undoubtedly contributed to the running dispute over the scheduling of the 24th Party Congress. According to the party statutes, the Congress was due to occur in 1970, but at the start of the year one party official implied that it would not be held until 1971. In April Brezhnev made the first of several assertions that it would be held in 1970, but no further steps were taken. Finally the Central Committee met twice in July and decided that the Congress would not be held until 1971.[59] This disagreement over scheduling betokened a clash over more fundamental questions.

About a week after Shelepin and Demichev had touted the USSR's superior economic dynamism vis-à-vis the United States, Brezhnev mounted a defense of his more pessimistic outlook. An essential element was his justification of criticism and self-criticism in economic policy. It was "natural," he said, "that we ourselves try to uncover existing inadequacies and errors and speak frankly about them." Self-criticism "is after all the law of development of our society," and it could not be abandoned for reasons of expediency:

We know, of course, that enemies of the Soviet Union . . . try to use our self-criticism to slander the socialist system. It was that way fifty years ago. It also happens today, when . . . the bourgeois press now and then bristles with sensational headlines about some sort of economic "crisis" in the Soviet Union. . . . The ideologists and politicians of imperialism have already many times thought up all sorts of "crises" of the Soviet system. But our country became more and more powerful. And the fact that enemies try to use our self-criticism for their purposes cannot weaken our determination to eliminate from our path everything which is . . . braking our movement forward.[60]

The defensive tone was unmistakable. Brezhnev was not rebutting bourgeois enemies of the USSR, but rather the most orthodox Soviet thinkers who felt it wrong to give any intellectual ground to such adversaries.

A major issue at stake in the fencing over self-criticism was whether to seek improved relations with the West—and if so, on what terms. In mid-1970 the USSR was engaged in hard diplomatic bargaining with the United States over a SALT agreement, and with the Federal Republic over a German settlement. Interpretations of Western motives and of the USSR's comparative economic dynamism had basic implications for Soviet policies in these two areas. They affected appraisals of whether the USSR could bear heavy military expenditures more easily than the United States, and they influenced judgments about whether concessions should be made in exchange for Western technology. Moreover, public self-criticism could affect Western policies on these same subjects. The image of the Soviet economy prevailing in Washington and Bonn could harden the American position in the SALT talks and encourage Western negotiators to extract a high diplomatic price in a German settlement.

This interpretation of the dispute between nontraditionalists and traditionalists is supported by evidence that in 1970 the leadership was sharply divided over policy toward the West. Brezhnev had begun to allude to moderate circles in imperialist countries and was now defending the wisdom of arms control in a way that suggested he was trying to persuade critics who believed it was impossible.[61] But Shelest, carefully avoiding any mention of moderate imperialist circles, again underscored the West's belligerence, while Shelepin insisted that its belligerence was increasing.[62] Shelest's remarks on the theme of American imperialism were so inflammatory that *Pravda* excised them from its version of one of his speeches.[63] Meanwhile, other polemics revealed high-level divisions over policy toward West Germany. One exercise in esoteric communication involved a vitriolic exchange about Lenin's drive to overcome opposition within the Bolshevik leadership to concluding a peace treaty with Imperial Germany in 1918. Given the explicit parallels drawn to present policy, the real issues were undoubtedly whether to sign a treaty with the Federal Republic in 1970 and whether the economic benefits would outweigh the political costs.[64] In view of these facts it is quite certain that the dispute between Brezhnev and his opponents over comparative economics also concerned foreign policy, and that his discourses on the weaknesses of the Soviet economy struck some Politburo members as an unwarranted gift of leverage to the "politicians of imperialism" with whom the USSR was negotiating. In June, when Brezhnev won a softening of the Soviet positions on SALT and the German question,[65] he won it over the resistance of these top-level critics.

Brezhnev's stress on self-criticism in economics probably irritated his opponents for another reason as well. Judging by the timing, dissenters apparently seized on the party's admission of serious economic problems as a further argument for radical political liberalization. Roughly a month after the authorities published their February declaration on the economy, Andrei Sakharov and two other dissenters stressed this argument in an unsolicited memorandum to Brezhnev, Kosygin, and Podgornyi. Painting a bleak picture of declining Soviet growth rates and incipient technological stagnation, the three authors asserted that without far-reaching democratization the USSR would fall further behind the West and ultimately revert "to the status of a second-rate provincial power."[66] The memorandum, which was quickly published in the West, typified a growing tendency among dissenters concerned about Soviet technological performance to link the question to broader issues like democracy and human rights.[67] Soon afterward a document purporting to be Varga's political testament also appeared in the Western press. Condemning the economic secrecy that masked the hardships borne by the Soviet populace, it assailed the ineptitude and privileges of the administrative elite and called for "a radical change of leadership" to establish genuine socialist democracy.[68] In all probability, events of this kind forced Brezhnev and his supporters to defend themselves against the charge that they were strengthening the regime's domestic opponents, although in actuality they felt anything but sympathy for dissenters like Sakharov.

Brezhnev's attempt to heighten the personal responsibility of Soviet officials also provoked opposition. In June 1970 Brezhnev sought to quiet the fears aroused by his earlier threats against officials guilty of poor economic performance. He repeated that achievement of the goals set by the December 1969 plenum would be the criterion for judging economic institutions and cadres. But he also commented that the party's existing political style was one of "trust and a respectful attitude to people," and he promised that "we do not intend to . . . return to the methods of administrative fiat decisively condemned by the party."[69] These reassurances suggested that a substantial number of persons feared Brezhnev's harsh demands for faster technological progress, and that some had perhaps accused him of reversing the policy of "trust in cadres" adopted by the oligarchs when they ousted Khrushchev in 1964.

Brezhnev's reassurances failed to placate his Politburo critics. In October 1970, Shelest expressed vehement disagreement with Brezhnev over Soviet economic performance. Describing the USSR's recent achievements as "great," he acknowledged that bold criticism was in order in fields where the country was encountering difficulties. But, he added, it was impermissible to allow the slogan of criticism and self-criticism to degenerate into "groundless, spiteful criticism, into cheap sensationalism." Unfortunately there were "cases where various kinds of fault-finders and Philistines use our individual difficulties . . . to fan and inflame fears." This occurred "where the party, Soviet and economic organs . . . do not give a decisive rebuff to various kinds of fault-finders and slanderers."[70] Four months after Brezhnev had defended frank economic criticism as valuable, Shelest was publicly rejecting his argument. Moreover, Shelest's words implied that the "party organs" which had allowed such conduct should act to stop it. In a scarcely veiled way, he was calling on the Politburo to deliver a "decisive rebuff" to Brezhnev's campaign for better economic performance, and he was soliciting support from the economic agencies threatened by that campaign.

Shelest plainly disgreed with Brezhnev's attempt to blame economic inefficiency and technological inertia on Soviet officialdom. He granted that the task of promoting faster innovation and greater economic efficiency made "fundamentally new demands" on the skills of officials. But he remarked that since 1966 the party had achieved its goal of developing the cadres to meet these needs. "Our cadres have turned out to be on the level of these requirements and have shown the capacity successfully to solve more complex tasks," he concluded.[71] Shelest's message was that cadres should not be punished for the malfunctioning of the economy because on the whole it was not their fault—a view directly at odds with the position taken by Brezhnev at the December 1969 plenum.

Nontraditionalist Specialists on the Offensive

While the top leaders squared off over these issues, academic commentators continued to argue about the USSR's comparative economic performance. Encouraged by Brezhnev's forceful speech to the December plenum, the nontraditionalists became more assertive in early 1970, accusing their opponents of dogmatism and the selective use of empirical evidence.[72] Inozemtsev warned it would be "deeply erroneous" to underestimate Western technological dynamism, as some past Soviet observers had done. The scientific-technological revolution was occurring "in all countries without exception, regardless of their socioeconomic structure." Led by the United States, contemporary imperialism had adopted a "more flexible and multifaceted strategy," augmenting its customary accent on military power with "increasingly more attention" to economic and technological forms of international competition.[73]

Scholars from the Academy's new U.S.A. Institute also stressed the changes in capitalist society and foreign policy.[74] At its first public conference some participants commented that American imperialism was increasing the spin-off from military R & D to civilian pursuits, while others noted the "extremely substantial changes" being made in capitalist economic administration under the influence of the scientific-technological revolution.[75] G. A. Arbatov, Director of the Institute, criticized unnamed writers who had underestimated Western economic dynamism and had thereby provided ammunition to the apologists of capitalism; he added that Marxism-Leninism was incompatible with "efforts to force the whole richness of actual political reality [sic] into a Procrustean bed of frozen dogmas."[76] Arbatov made an especially revealing comment at a General Meeting of the Academy called about the same time to discuss how to accelerate innovation. While the central reason for dealing with the issue was to meet domestic needs, he said, "the foreign-policy aspects of this problem must also be kept in mind." American imperialism was pursuing technological advance not only for military purposes: "In the plans for a new East European policy which are being intensively developed in the U.S.A., a large place is occupied by the use of scientific-technological successes to attempt to include the socialist countries of Europe in the sphere of economic influence of the West."[77] The implication was that the Soviet Union should respond by accelerating the development of its own technology—particularly nonmilitary technology—as a countermeasure.

Traditionalists struck back with their own accusations. An editorial in a major ideological journal was typical. "In their attacks on Marxism," it said, "the theorists of opportunism usually cite new phenomena . . . connected with the unfolding scientific-technological revolution . . . to prove that capitalism has allegedly become 'organized', 'orderly', 'regulated', and so on, and that therefore the necessity for revolutionary change . . . has fallen away." Such opportunists "are not averse to accusing of dogmatism those who do not agree with this, that is, those who remain true to the scientific principles of Marxism-Leninism."[78] In reality, argued the editorial, the opportunists were trying to resurrect untenable reformist ideas.[79] Other traditionalists claimed that imperialism was doing "enormous work" disguised as scholarship to discredit the Soviet economy, and that this work was influencing "some economists" in the socialist bloc.[80] More specifically, they condemned the "bourgeois" notion that capitalism was growing mostly by intensive means, whereas socialism was growing mostly by extensive means. This claim, they maintained, was misleading and distorted the facts by utilizing doctored statistics.[81]

Despite these traditionalist rebuttals, some officials and social scientists continued to endorse the nontraditionalist view. Trapeznikov, of the State Committee for Science and Technology, told an Academy meeting on innovation that while the USSR held "advanced positions" in managing certain systems, it was unfortunately "impossible to say this about the management of economic systems."[82] Another State Committee official admonished that it was "extremely dangerous" to lag "in using the enormous advantages of socialism over capitalism in the . . . accelerated development of science and technology [and in] the improvement of administration."[83] Dzherman Gvishiani, a deputy chairman of the State Committee, likewise warned against neglecting "the extremely substantial" changes in capitalism caused by governmental involvement in R & D management. He also sounded the theme of a Western shift of emphasis from military to nonmilitary technology. "Of course," he conceded, "the basic part of state means is directed to research pursuing military goals; however, American industrial firms broadly use the results of this research for . . . nonmilitary production . . . American governmental organs are expending great efforts on the improvement of the existing system."[84] Meanwhile, some economists asserted that "the internal economic mechanism of capitalist society has succeeded on the whole in adapting to the [economic] requirements of the contemporary scientific-technological revolution," and that this had "greatly increased the necessity" for faster Soviet innovation.[85]

Further Disputes among Leaders and Specialists

During the rest of 1970 the debate between the two schools of thought simmered. In a speech on the anniversary of the Bolshevik takeover, Suslov granted that stepping up innovation was a vital problem. However, he avoided any reference to the "scientific-technological revolution," which figured so prominently in the thinking of nontraditionalists, as a significant factor in the competition between East and West. He used the term only when condemning "the ideologists of imperialism and its revisionist underlings" for false assertions that accelerated innovation could reduce exploitation and prevent socialist revolutions under capitalism.[86] Apparently he objected to the concept of a worldwide technological revolution because some economists were suggesting that Western technological change could raise popular consumption without harming growth, while others were using Western experience to argue that faster Soviet technological progress required a diversion of resources from heavy industry to consumption.[87] The latter notion was probably especially distasteful to Suslov, who cautioned against exaggerating the importance of "economic incentives," that is, consumer goods, in the USSR. On the whole, Suslov preferred to stick to established priorities and established claims of Soviet superiority. He summed up this way: "The correlation of forces in the struggle between socialism and capitalism does not always change as rapidly as is desirable, but it is changing unswervingly in favor of socialism."[88]

Suslov's speech provoked a quick rejoinder. The author of record was Inozemtsev, but the article's appearance in *Pravda* just a few days after Suslov's speech indicates that it had high-level backing, probably from Brezhnev. Inozemtsev allowed that Western adaptation to the scientific-technological revolution could proceed "only to certain limits." But he sounded the familiar warning that to underestimate Western attempts at adaptation would be "incorrect," and he pointedly repeated that "in the most developed capitalist countries the transition to a predominantly intensive type of [economic] reproduction has been realized."[89] Since Brezhnev and some other leaders were arguing that the USSR had yet to make this transition, the message was that the traditionalists were still overestimating the strengths of the Soviet economy.

Brezhnev's report to the 24th Party Congress in March 1971 reflected the controversy over technological progress. While he was apparently compelled to give some ground on the issue of the USSR's performance vis-à-vis the West, he also voiced dissatisfaction with the current treatment of this subject. As a concession, Brezhnev said the general crisis of capitalism was deepening, and he gave considerable space to the evidence of unemployment and inflation that helped buttress this assertion. On the other hand, he claimed that some Western regimes had embarked upon "partial reforms" in order to maintain control of the masses, and he noted that Western monopolies were "widely using the results of scientific-technological progress to . . . heighten the effectiveness and rates of development of production."[90] In the same vein, Brezhnev stressed the need to study "new processes in the capitalist economy which are occurring, in particular, under the influence of the scientific-technological revolution."[91] Obviously he felt such changes were not yet receiving sufficient attention, since he remarked in this connection that "repeating old formulas in situations where they have outlived themselves . . . harms the cause, and creates additional possibilities of revisionist forgeries of Marxism-Leninism." The treatment of such issues in the party's "theoretical work," he concluded, was not entirely satisfactory.[92] Very likely one purpose of this argument was to parry opponents' charges that Brezhnev's unorthodox views of capitalism and the scientific-technological revolution were encouraging the current of dissent typified by the 1970 Sakharov memorandum. Another motive was to support his contention that the significance of Soviet economic and technological competition with capitalism had greatly increased.[93]

According to Brezhnev, the scientific-technological revolution demanded a dramatic improvement in the functioning of Soviet institutions. Since the 1930s the economy had assumed "completely new dimensions," which in turn required better administration.[94] Calling for a transition from an uneven pattern of innovation to uniformly effective innovation in all sectors, he summoned the party *"organically to unite the achievements of the scientific-technological revolution with the advantages of the socialist system of economy."*[95] As steps toward this goal Brezhnev advocated more computerization and long-term planning, the creation of production associations, and modified incentives for industrial enterprises.[96] Not least significant, he called on heavy industry to produce more consumer goods and urged increased diffusion of defense technology to the rest of the economy. In view of the high technological level of defense industry, he said, the achievement of such spin-offs "is acquiring paramount importance."[97]

Brezhnev also showed an unambiguous interest in expanded technological ties to the West. He commented favorably on several specific industrial projects that the USSR had undertaken with Western corporations.[98] Moreover, he endorsed an expansion of foreign economic relations as "a major reserve for heightening the effectiveness of the economy" that could improve "all our industry."[99] He thus treated intra-CEMA and East-West trade together, without indicating his past preference for CEMA, and suggested that all industrial sectors, rather than only a few lagging ones, would benefit from closer Western ties.

Brezhnev's forceful campaign for increased innovation and economic efficiency discomfited some of those who in principle should have agreed with him—especially Kosygin. At the 24th Congress Kosygin moved back toward a traditionalist view. Only socialism, he argued, permitted the "full and comprehensive development" of the scientific-technological revolution. Citing comparative statistics on recent growth trends, Kosygin claimed that the Soviet economy "at all stages of its development has always clearly demonstrated indisputable advantages over the capitalist economy."[100] These words marked a shift of emphasis from Kosygin's past statements about capitalism's "adequately high and stable rates" of growth and its "further development of scientific-technological progress."

What accounts for Kosygin's surprising change of stance? Part of the explanation may be that the comparative economic performance of the USSR did in fact look somewhat better in 1971 than in 1969. Growth rates had declined in the West since 1969, particularly in the United States.[101] A more important motive, however, was that Kosygin stood to lose politically if the issue of Soviet economic performance was appropriated by a Politburo rival. Since 1965 Kosygin had been the Politburo's chief economic expert and the leading spokesman of industrial reform. Brezhnev's onslaught in December 1969 had threatened both the reform program—since Brezhnev expounded the virtues of highly centralized economic management—and Kosygin personally.[102] Under these circumstances Kosygin naturally acquired an interest in defending the recent economic record of the USSR, and his new accent on the defects of capitalism was stimulated partly by a wish to make that record look better against a backdrop of Western economic crisis. Kosygin's shift of stance did not put him squarely in the camp of the traditionalists.[103] The fact that he still pushed for an expansion of Western economic ties[104] indicates that his new treatment of capitalism was primarily tactical in nature. Nevertheless, the traditionalist ideologists and the economic administrators in the Politburo appeared to be forming a tacit alliance to check the political momentum of Brezhnev's economic campaign.[105]

For the moment the alliance succeeded in making Brezhnev accommodate himself to the dominant Soviet image of capitalism. In July 1971 he remarked the "striking contrast" between the socialist countries, which were confidently making social progress, and the West, where "the noose of the general crisis of capitalism is drawing up . . . tighter and tighter." Afflicted by a deep political crisis and "the constantly feverish condition of the economy," capitalism faced mounting difficulties which no reformer could remedy.[106] The themes of partial reform, heightened economic effectiveness, and new phenomena in capitalist development were missing entirely from the speech. For the time being, Brezhnev's attempt to employ an altered image of the West in support of his drive for greater Soviet economic effectiveness had been stymied.

The controversy over the USSR's ability to match the West technologically was hardly over, however. Demichev soon returned to the issue in an article in *Kommunist*. Cautioning that the struggle against revisionist views of socialist economic development required "the most serious attention," he repeated that socialism's advantages over capitalism in innovation were "indisputable."[107] Yet some revisionists were arguing that the socialist economy was incompatible with the requirements of technological progress. "We do not deny that a few highly developed capitalist countries still surpass the Soviet Union in particular branches of science and technology," Demichev said defensively, "but this hardly demonstrates that the capitalist system is superior in any way."[108] As if to buttress this claim, Kosygin asserted that socialism's economic advantages over capitalism related "to the whole period" in which the two systems had coexisted, and he cited a recent U.S. economic downturn as proof.[109] Shortly afterward, however, *Kommunist* published another article sharply at odds with these statements. The article proclaimed that it would be "deeply erroneous" to underestimate the West's technological dynamism, and it denounced the "infamous 'theory' of the stagnation of productive forces" in capitalist society.[110] Together with evidence of conflict over the forthcoming five-year plan,[111] these words indicated that the Politburo dispute between Brezhnev and his supporters, on the one hand, and the more traditionalist coalition of ideologists and economic officials, on the other, was still smoldering.

Because the controversy had major implications for Soviet external relations, it must have contributed to the new tensions over foreign policy that surfaced about this time. Seeking to reach a German settlement, Soviet negotiators had entered into talks on the status of Berlin in 1971, and their interest in bringing the talks to fruition was heightened by signs of a rapprochement between Washington and Peking.[112] But relations with the West were still a source of disagreement. Late in November the Central Committee heard a report from Brezhnev on foreign policy since the 24th Party Congress. While the Central Committee adopted a resolution on the topic unanimously, it approved "the work done by the Politburo" to implement the foreign policy line of the Congress—not Brezhnev's report to the plenum.[113] Given the conflicts in the Politburo, together with Brezhnev's effort to downplay that body's policymaking role at the 24th Congress,[114] the distinction was significant. It was all the more so because the plenary debate on Brezhnev's report included persons who had previously expressed strikingly divergent views. The participants ranged from Shelest to Inozemtsev, now a new candidate member of the Central Committee.[115] In reviewing the foreign policy work of party and state organs, the plenum's final resolution hinted at disagreement. The "character of the forthcoming tasks in this area," it said, "requires constant improvement in all their activity."[116] This note, unusual in a foreign policy resolution, implied some deficiency in the recent conduct of foreign policy.

A central issue connecting the disputes over foreign policy and the scientific-technological revolution was whether the USSR still faced the familiar imperialist military threat or now had to meet a Western technological challenge of a new kind. Shortly after the November plenum, Inozemtsev expounded the changing roles of economic power and military force in Western foreign policy. "In certain measure," he said, "the political strategy of imperialism is shifting to the plane of the economy, science, and technology."[117] Although Inozemtsev hastened to add that imperialism had not rejected military means, he clearly thought the shift extremely important. Today the military power of "any state" depended closely on its economic and scientific-technological potential, and the capitalist states were taking this into account. The United States was giving priority not to "current military measures connected with the production . . . and deployment of existing weapons systems," but to "measures of a long-term nature directed toward acceleration of the rates of growth of the whole economy, [and] toward further development of scientific-technological progress (which, incidentally, is the basis of research and development for new, more effective types of weapons)." In the last decade and a half total U.S. military expenditures had not even doubled in absolute terms and had declined by one-third as a share of GNP. Meanwhile U.S. overall expenditures on R & D had quintupled.[118] Strikingly, Inozemtsev was arguing that general technological development had so increased in strategic importance that the United States was reducing the military share of GNP and diverting R & D resources into sectors more likely to generate economic growth.

These trends, Inozemtsev remarked, demonstrated the importance of ensuring the USSR's rapid economic growth and "leading role in the development of science and technology. . . . Both in the present and in the future—the immediate as well as the comparatively more distant [future]—this is the key to the solution of our basic internal and foreign tasks, the guarantee of new successes . . . in the competition with capitalism."[119] The Soviet Union, in short, should emulate America by reducing the share of resources going to the military—and it should do so in the "immediate" future, rather than paying lip service to this goal and then postponing it. In Inozemtsev's eyes, Brezhnev's call for better spin-off from military industry was "very important." "Such a framing of the question," he pointedly told the members of the Academy, "poses a series of new problems not only for our economic science, but also for scientists in the most varied fields of knowledge."[120] Scientists working on military R & D, in other words, ought to make a greater contribution to nonmilitary needs. Some science planners shared this view.[121]

Other observers, however, disagreed that such changes had occurred in the West or were desirable in the USSR. As one military expert on R & D put it, "under imperialism the best achievements of human genius are subordinated to . . . aggressive forces and the military-industrial complex, which apply the majority of scientific discoveries to military purposes. An active process of the militarization of science is thereby taking place. In the United States . . . allocations for military research make up 60 percent of all allocations in science."[122] Whereas Inozemtsev emphasized the declining military share of American GNP, this writer stressed the large military share of American spending on science. Turning to Soviet defense policy, he highlighted the concept of national scientific potential, giving it a more military slant than did Inozemtsev. "In military theory this concept implies primarily science's capacity for immediate . . . and effective resolution of the problems posed by the development of warfare."[123] Soviet science had to be harnessed to the goals of the military establishment because "top-flight science does not by itself guarantee the successful resolution of the problems of the country's defense. It also needs to be ready for and capable of resolving military tasks in particular."[124] About the same time, other commentators implicitly rejected Inozemtsev's contention that better civilian spin-offs were still needed from military R & D and that it might be wise to limit military research.[125] These different outlooks had sharply divergent implications for domestic priorities and for foreign policy, particularly policy toward the ongoing SALT negotiations with the United States.

In view of such disagreements, it is not surprising that differences over the scientific-technological revolution continued to appear. These differences cropped up in discussions of the 24th Party Congress's slogan of uniting the technological revolution with socialism. At the end of 1971 Suslov paid lip service to this goal, but he again emphasized the superiority of Soviet over capitalist growth rates and remarked that the scientific-technological revolution and socialism were now being united "in the very best way."[126] Less than two months later, Brezhnev said the 24th Congress's prescriptions for economic improvement must be realized "more rapidly and more energetically than has been done until now," and he promised a Central Committee plenum on the theme of technological progress.[127] During 1972, however, no such plenum was held.[128] At the end of the year Brezhnev voiced frustration with earlier reforms. "In some ways," he said, "the measures adopted have not justified themselves."[129] Claiming that a "turnabout" in economic efficiency was needed to meet the competing demands of consumption, investment, and defense, Brezhnev remarked that the Politburo would now take up the improvement of economic planning and administration.[130] He added pointedly that Lenin's strictures against dogmatism in party theory "must be the motto of every Marxist."[131] Undeterred, Suslov replied that "the decisions of the 24th Congress are being successfully realized; the course defined by the Congress— toward the further strengthening of the country's economy, the acceleration of scientific-technological progress, and the improvement of the material condition of the people—is being introduced in life."[132] He lauded Brezhnev's "remarkable" discussion of the role of party theory, but he excised Brezhnev's references to "dogmatism" and quoted only the bland assertion that the party "has always and will always support an innovative, Leninist approach" to problems.[133]

The significance of the scientific-technological revolution for foreign policy also remained a source of tension. Not long before the conclusion of a four-power agreement on Berlin in November 1971, Shelest softened his public position on the German question, but the change was so drastic that one Western expert has plausibly suggested it "represented a command performance and did not express any underlying change of sentiment."[134] Shelepin, too, gave ground by stating that the Berlin agreement signified a relaxation of tension in Europe. But Shelepin refused to extend this generalization to East-West relations as a whole, saying that the "reactionary, aggressive policy of imperialism, and first of all of the U.S.A., is a permanent source of tension in international relations."[135] In March 1972, two months before President Richard Nixon was due in Moscow to discuss SALT, Vietnam, and Soviet-American economic relations, Shelepin publicly analyzed the interaction between capitalism and the scientific-technological revolution. His message was that despite some "new phenomena" in capitalism, its "exploitative, reactionary" essence was unchanged; it was still afflicted by "serious economic shocks" and was losing the race for higher labor productivity to socialism.[136] Accordingly, Shelepin slighted Brezhnev's "peace program" and, in a bid for support from the military and heavy industrial lobbies, warned that the priority growth of consumer goods proclaimed by the 24th Party Congress should not be interpreted "simplistically."[137] Clearly he did not believe in the economic necessity or the political advisability of a rapprochement with the United States.

Brezhnev's Victories

Despite such resistance, Brezhnev achieved important victories on the issues of technological progress and arms control. The November agreement on Berlin promised to widen access to Western technology by relaxing tensions with Western Europe. The debate over Soviet-American economic ties and SALT peaked in May 1972, when Brezhnev won a Politburo battle over whether Nixon's visit to Moscow should be cancelled because of stepped-up American military action against North Vietnam. From the beginning, the decision to invite Nixon had been distasteful. Never before had the USSR pursued major bilateral agreements with a power at war with another socialist state—and now, to make matters worse, the United States appeared to be escalating rather than reducing its role in Vietnam. On the eve of the visit Brezhnev made a key foreign policy speech to the Central Committee and won its approval for his report, rather than simply approval for the policy of the Politburo as a whole.[138] That this was a defeat for those who opposed Nixon's visit is shown by the fact that Shelest, who had long taken a hard line toward the United States, was simultaneously transferred from his powerful position as head of the Ukrainian party to a lesser economic post. The May Central Committee plenum cleared the way for Nixon to come to Moscow and gave Brezhnev the latitude to conclude a SALT accord and agreements for Soviet-American cooperation in space exploration, science, and technology. In October the two sides signed an agreement that promised to expand bilateral trade (and that required the approval of the U.S. Congress).[139]

In the spring of 1973 Brezhnev made further gains. In March he pushed through a timetable for amalgamating research institutions and industrial enterprises into multi-unit associations that were expected to speed up indigenous technological innovation. He also firmed up support for his policy of seeking Western technology and credits. At a Central Committee plenum in April Brezhnev engineered the removal of Shelest from the Politburo, along with G. I. Voronov, another member who had reputedly opposed an easing of tensions with the United States.[140] While the plenum called for vigilance against imperialist probes, it emphasized that the Cold War was giving way to improved relations. Praising Brezhnev's "great personal contribution" to the solution of foreign policy issues, it instructed the Politburo to follow the guidelines set forth in his plenary speech and endorsed "an activation of mutually beneficial foreign economic ties" with the West.[141] It was still unclear how far the Politburo would carry this policy, since two of the three new Politburo members elected by the April plenum headed institutions with a vested interest in curbing a broad liberalization of relations with the West. Whether A. Grechko (Minister of Defense) and Iu. Andropov (Chairman of the KGB) would support Brezhnev on this score remained to be seen.[142] Nevertheless, Brezhnev's hand was strengthened. The significance of the fate of Shelest and Voronov would be lost on neither the old nor the new members of the Politburo.

Bolstered by these victories, Brezhnev pressed harder for greater reliance on Western technology. During a visit to the United States in June 1973, he stated that there were now "enormous possibilities" for increased scientific and industrial ties with America, which could be formalized for periods up to twenty years.[143] Later in the year Brezhnev observed that it was unreasonable to limit economic cooperation to trade alone. The scientific-technological revolution could now be fostered "only by depending on the broad international division of labor" and on "large-scale economic cooperation—both bilateral and multilateral." This requirement, he said, "relates not only to Europe but to all continents." Moreover, such cooperation would also strengthen peaceful relations between states.[144] These words showed that Brezhnev viewed a new economic relationship with the West as essential for promoting technological innovation in the USSR, rather than simply as a means to attain the noneconomic goals of Soviet diplomacy. Equally important, the passage showed that he felt such a relationship should include the United States. Recently Shelepin had tried to draw a distinction between détente with Western Europe and persisting hostility toward the United States, and some Politburo members had mentioned Europe more frequently than America as a possible economic partner.[145] Brezhnev's message was that economic ties with the United States were an essential part of the reorientation of foreign relations which he favored.

Specialists concerned about the comparative performance of the economy backed Brezhnev's stress on Western ties and his claim that the country faced new technological requirements. G. A. Arbatov, head of the Academy's U.S.A. Institute and a full member of the Central Committee, argued for "the broad development of mutually beneficial [economic] cooperation" with the West.[146] Echoing the attitude of Inozemtsev, he suggested that nonmilitary "factors of strength" were playing a growing role in the competition between East and West.[147] The United States, Arbatov maintained, was shifting many R & D resources to nonmilitary uses with greater foreign policy utility, and American economic penetration of Western Europe and Canada showed the results.[148] Like Inozemtsev, Arbatov gave the impression that he felt the USSR should undertake a similar reorientation of its R & D effort.[149]

Other commentators, however, rejected this view of Western economic ties and Soviet needs. Marshal Andrei Grechko insisted that the military threat posed by imperialism remained strong, and, while making a brief gesture toward the "peace program" inaugurated at the 24th Party Congress, carefully refrained from mentioning its international economic dimension.[150] Instead Grechko interpreted the program as a formula for avoiding war by matching imperialism's military might.[151] Soviet military observers clearly felt that this task remained central. One observer asserted that despite the growing skepticism of Western scientists toward military research, "as before, militarization is the characteristic feature of the development of the sciences there," especially in the United States.[152] Grechko pointed the moral: "The interests of a reliable defense of the Soviet Motherland require [us] not to weaken the front of scientific research, to continue . . . to use the results of scientific-technological progress for creating prospective types of weaponry and military technology."[153] Any transfer of military R & D resources to civilian needs, in other words, was out of the question.

Suslov, too, remained skeptical of Brezhnev's enthusiasm for Western technology. A month after Brezhnev visited the United States, Suslov registered his reservations. He gave the Politburo more credit, and Brezhnev less, for recent foreign policy initiatives than had the April plenum.[154] Moreover, he avoided Brezhnev's association of the scientific-technological revolution with East-West cooperation. While Suslov noted that the April plenum had made a beginning on beneficial East-West economic ties, he emphasized that "colossal reserves" were available inside the Soviet system to meet the demands of the technological revolution. He added: "Finally, exceptionally large and long-term sources for the growth of socialist production lie in the international division of labor, in socialist economic integration on the basis of broad scientific-technological cooperation."[155] This implied that many of the USSR's needs for foreign technology could still be met within the boundaries of CEMA. Although Suslov may have been willing to grant some validity to the idea of slowing the military race with the United States,[156] he viewed the prospective role of Western economic ties far more dubiously than did Brezhnev.

Confronted with such resistance, Brezhnev again emphasized the need to upgrade Soviet economic performance and make good use of economic ties with the capitalist world. In July 1973 he demanded "genuinely revolutionary measures" to step up innovation and again called for special consideration of this problem at a Central Committee plenum.[157] Five months later he followed up with a very forceful speech to the annual budgetary meeting of the Central Committee. The measures taken thus far by the Politburo, Central Committee, and Council of Ministers to improve economic administration were, Brezhnev said, "insufficient." "Whether we desire it or not," the task of improving economic administration "is being raised by life itself."[158] It could not be approached from "narrowly economic, still less from technocratic viewpoints," and the "ossification of organizational forms" could not be permitted.[159] The regime must improve economic planning, change the powers of the ministries, strengthen economic incentives, and step up the growth of consumer goods to compensate for unjustified departures from the preferential rate for these goods decreed by the 24th Party Congress.[160] Brezhnev also remarked that economic links with the West, which were "extremely useful," made "new demands" for the rapid construction of vast projects, and that this required the "unremitting attention" of the central planning organs.[161] These words contained a hint of displeasure, but an elision in the published text prevents us from knowing whether he made any stronger comments on the subject. The excerpts completely avoid the broad foreign policy questions that were reportedly discussed at the plenum.[162]

Persisting Opposition to Brezhnev's Program

Brezhnev's new offensive provoked strong resistance. One leader, Romanov, praised his speech to the plenum for its frank discussion of economic difficulties and firmly endorsed his leadership in foreign policy.[163] Suslov, on the other hand, asserted that socialist economic achievements were especially striking in view of the deep economic and political crisis of the West, and he described recent Soviet technological improvements in glowing terms that implicitly contradicted Brezhnev's call for new policies.[164]

In early 1974, two major party journals ran editorials that reflected sharp tensions over the economy's performance and the distribution of power in the leadership. The editorial in *Kommunist*, the party's main theoretical organ, first threatened to remove economic officials who refused to heed criticism of their work, yet then reversed itself by threatening to remove any presumptuous leader who overreached himself in proposing solutions to current problems. It clearly embodied an exchange of warnings between Brezhnev and his opponents.[165] The second editorial, in the major journal on party history, was even more self-contradictory. On the one hand, it cautioned that collective leadership could not always protect the party from errors, and it flatly asserted that self-criticism helped improve socialist society. It rejected the charge that criticism should be muted because it aided the enemies of socialism—a theme that echoed the earlier controversy between Brezhnev and his opponents over the impact of self-criticism on Soviet diplomacy. Attacking the tendency to brand conscientious critics as "slanderers" and "blackeners," the editorial also quoted Brezhnev's statement that "some leaders" did not respond to criticism properly.[166] Yet, on the other hand, the same editorial lauded the October 1964 plenum that had removed Khrushchev. It warned of many historical cases when antisocialist forces had used "revisionist, demagogic, and careerist elements within this or that communist party, [and] under the slogan of criticism and self-criticism tried to break down the party, to undermine its authority in the eyes of the toilers." These elements misrepresented "miscalculations in the direction of the economy as the unsoundness of the economic system of socialism." The party, said the editorial, "comes out decisively against spiteful fault-finding [and] blackening."[167] The editorial, in short, condemned both "blackening" and resistance to "blackening." Brezhnev and his opponents were obviously still at odds over the propriety of economic self-criticism.

The evidence suggests that this conflict concerned foreign policy as well as domestic economics. In the preceding months the leaders had differed over the value of Western technology, and it was becoming increasingly clear that influential U.S. political circles were determined to use economic relations to force a relaxation of the restrictions on the emigration of Soviet Jews.[168] It is thus especially noteworthy that at the same time as Soviet editorialists began to polemicize about economic self-criticism, other commentators focused unusual attention on the relationship between domestic and foreign policy. The striking thing is that although these commentators endorsed economic détente in principle, they disagreed on the importance of domestic needs, and particularly economic needs, in determining the Soviet line in international affairs.

Perhaps most revealing was an exposition in *Kommunist* of Lenin's ideas on Soviet foreign policy. Lenin, said the author, had underscored the class character of the regime's foreign line. "The very deepest roots of both the domestic and foreign policy of our state," Lenin had said, "are determined by the economic interests and the economic position of the dominant classes of our state. These theses . . . must not be lost from view for a minute, in order not to lose ourselves in the thickets and the labyrinth of diplomatic contrivances."[169] Foreign and domestic policy, urged the writer, were "indissoluble parts of a single whole. The separation of foreign policy from policy in general or the opposition of foreign policy to domestic policy would be a contradiction with the scientific approach to the analysis of politics and with Marxism." This note was unusual; Soviet writers rarely raise the possibility that calculations of domestic and foreign policy might point in opposite directions. The writer then drew an implicit line between this lesson and Politburo deliberations by quoting Lenin again. Lenin had noted that the Politburo weighed "many small and large questions about 'moves' from our side to answer 'moves' of foreign powers, to prevent their, shall we say, tricks," and the Leninist approach to foreign policy was irreconcilable with "dogmatism, subjectivism."[170] The moral seemed to be that domestic economic needs should play a large part in the formulation of foreign policy, and that any effort to minimize their importance was wrong, notwithstanding the "diplomatic tricks" of foreign powers.

Shortly afterward, in an article written as a gloss on the "peace program," another spokesman offered a different interpretation of the relation between domestic economic interests and foreign policy. Although nonstate relations, such as those involving corporations, could sometimes outpace the foreign acts of states, the essence of international relations was still the foreign policies of states themselves. "The principle of the *primacy* of internal [social] relations and their unity with international relations . . . does not give grounds to negate the *relative in-*

dependence of international relations." Indeed, "the relatively independent role of foreign policy is growing stronger."[171] Having rejected the connection between internal and foreign policy posited in the *Kommunist* article, the writer challenged its view of the relationship between economics and politics. In "foreign relations more than anywhere else," he said, "the precedence of politics over economics manifests itself, being expressed, for example, in the initiating role of states in the arrangement of economic ties."[172] Probably these words were meant to counter the belief that Western economic interests had given détente an irresistible momentum which permitted the USSR to seek more Western technology without political risk. Later the author pointedly observed that for commercial reasons socialist countries sometimes preferred capitalist to other socialist trading partners. But, he warned, "socialist states must be guided in external economic ties *by the ... priority of the basic interests of socialism over the interests of immediate benefit.* ... [It] is doubtful that a conscious revolutionary-internationalist can approve cases when the contacts of socialist countries with capitalist ones conflict with the interests of [socialist] cooperation as a whole."[173]

Tension over the economic dimensions of domestic and foreign policy sharpened further in late 1974. In September *Pravda* published an article clearly meant to undermine the political position of Suslov.[174] But Suslov was not the only leader under fire. In October *Kommunist* ran an editorial that strongly underlined the dangers of departing from collective leadership. Warning against ill-conceived changes in domestic economic policy, it stated that the party "rejects any efforts artificially to push socioeconomic development." In his time, Lenin "spoke out decisively against rushing forward, against adventuristic plans for the 'instant introduction' of socialism and communism, warning that there is nothing more dangerous than to design policy ... on the basis of wishes alone, without taking account of objective ... possibilities." Soviet economic growth would inevitably be braked by voluntaristic attempts to force the transition from a socialist to a communist economy and by "the premature liquidation of socioeconomic forms which are necessary in the current stage."[175] The editorial then drew a clear connection between such economic mistakes and the sort of deviation from collective leadership that had brought down Khrushchev in 1964. The October 1964 meeting of the Central Committee had confirmed collective leadership as the highest principle of party guidance. Collective leadership "is a guarantee against one-sidedness, against the adoption of subjective decisions. ... Collectivity is characteristic of political leadership at all levels, from primary party organizations . . . to the party Central Committee, which defines the political line that directs the development . . . of the country in the period between CPSU Congresses."[176]

The editorial concluded with a veiled warning about foreign policy. Although it quoted from the Central Committee's April 1973 foreign policy resolution, it omitted some of the most positive phrases and added an unusual caveat: "Coming back to the tenth anniversary of the October [1964] plenum of the CPSU Central Committee, it is impossible not to mention that it also marked a sharp turn to the genuinely scientific development of the *foreign policy line* of the party and state."[177] This sentence was highly significant. The editorial had not actually mentioned the anniversary of the plenum. Rather it had dwelt on the policy dangers of one-man rule and the steps taken at the plenum to counter them; if it was coming back to any point, this was it. Nor was the connection of the October plenum with foreign policy simply an editorial convention. To my knowledge, this is the only time a Soviet editorial linked the two. The aberration was especially striking because the article had discussed the October plenum so pointedly, and because the evidence suggests that Khrushchev's inclination to exchange diplomatic concessions for Western technology had contributed to his ouster at the plenum. In short, the passage was almost certainly meant to signal that just as in domestic affairs, any attempt by some leader to "rush forward" in foreign policy was impermissible.

Brezhnev was the main target of these strictures. Of the top leaders, he had been the most outspoken about the need to enter a new stage of intensive economic growth, whereas some others had shown considerable skepticism about the urgency of this transition. Moreover, as part of his drive for intensive growth, Brezhnev had warned against the "ossification" of economic institutions, which probably made him guilty in the eyes of his critics of seeking the "premature liquidation of socioeconomic forms." Just as important, he had tried to speed up production of industrial consumer goods and to incorporate this goal in the fifteen-year economic plan for 1976–90 whose formulation he was pushing hard.[178] This made him vulnerable to *Kommunist*'s attack on persons who wished prematurely to abandon the restrained consumption policies of socialism in favor of the abundance promised by full communism; and his subsequent effort to ram through higher priority for consumer goods was defeated by powerful opposition at the top.[179] At the 25th Party Congress in early 1976, no fifteen-year plan was approved, despite Brezhnev's earlier promises that it would be, and the delegates endorsed the primacy of producer goods against his wishes.[180]

In all probability, Brezhnev's enthusiasm for Western technology caused added dissatisfaction among his critics at this time. He had connected his campaign for intensive growth with the prospect of enlarged Western trade and credits, and his emphasis on long-term economic agreements with the West was probably intended to mesh with the time frame of the fifteen-year economic plan.[181] Yet the United States, the capitalist country with which Brezhnev was most eager to build up economic relations, was using the lure of advanced technology to press for concessions on Jewish emigration that the party authorities found politically offensive. It is striking that *Kommunist*'s editorial stressing collective leadership in foreign policy went to press on 25 October 1974—one day before Secretary of State Kissinger, then in Moscow, received a Soviet letter denying that the USSR had given the Nixon administration a private promise to expand Jewish emigration in exchange for most-favored-nation status and U.S. government-backed credits.[182] The fact that the Soviet press did not publish the letter, but allowed Kissinger to continue to bargain on this issue with critics of the Soviet Union inside the U.S. Congress, indicates that the Soviet leadership had not swung decisively against the idea of expanding economic relations with the United States. But the decision to send the letter does show the highly sensitive nature of this question.

Some additional evidence suggests that the top leaders were divided over exchanging political concessions for U.S. trade and credits. Three days after Kissinger received the Soviet letter, a major party journal ran an important lead article defending the Brezhnev line on relations with the West.[183] The subject was the Leninist approach to political compromise—not with other communists but with class enemies. The article stressed that the Leninist approach had "enormous" significance "in current conditions" and was useful in international as well as domestic politics.[184] Compromises could be either "formal or informal."[185] They required, in Lenin's words, "the *obligatory* . . . use" of "any sort of 'split', even the smallest, between enemies, of any sort of opposition of interests between the bourgeoisie of different countries, between different groups or sorts of bourgeoisie within particular countries."[186] This prescription dovetailed with the use of tacit compromises over emigration to neutralize the more hostile elements of the U.S. political establishment and gain economic benefits for the USSR. Next the article denounced "the phrase-mongering of left doctrinaires . . . who recognize only 'the direct' path to socialism"—that is, those who rejected even tactical compromises with imperialists—and it cited the Treaty of Brest-Litovsk to justify its position.[187] It also gave favorable mention to Lenin's policy of setting up concesssionary enterprises with Western companies, and while it said that he had firmly defended Soviet sovereignty in this matter, it noted that he had made "partial [diplomatic] concessions" at the Genoa economic conference in 1922.[188] The concluding section praised Soviet agreements since 1969 with West Germany and the United States. By stating that the socialist countries, while they would pursue intelligent compromises with imperialists, would strive to increase their own military might, it did signal that détente would not be carried to lengths that seriously hampered the growth of the military establishment.[189] But the article was otherwise conciliatory and uncharacteristically refrained from denouncing recent Western efforts to influence Soviet internal affairs. In short, it constituted a defense of the quest for further Western (especially U.S.) economic ties against the unnamed "left doctrinaires" who questioned this policy.

About three months later the Soviet authorities renounced the trade agreement of 1972, in which the Nixon administration had pledged to obtain Congressional approval of most-favored-nation status for Soviet exports. The sequence of events suggests that this move was triggered by two new considerations: the low ceiling on government-backed credits that the Congress enacted in December, and the added linkages that the Congress drew between a relaxation of the ceiling and Soviet moderation on matters such as the Middle East and arms control.[190] These additional restrictions must have tipped the balance in favor of the arguments of Brezhnev's critics that American technology— especially so little of it—was not worth the political price. It is even conceivable that Brezhnev and his supporters may have concurred, although other signs of internal tension within the elite[191] make this seem improbable.

At any rate, this episode did not resolve the broader questions of the Soviet economy's performance and the importance of access to Western technology. In April 1975 Brezhnev obtained the removal of Shelepin, one of his strongest critics, from the Politburo. However, leaders continued to express divided sentiments on economic relations with the West. Some emphasized the relative strengths of the Soviet economy and treated East-West relations as a political or strategic matter in which economic ties had little place; others still emphasized the need for more economic interchange.[192] Moreover, longer expositions of party policy also contradicted one another. In mid-1975 *Kommunist* ran an article on international economic relations that defended closer ties with the West. The author went out of his way to quote Lenin's rebuff to the "ultrarevolutionaries" who opposed beneficial relations with capitalist countries after the Revolution on the grounds that "one of the systems must 'be removed.' "[193] A few months later *Pravda*, commemorating the anniversary of the 14th Party Congress, responded with an article that interpreted the Leninist legacy quite differently. Heavily stressing the danger of capitalist encirclement during the 1920s, the writer praised the 1925 Congress for selecting an autarkic path that saved the Soviet Union from becoming an appendage of capitalism and strengthened its ability to revolutionize the workers of other lands. The critics of this program at the 14th Congress, he maintained, had advocated a policy that "aimed at . . . strengthening the country's dependence on the world capitalist system" and that "signified, in essence, the rejection of the construction of socialism." The author, associated with the party's conservative Academy of Social Sciences, manifestly felt that a widening of economic ties with the West might still endanger the Soviet system. "The Leninist teaching . . . on which the 14th Congress was based," he said, "is also relevant for the contemporary period—for the stage of developed socialism." Not long afterward, however, Vadim Zagladin, a central party specialist on foreign relations, explicitly denied that this article had any bearing on current policy.[194]

Technological Strategy and Western Economic Ties

The question of enlarging economic relations with the West was one of several decisions about technological strategy. How much technology should the country create through its own R & D? How much should it acquire abroad, and how should it obtain this know-how? Obviously these choices had great political significance, and after 1968 a number of elite groups indicated their policy preferences, either directly or obliquely.

Attitudes of Secondary Members of the Elite

Many social scientists favored turning to the West for more technology. Arbatov, for example, backed this policy. In early 1970 he quoted Lenin's emphatic endorsement of economic ties "with all countries, but *especially* with America," adding that these ideas constituted "long-term principles of Soviet policy, having enduring significance."[195] Later, as the balance of opinion among the top leaders swung toward détente, Arbatov voiced this preference much more openly. In mid-1972 he praised the agreements concluded at the recent summit between Brezhnev and Nixon for contributing to faster economic development, higher living standards, and the construction of communism in the USSR.[196] Inozemtsev, too, backed wider Western ties.[197] Other social scientists, advocating the active use of world science and technology, argued that the scientific-technological revolution demanded complex R & D undertakings which exceeded "the capacities of one country, regardless of its might" and regardless of its social system.[198] R & D specialization between East and West, they predicted, would allow all participants to economize on expenditures and produce better technology.[199]

But, critics might ask, was such cooperation compatible with the long-term struggle for survival between the socialist and capitalist worlds? Arbatov claimed it was. He granted that economic and technological competition were forms of international class struggle but contended that they had a "unique aspect. . . . Peaceful competition in these spheres not only does not exclude, but on the contrary presupposes the broad development of mutually beneficial cooperation. . . . We can with full justification assert that along with new bridgeheads of struggle, the scientific-technological revolution also creates new bridgeheads of peaceful collaboration."[200] It was difficult to make this argument persuasively, particularly since Arbatov and his colleagues were trying simultaneously to convince skeptical Westerners that economic ties helped the West as well as the East. How, then, the Soviet skeptics demanded to know, could these ties also aid the USSR against the West?

The proponents of greater reliance on Western technology answered that although the West would benefit, the USSR would benefit more, and that such ties therefore would promote the ultimate victory of socialism.[201] In 1974, when Soviet hopes for a breakthrough in trade with the United States were threatened by Congressional amendments to the Trade Act, one social scientist asserted that the advantages of such ties for the USSR were "indisputable." It was necessary, she said, "to remember that here we are talking first of all about . . . which system more fully opens the way to stormy progress of science and technology."[202] Her implied premise was that the USSR would gain more from such links than would the West. Proponents also tried to neutralize the view that by helping the capitalist economies, East-West ties would reduce the strains within these systems. Arbatov dismissed the idea that the USSR should not expand trade with the United States because this would reduce U.S. unemployment, supply America with needed raw materials, and increase its military potential.[203] Another spokesman contended that more trade would improve the economic situation of Western workers and thereby "strengthen the politico-ideological position of socialism in the world."[204] He glossed over the possibility that closer economic links might solidify worker support for capitalism by bolstering the capitalist economy.

These social scientists also sought to counter the claim that wider economic exchange would invite ideological erosion and Western interference in Soviet affairs. Arbatov acknowledged that some American circles wished to use economic and technological relations as a political lever against the USSR, but he maintained that this was not a significant danger. By simultaneously strengthening indigenous innovation and cooperation with other socialist countries, he argued, the socialist camp would be able "successfully to oppose any probes by its enemies." Moreover, long-term economic ties between East and West created "mutual dependence," that is, mutual vulnerability, and this could be regarded "only as a positive fact."[205] Nor did Arbatov accept the theory that the USSR was susceptible to ideological subversion through such channels. Citing the 1930s as a period in which such ties exerted no appreciable political influence on the USSR, he asserted that the country was even more impervious to such influences in the 1970s, given its present industrial might.[206] At times, indeed, Arbatov seemed more worried that the very durability of Soviet ideology would obstruct wider economic ties. Appealing for a "rejection of war propaganda and of hatred for other countries," he contended that the ideological struggle between the two systems should be conducted so as not to harm the "wholesome processes" currently improving international relations.[207] The striking thing about this statement was that it could easily be read as applying to Soviet propagandists as well as to their Western counterparts. Ordinarily Soviet writers stuck to the stock formula that improved relations between East and West required a simultaneous intensification of the ideological struggle.

Not all social scientists advocated an economic opening to the West, however. In 1969 the director of the East German Institute of Social Sciences argued in a Soviet social science journal that although some ties with the West were necessary, "not one Marxist can seriously expect that the monopolies and imperialist states will make a contribution which would aid the achievement of the superiority of socialism in the . . . decisive areas of contemporary science and technology." Moreover, the "effort in particular socialist countries to solve the problems of the scientific-technological revolution with the help of imperialist groups of monopolies or from a position of some sort of 'independence' is in fact pregnant with the threat of dependence on the imperialists."[208] The article was probably published on the initiative of Soviet politicians who agreed with it,[209] but other social scientists also agreed. The following year a Soviet specialist on CEMA remarked pointedly: "It is impossible seriously to expect that the countries of CEMA will be able to solve the problems of technological progress by importing technology from Western countries."[210] After 1970 social scientists apparently ceased to air this view, but it would be surprising if some did not still adhere to it privately.

Like most social scientists, many natural scientists and science administrators wished to draw more fully on Western R & D. As before, representatives of the science establishment continued to push for a sharp expansion of the budget for domestic R & D, and some economists supported their demands.[211] But on this condition many scientists were eager for wider international cooperation. In 1970 Academy Vice-President Millionshchikov asserted that the range of potential cooperation between the USSR and the United States was "extremely broad."[212] In 1973–74 the Acting Chief Academic Secretary called for sharply increased travel to the West by scientists, who should spend "long periods . . . in the best foreign laboratories," and indicated that in the future the Academy would put much greater emphasis on joint research abroad.[213] Another high Academy official implicitly defended international scientific ties from the charge that they invited foreign subversion. Soviet scientists were aware that such ties "are often used for purposes alien to the interests of peace and social progress," but participation in foreign scientific relations, he assured his readers, facilitated resistance "to such an inhumane practice."[214] Meanwhile a top official of the State Committee for Science and Technology argued for long-term technological cooperation. Peaceful coexistence, he said, involved not only mutual understanding but also "perhaps, even interdependence," and "we are decisive opponents of the ideology of autarky." U.S.–Soviet economic ventures were likened to a large jet plane that after a long takeoff "easily gains altitude and builds up tremendous speed."[215]

Some resistance to enlarged Western cooperation, however, persisted in the Academy. In 1971 N. N. Semenov, a member of the Academy Presidium, underscored the importance of developing Soviet chemical technologies superior to those of the West: "Even when using the best foreign processes, when buying factories abroad, we receive yesterday's technology—or at best, today's. Whereas study of . . . chemical processes gives inexhaustible possibilities for their radical improvement."[216] Although the USSR had already purchased a sizable amount of Western chemical equipment and was moving toward further purchases, Semenov opposed the practice. Indirect confirmation of such attitudes later appeared in an article printed "for discussion." The author, an Academy member, criticized "the organs planning science" for "major miscalculations" in R & D plans. Maintaining that the cause was an inadequate appreciation of international R & D specialization, he advocated more cooperation with other CEMA members and with the West.[217] Obviously he felt that some persons in the agencies overseeing research still did not fully appreciate the value of Western R & D.

Foreign trade officials expressed vigorous support for wider Western ties. Praising the "large prospects" for compensation agreements with the West, the Ministry of Foreign Trade endorsed increased East-West trade as meeting the needs of the scientific-technological revolution.[218] Writers in the Ministry's journal still occasionally raised the specter of damage to the economy by the forces of the world market, but they did this only to protect the Ministry's power against persisting pressures for reform of the trade apparatus.[219] In late 1974, when tensions over exchanging political concessions for U.S. credits peaked, the Ministry journal printed an article that championed economic dealings with the West. The author cited Lenin's strictures against those who thought "the interests of the international revolution allegedly prohibit any sort of peace with the imperialists." Lenin, he said, understood that peaceful coexistence would not be "idyllic," and his teachings in favor of trade with the imperialists were especially timely in the present day.[220]

Another protrade group consisted of industrialists who believed that imported producer goods would help them meet the party's demand that they raise their output to world technological levels. This demand was most painful for officials from sectors that had received low priority in the past and had lagged behind Western technology. Branches like oil production, chemicals, and consumer goods fell within this category. On the heels of Brezhnev's 1973 visit to the United States, for example, the Minister of Oil Industry predicted that a series of major contracts would soon be concluded with American oil and engineering concerns. The tone of his comments was distinctly favorable.[221] During subsequent negotiations, the same official reportedly said that such ties were not necessary for Soviet industry, but these words were almost certainly part of an effort to drive a favorable bargain with prospective Western partners.[222] In 1974 the Ministry's journal again painted a positive picture of possible ties with the U.S. oil industry.[223] The Ministry of Cellulose and Paper Industry was also reportedly enthusiastic about acquiring Western equipment because domestic producers were not meeting its technological needs.[224] A similar enthusiasm probably characterized the officials of other lagging industrial sectors.

On the other hand, some administrators of the producer-goods branches opposed greater reliance on Western technology. These branches had traditionally enjoyed an overriding claim on the budget and had been shielded from foreign competition through the state monopoly of foreign trade. Because large-scale imports of Western technology would reduce their share of the Soviet market, their leaders must have had reservations about this prospect. An economic journalist who later emigrated states that in the early 1970s some of the most energetic managers were ashamed of the degree to which industry already depended on foreign technology.[225] In 1975 A. I. Tselikov, head of a major industrial research establishment and also a member of the Academy, published an article in *Kommunist* that questioned the growing Soviet interest in Western technology. Tselikov argued that the USSR should concentrate on producing machines for export, because they were more profitable than raw-materials exports, and should step up the expansion of the machine-building branches. Implicitly this argument contradicted the idea of paying foreign companies for imported turnkey plants with raw materials exports. Disputing the value of foreign licenses, Tselikov also asserted that by the time licensed production began, the goods were often outdated. He went on to endorse machinery imports, but "only within general limits" and only in sectors where the USSR lacked experience—for example, in automated lines for producing consumer goods. In sectors where Soviet industry had manufacturing experience, he contended, it could produce machines surpassing the output of the best foreign firms. Here Tselikov listed several machine-building branches and cited three recent cases in which Western installations in the USSR had proven unsatisfactory and been replaced with Soviet equipment.[226]

The expression of such views provoked a rebuttal from the Gosplan journal, which maintained that international cooperation, including the purchase of foreign licenses, was a "major reserve" for accelerating Soviet innovation. Noting that many ministries were still trying to solve technological problems long since solved abroad, the journal remarked that "in a series of instances" R & D specialists viewed the purchase of foreign licenses as a derogation of their own scientific achievements. There can be no doubt that this remark was directed as much at ministers as at R & D personnel. Rejecting such hostility toward foreign technology, the journal called for wider use of foreign know-how.[227] The airing of such contradictory views indicated that the national and professional pride of many industrial officials was still an obstacle to large-scale technology transfers from the West.

A few military writers, encouraged by the strength of "realistic circles" in the imperialist countries, called for much wider trade and economic ties with the West.[228] In terms of their rank and the frequency of their appearances in the press, however, the exponents of this view constituted a definite minority within the military establishment.[229] Like Marshal Grechko, leading military figures emphasized imperialism's aggressive foreign policy and the persisting danger of war, thereby implying strong doubts about the wisdom of significant economic dependence on the West.[230] When they spoke of international economic ties at all, they concentrated on the virtues of tight economic integration within CEMA. In an article on the economy and the armed forces published in early 1973, a Deputy Minister of Defense dwelt on the new steps taken by NATO countries to upgrade the convertibility of their economies for war. He strongly emphasized the value of CEMA for the USSR but made no reference to the economic agreements recently concluded with the United States.[231] The importance of economic independence from the West was also stressed by lower-level officers. One writer, concluding that Soviet independence from the world economy had been "one of our great advantages" against Nazi Germany, contended that in current circumstances improving the economy's adaptability for war was "a most important task."[232] Like military officials in most powerful states, these officers feared that broad economic relations with rival regimes would undermine their country's freedom of strategic maneuver. They took little comfort from Arbatov's reassurances that such relations would create mutual, rather than only Soviet, dependence.

The military skeptics were also worried that détente might undermine the military's preeminent claim to budgetary and research resources. No doubt they would have been happy to acquire the Western strategic equipment, such as computers, which Soviet industry could not produce on the same level. This is one reason why some Western observers suggested that the military backed Brezhnev's vision of détente.[233] But there was a more compelling reason for the military to resist that vision. During the early 1970s Brezhnev tried to upgrade the standing of consumer goods in relation to producer goods, and other exponents of economic détente argued that the military budget should be curbed and a greater share of R & D resources be given to nonmilitary ends. This meant that, on balance, a comprehensive détente with the West might hurt the armed forces technologically, rather than help them.

Military officials were very sensitive to this danger. In the early 1970s they obliquely disputed the idea of revising R & D priorities, but by mid-decade they were openly resisting. In a book published by the Ministry of Defense, one military specialist on R & D remarked that the development of military programs "cannot always correspond to the general direction of the development of research stipulated by the [internal] requirements of the further development of socialist society." Socialism, he said, "has the possibility" to establish the correct balance between military and nonmilitary research programs—thereby implying that the current balance was incorrect. Clearly he thought the military was receiving an insufficient share of R & D, since he attacked unnamed Soviet critics who had asserted that civilian spin-offs were inadequate and that "military requirements brake the development of science or are the cause of its one-sided development."[234] The same writer was unusually outspoken about the liabilities of dependence on the West. "Of course," he said, "one . . . must use the results of fundamental research obtained in other countries. However, from the point of view of strengthening the defense capacity of the country, dependence . . . on the development of science in the countries of the opposition coalition is extremely undesirable and dangerous."[235] At a time when greater East-West cooperation was the dominant party line, these were strong words.

Most party ideologists were also suspicious of expanded Western ties. One illustrative article was published in late 1972, a few months after Nixon had visited the USSR and signed an agreement for U.S.–Soviet space cooperation. In the writer's view the purposes of socialism and capitalism in exploring space were fundamentally different. Moreover, the structure of the socialist system gave it "immeasurably broader possibilities" for conquering space than capitalism possessed, and U.S. achievements like the moon landing would have been impossible "without the many results that we obtained in the Soviet Union in the course of the exploration of the universe." If these statements had any policy implication, it was that the USSR should keep more of its space research secret, rather than cooperate with the United States. The author also contended that space exploration influenced "ideology above all"—a view that tacitly defined space as a realm in which, because of the standard prohibition against accommodation with bourgeois ideology, there could be no East-West cooperation. Working from these premises, the author criticized unnamed persons who wished to "unite all 'space' countries into a single entity without seeing that . . . mutually fruitful international cooperation is possible only with a socialist transformation of society."[236] All in all, it was a remarkably forceful attack on the policy of East-West technological cooperation that Brezhnev and his supporters were advocating.

Party ideologists remained especially distrustful of technology transfers that required extensive personal contacts between Soviet experts and Westerners.[237] As the official drive for Western technology got under way in the 1970s, ideologists began to modulate their positions on this issue,[238] but serious tensions over the choice between ideological purity and stepped-up technological borrowing remained. Perhaps the best example came in early 1974, when V. N. Iagodkin, the ideological secretary of the Moscow city party committee, published an article in *Kommunist*. Iagodkin, who had already shown deep concern about the ideological dangers raised by détente,[239] emphasized that "the development of business, scientific, and cultural ties and other contacts between socialist and capitalist countries . . . increases the possibilities for the penetration into our environment of alien ideas. *This poses new tasks of ideological-theoretical and educational work, requiring from each communist ideological maturity, high political consciousness, and communist conviction.*"[240] Iagodkin's stress on the ideological liabilities of foreign technological contacts signaled serious disagreement. Shortly after this issue of *Kommunist* was distributed, the authorities, in an extraordinary act, recalled the issue and destroyed hundreds of thousands of copies.[241] When the new version of the same issue appeared a few days later, Iagodkin's article bore a title different from the one *Pravda* had previously announced—it was the only article whose title had been changed—and his treatment of contacts with the West had been altered.[242] In addition to citing the negative consequences of such contacts, the revised text stipulated that their extension "favors the peaceful advance of world socialism and wider dissemination of communist ideas in the whole world."[243] The article thus accented the benefits of contacts, rather than dwelling only on their drawbacks.

Shortly afterward, another ideologist presented an assessment of foreign contacts that was strikingly different from the view Iagodkin had tried to advance. One thought recently expressed by Brezhnev, said the writer, was of special importance: the broadening of contacts between East and West facilitates "the dissemination of the truth about socialism and of the idea of scientific communism." While this official acknowledged the subversive intent of some Western circles, he did not say that Soviet intellectuals were susceptible to foreign ideology. Instead he emphasized that exchanges of travelers were a useful way of refuting Western misrepresentations of socialism. Moreover, he hinted that the country's current foreign line was being poorly treated in party propaganda, which was still giving only a "modest place" to major international developments. "Not everything is all right" in expositions of Soviet foreign policy, he said, because "unfortunately" some central propaganda materials explaining that policy "are being delayed."[244] These delays were presumably due to high-level disagreements over the interpretation of recent changes in foreign relations. Despite this official's willingness to take a more positive view of personal contacts, most party ideologists remained highly skeptical of the idea.[245]

The political police, although they occasionally paid lip service to wider scientific and technological exchanges with the West,[246] likewise distrusted the notion. They bore responsibility for screening out potentially subversive foreign ideas, and a large part of their power rested on their charter to prevent state secrets from falling into foreign hands. This gave them a strong bureaucratic motive for asserting the technological superiority of the Soviet system and for emphasizing the threat of foreign espionage. In 1970, one KGB spokesman described a massive American intelligence effort to gain valuable Soviet information through open sources such as scientific exchanges and publications, and he called on Soviet specialists to combat this "serious danger" by being more circumspect in publishing their work and not being "superfluously frank" in personal contacts.[247] In 1971 a Deputy Chairman of the KGB claimed that the USSR's growing scientific and military might had caused the imperialists to pursue a policy of "total economic espionage." In this effort Western intelligence services were using "scientific and other organizations, specialists in the fields of science, technology, military affairs, persons in culture and art, journalists, representatives of business circles." Indeed, cases had been noted "in which some foreign specialists working on the construction of a series of enterprises in the USSR were used for subversive purposes."[248] The last sentence showed a distinct lack of enthusiasm for even the most tangible fruits of foreign exchanges—the massive industrial complexes, such as the Volga Auto Plant, which were being built with the aid of numerous Western advisers.

But if the USSR was as superior technologically as the police claimed, how could it be in danger of "total" Western subversion? The advocates of more technological borrowing used a weaker variant of the same premise to support the opposite conclusion; the country's economic strengths, they maintained, made it impervious to subversion. This counterargument rankled, and the growing internal tensions over exchanging political concessions for U.S. trade provoked the police to denounce it directly. The third issue of *Kommunist* for 1974—the same issue in which the ideologist Iagodkin made his ill-fated attempt to highlight the danger of closer contacts with the West—carried a review of a book written by the KGB official just quoted. Paraphrasing sympathetically, the reviewer said: "And if sometimes it is necessary to encounter moods of . . . carelessness, if some people are still inclined to consider that the Soviet Union and the other countries of the socialist commonwealth, because they have achieved enormous successes in all areas of economic and cultural life, are no longer at all endangered by the intrigues of the imperialist intelligence services, then this is a deep error. . . . The enemy is insidious and resourceful."[249] The implication was that the existing barriers to contacts between Soviet and Western specialists should be kept in place. The chief danger was Western subversion, not the threat of being left behind by Western technology.

Changes in Technological Strategy

In the late 1960s, when Brezhnev began to push for more rapid technological innovation, Soviet R & D spending climbed sharply. The annual increase in science expenditures reached 11.1 percent in 1969 and an extraordinary 17 percent in 1970, following Brezhnev's dramatic speech to the December 1969 plenum.[250] But as Soviet interest in Western technology became stronger, the growth of spending on science slowed, dropping to 5.1 percent in 1974 and 6.1 percent in 1975.[251] Although these rates were still substantial, they were the lowest in at least two decades. In part they signified the regime's desire to get more "intensive" results from its scientists, rather than to keep upping science expenditures at a rapid pace.[252] But they also manifested the determination of a strong group within the leadership to acquire more technology from the West. The unusually low increase in R & D for 1974 was planned in 1973, when hopes for major technological benefits from détente were at their peak.

The desire for more Western technology also showed up in the pattern of trade. From 1971 to 1975 total Soviet trade grew at more than twice the rate projected for those years by the Five-Year Plan, and the Western share increased markedly. In 1970 only 21.3 percent of the total was with the industrial West. By 1974 the proportion had risen to 31.3 percent.[253] Trade with other socialist countries, on the other hand, declined from 64.5 percent of the total in 1971 to 56.3 percent in 1974.[254] At the same time, despite its growing appetite for Western technology, the regime sought to maintain its economic hold over Eastern Europe. Thus in July 1971 it pushed through an ambitious CEMA program that called for closer intra-CEMA cooperation in R & D and economic planning.[255] It also avoided sharply reducing its energy exports to Eastern Europe, although this would have provided more hard currency for the purchase of Western technology from energy sales in the West.[256]

Important changes occurred not only in the scale but in the modes of acquiring technology from the West. More vigorously than in the past, the regime sought long-term ties with Western suppliers. By the late 1960s it had signed a number of accords with Western governments for scientific, technological, and economic cooperation. Building on this foundation, in the late 1960s and the 1970s it established extensive technological relationships with Western firms.[257] At first the regime selected most corporate partners from West Germany; it then expanded the circle to other West European countries and Japan, and finally to the United States. By the mid-1970s these agreements numbered about 160.[258] They ranged from licensing deals and the construction of turnkey plants to arrangements for coproduction and specialization by the national participants.[259] Although Soviet negotiators were more reluctant than the East Europeans to accept the high measure of interdependence with the West required by coproduction schemes,[260] the agreements demonstrated a strong new desire to open channels that would help the USSR stay abreast of the latest Western advances.

While the figures cannot be calculated with precision, it seems certain that the number of Soviet specialists traveling to nonsocialist countries also increased. Substantial contingents of engineers and managers went abroad in connection with the major turnkey projects negotiated after 1968.[261] The same trend was visible in the number of scientists going abroad under the auspices of the Academy. Immediately after the invasion of Czechoslovakia, the number of scientists traveling outside the USSR declined slightly, to 4,225 in 1970.[262] It is reasonable to assume that this dip resulted from the regime's efforts to reduce the risk of foreign ideological contamination. But after 1970 the number of scientists abroad again began to climb, reaching 7,204 in 1976.[263] By mid-decade the contingent traveling to nonsocialist countries had grown noticeably, albeit not so rapidly as the group traveling to other socialist countries. For 1974–6, the number journeying outside the socialist bloc averaged 2,530 per year, as compared to 1,800 in 1966.[264] Thus the regime continued to ease the restrictions on travel in order to gain greater benefits from Western R & D. It is equally true, however, that the magnitude of these changes did not match the fundamental change in technological strategy which some commentators advocated.

Partly for this reason, serious barriers persisted to the acquisition of Western technology. The opening to the West did produce some economic benefits, in the form of extra capital construction financed via Western credits, up-to-date turnkey plants, and greater knowledge of Western industrial methods. But the transfer of foreign know-how continued to be impeded. One of the biggest obstacles was the restriction on personal contacts and communications with Western specialists. A follow-up study of the 1972 Soviet-American agreement on cooperation in science and technology concluded that scheduled travel to the United States by Soviet participants "is often delayed or simply never undertaken," and that communications with Soviet R & D institutions involved long delays.[265] Similarly, Soviet security restrictions frequently prevented Western visitors to the USSR from obtaining information needed to expedite technology transfers, including the names and addresses of many of the specialists with whom they met to discuss their company's technology.[266] Despite its increased enthusiasm for Western technology, the regime was still trying to acquire it "in a mechanical 'stand-off' fashion without establishing true organic relationships" with Western researchers and managers.[267]

The problem was compounded by difficulties in providing back-up supplies and management for the introduction of Western technology.

The introduction of some processes acquired under Western license took an unusually long time by Western standards.[268] Similarly, a study of large industrial installations imported from Great Britain shows that the elapsed time between conclusion of the import agreement and commissioning of the installation was two to three times the interval required for comparable British exports to other Western countries.[269] Moreover, there was a marked tendency not to update installations that had been highly advanced at the time of commissioning. For example, one report indicated that the giant Toliatti plant constructed with the aid of Fiat had not been modernized in line with the Western technological developments of the 1970s, and the advanced methods it incorporated had not been widely disseminated to the rest of the Soviet automobile industry.[270] The study of British installations in the USSR also showed that even after commissioning, the size of the labor force at new chemical works exceeded West European levels by 50 to 70 percent, although machine-tool works did not show similar shortfalls in productivity.[271] These facts indicate that the effective use of Western technology required the regime to overcome some of the same impediments which inhibited the introduction of indigenous technology. The opening to the West, while economically beneficial, had not eliminated the need to upgrade the technological performance of domestic scientific and industrial institutions.

Innovation and the Politics of Bureaucratic Change

After 1968, the elite canvassed three possible methods of improving domestic research and innovation. One was to promote further autonomy and competition among institutes and enterprises—an idea that encountered mounting opposition. Another notion, much in vogue, was to better the central planning of technological progress. A third approach was to amalgamate research and production units in such a way as to reduce the organizational barriers to innovation.[272]

In late 1968 the regime adopted a package of decrees that included elements of each approach. The more comprehensive decree, which addressed the problem on a nationwide basis, observed that the gap between research and application was "excessively great" and that the time required to assimilate scientific discoveries was "extremely large."[273] A second decree, which dealt only with the electrical engineering industry, gave more specific reasons for reorganizing R & D in that branch.

Among them it listed the need to boost exports and, intriguingly, the need to strengthen national defense.[274]

Arguments for and against Greater Technological Competition

The package of decrees called for "broad competition," which was "weakly developed" at present, in creating new technology. Although it aimed to simplify research finance and strengthen the economic benefits accruing to applied researchers from successful innovations, it sought to buttress these inducements with competitive pressure. In order to prevent monopolies in important research fields, it said, the Academy and the ministries should direct several R & D organizations with different approaches to work on the same problems. Competition of this kind was to be promoted "in necessary cases" and was ordinarily to stop short of the creation of prototypes, but in especially important projects competitive prototypes were to be produced as well.[275] This measure applied to institutes and design bureaus, not to industrial enterprises; but it implicitly raised the question of whether the technological monopolies of industrial enterprises might not be equally harmful.

Some high-level officials regarded this provision as very important. Trapeznikov, the First Deputy Chairman of the State Committee for Science and Technology who had earlier underscored the defects of central control, emphasized that the results of R & D competition would affect decisions about expanding a research unit's resources, changing its profile, or "even its closure."[276] Competition, in other words, would provide the technological discipline which administrative instructions could not.

Other commentators supported the idea of finding an effective substitute for administrative controls. One criticized the prevailing desire to fit research into a comprehensive organizational framework. Forms of scientific activity were possible, he said, in which scientific "benefits result not from order, but from chaos, disorder." The failure to appreciate this was due to mistaken definitions of order and disorder, rather than to the real nature of science. The example of molecules inside the boiler of a steam engine, the writer suggested, showed that an allegedly "irrational" chaos could produce very useful results; and the relation between structure and chaos in the engine was directly analogous to the proper relation between formal organization and scientific activity.[277]

The article was a fundamental critique of the dominant notion that R & D could best be organized through central regulation.

Another writer argued more concretely for greater autonomy and competition among institutes and enterprises. At least two institutes should compete for each governmental research contract. Such an arrangement was not "parallelism" but "creative competition," which would prevent the monopolization of scientific fields.[278] Likewise, the value of competition among industrial producers was "growing immeasurably" in the conditions of the scientific-technological revolution. Economic competition would create powerful compulsory pressures, rather than only moral exhortation, for lagging enterprises to absorb the technology being applied at more advanced enterprises.[279] The goal of reform, the writer concluded, should be "full commercial accounting of production associations competing for the best satisfaction of the demands of consumers."[280]

Most elite members, on the other hand, rejected competition, whether between institutes or enterprises. The Prague Spring had aroused new fears about the political implications of economic reform, and the enemies of decentralization seized the opportunity to argue their case. About the time Trapeznikov and others advocated more competition, the journal of Gosplan attacked the idea. Enterprises operating in competitive markets, it asserted, were "doomed to technological and economic stagnation." Moreover, the alleged freedom of capitalist enterprises was "illusory," being "limited by dependence on the social conditions of the market, the competition of other enterprises, and so forth." Real freedom came from central planning, which "gives each socialist enterprise a clear prospect for its development" and thereby "ensures it an incomparably greater degree of independence than that which any capitalist enterprise possesses." Under the plan "every production collective can be certain of the need for . . . its work, and the members of the collective [can be certain] of their material well-being." The editorial, in short, sought quite plausibly to persuade managers and workers that competition would bring hardship to many of them and that central planning protected them from such "anarchosyndicalist" dangers.[281] Where the proponents of competition saw technological irresponsibility in the current system, the editorial professed to see a laudable independence. Whereas the proponents envisioned competition as a necessary incentive for high performance, the central planners predicted it would cause only personal misfortune.

Given the disappointing results of the 1965 industrial reform, many centralizers wished to roll it back, even though it fell far short of market competition. In late 1969, when the December plenum highlighted the problem of innovation, this attitude became especially pronounced,[282] prompting one liberal economist to mount a defense of the reform. He granted the critics' point that the reform had not yet significantly improved growth and innovation. But he rejected their argument that "economic methods" of guiding the economy were the same as "scientific administrative methods." The critics were unreasonably demanding instant results, he said, when in fact the expansion of enterprise freedom had not been carried far enough. A lack of perspective, he added, was preventing "some of us" from appreciating that the 1965 reform was equal in importance to the NEP cutback in economic controls during the 1920s and the Stalinist economic reorganization in the early 1930s.[283]

Such arguments, however, carried little weight. Opponents continued to cite the defects of the 1965 reform to block further decentralization, although many of them had hamstrung the reform in the first place by arguing for great caution in its formulation and implementation. At the 24th Party Congress, where Brezhnev underlined the need for ways of speeding innovation that were "peculiar to socialism," some speakers attacked the 1965 reform. One powerful regional party chief labeled the reform's poor technological results "one of its major inadequacies" and urged a reduction of enterprise autonomy.[284] Calling the 1965 changes "premature," another regional party official said that the stress on profits and sales had led not to greater output but to "an unjustified inflation of wholesale prices" and slower innovation.[285] A few months later Demichev, a top ideological official, condemned recommendations to permit competition among enterprises, claiming that this would entail renunciation of central planning, fragmentation of industrial ownership, inflation, and budgetary waste.[286] The party also adopted a decree castigating Academy economists for laxity in unmasking revisionist economic theories.[287] Such pressures prevented any noteworthy increase of autonomy and competition among institutes and enterprises.[288] Instead the political pendulum swung toward other possible remedies for technological inertia.

The Search for Better Central Planning

Perhaps the most appealing idea was to improve central planning. Brezhnev, in particular, based his economic campaign on the assumption that better planning could give a major boost to innovation. The 1968 decrees ordered the compilation of long-term scientific-technological prognoses as guidelines for developing major industrial facilities and integrating R & D results into production plans.[289] The decrees also called for systematic planning of prices, as well as better circulation of technological information and more slack resources, as ways of accelerating technological change.[290] Just how to coordinate all this planning, however, was far from clear. At least five planning agencies were involved: the Academy, the State Committee for Science and Technology, the State Construction Committee, Gosplan, and the State Committee for Prices.[291] These agencies often worked at cross-purposes, and the sheer amount of calculation required was enormous.

The difficulty of coordinating this planning effort was reflected in the fate of the prognoses. Since effective technological forecasting required economic assumptions, mathematical economists in the Academy helped generate the prognoses required by the decrees. An account of the long-term economic forecasting being done pursuant to the decrees listed four participating institutes—three from the Academy and one from Gosplan. Supervision of this work was vested in the Academy's Scientific Council on Optimal Planning and Administration of the National Economy.[292] Near the close of 1969, about the time Brezhnev delivered his speech to the December plenum, the Academy Presidium conducted a special review of the research of its two largest economics institutes. In a rather critical appraisal, the Presidium's review panel called for heavier emphasis on mathematical economics and much closer ties between the institutes and the planning organs.[293] In late 1971, after Brezhnev had demanded better central planning and administration at the 24th Party Congress, a party decree pilloried one of the Academy's economics institutes for giving "extremely little attention" to improving forecasting, long-term planning, and technological progress.[294]

Even under the party's prompting, it was hard to divide responsibility for the forecasts that were to guide technological policy. The boundary between forecasting specific technologies, where the Academy and the State Committee for Science and Technology had strong expertise, and forecasting investment and production, where Gosplan clung to its prerogatives, was hazy. In mid-1971 President Keldysh remarked that the Academy required a "general socioeconomic prognosis" in order to make intelligent predictions about specific R & D fields, because the two types of forecasting were closely connected. "But," he added point-

edly, "there are evidently other opinions [about this matter]. It is necessary to discuss all of this."[295] The question was whether Academy
members projecting future technologies should leave the broad economic judgments to Gosplan or whether the academicians should actively influence the assumptions underlying future economic plans.
Only in the latter case might the plans accelerate innovation. Gosplan
had a well-established propensity to use outdated assumptions and
technological parameters in formulating plans.

In 1972 the regime tried to clarify the lines of authority in technological
forecasting. The guidelines for preparing the Fifteen-Year Plan instructed
the Academy and the State Committee for Science and Technology to
write a "Comprehensive Program of scientific-technological progress
and its socioeconomic consequences for 1976–1990 with justifications
and calculations."[296] This gave the Academy's natural scientists and
econometricians a broad mandate. On the basis of the Comprehensive
Program and submissions from the ministries, Gosplan was to oversee
formulation of the Fifteen-Year Plan and then present it to the top
leadership.[297] President Keldysh in turn pledged that the Academy
would take an active part in forecasting the development not only of
individual technologies but of the economy as a whole.[298]

Nevertheless, the attempt to incorporate the Academy's prognoses
into operational economic plans provoked controversy. Gosplan officials
were either unwilling or unable to take the forecasts into account. In
mid-1973 Gosplan's journal printed an article which warned that overestimating the significance of prognoses posed the "large danger" of
"not leaving, in essence, a place for planning."[299] A month later a
Deputy Chairman of Gosplan asserted that some economic studies
being done in the Academy ignored fundamental principles of Marxism-
Leninism; he added that "many assignments" which scientific organizations had received for the development of prognoses were as yet
unfulfilled.[300] The planners especially distrusted the ideas of Academy
econometricians on ways to optimize economic performance, and these
econometricians were deeply involved in preparing the Comprehensive
Program of scientific-technological progress.[301] Spokesmen for Gosplan
contended that application of such ideas to planning would cause
"enormous losses" to the state and could cause "the artificial obstruction
of the real rates of technological progress."[302]

These conflicts made it still harder to mesh plans with prognoses.
Many of the Academy specialists working on forecasts for the Fifteen-
Year Plan were promising that the country could achieve "a significant

speeding up of the rates of growth of labor productivity and a rise in the effectiveness of social production . . . as compared with the period of the Eighth and Ninth Five-Year Plans."[303] But Gosplan officials obviously felt these promises were false and acted accordingly. An editorial in *Kommunist* about this time called planning the "central link" in the economy, and it noted that the planning organs "are being subjected to justified criticism for some cardinal inadequacies in their work." One inadequacy was that planning "is often based 'on the level achieved', [and] does not rely on long-term norms and reliable prognoses."[304]

Nor were the prospects for an improvement very good. In March 1974 President Keldysh reported that the Academy and the State Committee for Science and Technology had finished work on the Comprehensive Program.[305] The implementation of these guidelines for technological change, however, was frozen by the same political deadlock that blocked the Fifteen-Year Plan to which they were supposed to give direction. More than two years later, the General Meeting of the Academy reportedly "analyzed in detail" the tasks that had to be done in "further work" on the Comprehensive Program.[306] As of 1979 the Fifteen-Year Plan was still in limbo, not yet having been published or approved,[307] and the Comprehensive Program was in the same state.

The Comprehensive Program, of course, represented a massive effort to formulate guidelines for innovation, and the delay did not prove that such guidelines were not being integrated into economic plans on a more limited basis. The actions of the political leadership, however, suggested that many ministries were still failing to achieve this sort of integration. In 1972 and 1973 *Pravda* sharply criticized several ministries for their half-hearted, "completely impermissible" efforts to fulfill their annual plans for technological innovation.[308] A Central Committee resolution published in August 1972 scathingly attacked poor technological planning by the Ministry of Ferrous Metallurgy, whose chief had expressed the view only a year before that his branch's technological achievements gave it "leading authority in the whole world," and demanded faster innovation during the current five-year plan.[309] Problems with ministerial innovation in the coal industry were the object of another Central Committee resolution published in April 1973.[310] By setting off a barrage of mutual recrimination between the several ministries managing and supplying the iron and steel sector, the first resolution provided new evidence of the difficulty of coordinating the technological activities of highly specialized bureaucratic subordinates in a noncompetitive economic setting.[311] Without market competition, it was hard both to motivate such subordinates and to apportion blame among them when a joint endeavor failed.

Equally striking, however, was the inability of the central agencies overseeing technological planning to agree about the existence of the problem. Six weeks after the party resolution on ferrous metallurgy, V. Loskutov, the Gosplan official responsible for planning production automation, discussed the pace of innovation in the iron and steel sector. In a telling omission, Loskutov avoided any mention of the party resolution, although it was pertinent in the most direct way imaginable. Because Soviet officials habitually cite the most recent party pronouncements on their subject, this silence manifestly indicated disapproval, and Loskutov shunned any criticism of Gosplan and the ministries attacked in the resolution. The existing state of affairs, he implied, was basically satisfactory.[312]

The outlook of the State Committee for Science and Technology, on the other hand, was far more critical. About a year later V. Miasnikov, Loskutov's counterpart on the State Committee, reported on ferrous metallurgy plants commissioned during the current five-year plan. Asserting that there was no substantial improvement over the facilities started up during the preceding plan, he concluded that "the designed level of the automation of these facilities does not correspond to today's requirements."[313] The Ministry of Ferrous Metallurgy, he argued, was doing a "poor job" of automating production. Its supplier ministries, though ordered by the party to produce more equipment to automate the sector, had not done so.[314] From this perspective, the situation in ferrous metallurgy was alarmingly bad. Wherever the truth lay, it seemed impossible that two central agencies whose views differed so sharply could integrate their planning efforts and exert sufficient pressure on the ministries to stimulate a major improvement in innovation.[315]

Similar difficulties plagued attempts to use centrally planned price changes to promote technological advance. Direct responsibility for the changes rested with the State Committee for Prices, whose powers were upgraded in 1969–70.[316] But the effort to incorporate price changes into current economic plans set the State Committee at odds with Gosplan and the Ministry of Finance, as well as with the industrial ministries. In writing the plans Gosplan and the Ministry of Finance tried to maintain the rates of profit growth achieved in previous years. They resisted adjusting plans for changes in product mix or prices, particularly when the adjustments meant less profit.[317] Partly for this reason, most machine-building ministries sought to preserve a high level of profitability in any way they could, including the continuation of high prices on obsolete products.[318] When criticized by the State Committee for such practices, ministerial officials were quick to claim that Gosplan and the Ministry of Finance had not adjusted past economic plans to take account of new prices, thereby putting the ministries in a painful economic bind.[319] No doubt the ministries would have taken advantage of their monopolistic position to engage in some behavior of this kind in any case. But the added pressure from plans ratcheted upward annually by Gosplan sharpened their impulse to misprice their products. It also helped defeat the State Committee's effort to put real financial teeth into the sanctions applied to offending ministries, and the ministries continued to engage in such conduct.[320] In price formation, just as in formulating technological prognoses, more rational planning proved far easier to advocate than to achieve.

Associations as a Means of Improving Innovation

The political authorities also tried to speed innovation by combining institutes, design bureaus, and enterprises into amalgamated associations that would reduce the administrative subdivision of the R & D cycle. The 1968 package of decrees recommended that the ministries set up associations and also establish full-fledged research institutes within large individual enterprises.[321] It did not, however, order the ministries to do so. It said rather that these steps should be taken "in necessary cases."[322] In only one branch, electrical engineering, were such changes made mandatory.[323] In all others, the future of the measures hinged on the politics of their implementation.

The politics of implementation revolved around two questions: how should the term "association" be defined, and was a shift to associations advisable? Commentators soon came to accept a rough distinction between production associations, which might include R & D units but were led by an enterprise and were geared primarily to mass production, and science-production associations, led by a research unit and oriented toward continuous innovation. But it was unclear just how large and how powerful the associations ought to be. Should they include only a few enterprises and R & D units? Associations of this size promised more flexible management, but diffusing power to them might resurrect the localism of the sovnarkhoz era—when, incidentally, the first production associations had been formed. Alternatively, associations might approach the scale of existing ministerial main administrations, the glavki. This would reduce the dangers of regional technological autarky, but such large units might be unable to improve R & D management.

Various bureaucratic groups differed sharply on these questions. Regional officials tended to favor the creation of relatively small associations. The earliest associations (firmy) were small combinations of enterprises formed under the auspices of the regional party organs in the early 1960s. After Khrushchev's fall, regional officials continued to push for more small-scale associations, including science-production associations, with the backing of the State Committee for Science and Technology and Gosplan.[324] Most central ministerial officials, on the other hand, strongly opposed this idea. Because associations were candidates to share ministerial powers and threatened to upset internal ministerial organization,[325] the ministries dissolved many of them after the 1965 reform. According to one report the dissolutions included associations that had raised the tempo of innovation at their constituent enterprises.[326] The wave of abolitions ended the rapid numerical growth of associations that had occurred before 1965. While some new associations were formed in 1965–8, considerably more appear to have been disbanded.[327] This bureaucratic interplay helps explain the tentativeness of the recommendations contained in the 1968 decrees.

Once the decrees appeared, regional and central planning officials began to press the ministries to adopt a new attitude.[328] This was particularly true after the December 1969 plenum made innovation a burning issue and raised associations as a possible answer.[329] In the ensuing campaign for more associations, party officials, especially those at the regional level, seized on the economic anxieties aired at the plenum to press for organizational change.

The campaign was epitomized by an August 1970 Leningrad conference at which party officials championed associations. In the opening address an industrial overseer from the Central Committee staff asserted that technological policy could best be managed through such units. Noting that some ministries were resisting this idea, he pointed out that in late 1969 the Central Committee had called on regional party organs to become more active in setting up associations. He concluded that it was necessary to consider forming "major production and science-production associations in all branches of industry."[330] G. V. Romanov, the Leningrad party leader who was later elected to the Politburo, reported that the associations in his region had sharply reduced the length of the R & D cycle, and he endorsed them unqualifiedly.[331]

Similar backing came from other party officials who dominated the conference.[332]

In the minds of these proponents, not only did new associations have to be created, but the powers of most associations, new and old, had to be broadened. After 1965 a few surviving associations had obtained the powers of ministerial glavki, but most had been cast into a state of limbo in which they had no more rights than did individual enterprises.[333] This did not suit party officials, who called for all associations to receive the powers of glavki and sometimes urged that the glavki themselves be abolished.[334] Enactment of this recommendation would have shifted substantial economic authority from the Moscow-based ministries to regional party and state agencies.

The efforts of the regional party organs provoked resistance both from below and from above. Prominent among the lower-level opponents of associations were enterprise directors, who apparently feared that a local association, because it could stay better informed about their activities than a ministerial glavk, would further circumscribe their autonomy. One party official remarked that some directors resisted associations out of a fear of losing their "social position," since they were presently members of the borough party committee and were subordinate directly to Moscow.[335] The local resistance also drew on some city and borough organs of the party and state bureaucracies, which feared a loss of influence over enterprises currently in their territorial jurisdictions—and sometimes a loss of revenue as well.[336]

The regional party authorities could have overcome this local opposition easily enough, except that it was allied with the central ministries. Time and time again at the Leningrad conference officials rebuked the ministries for obstructing the formation of associations. One official criticized "many" ministries for "standing on the sidelines" and attacked five ministries by name. Others did the same.[337] At the 24th Party Congress, Leningrad party secretary Romanov chastised those whose attachment to past managerial forms was slowing the establishment of associations, and he bluntly asserted that their conservatism was "braking the development of our economy." Another powerful proponent accused the ministries of trying "to cram the scientific-technological revolution into the framework of old methods and forms of organization," even when this impeded technological progress.[338]

Although the ministerial opponents generally kept silent, occasionally they argued their case in public. At the 1970 Leningrad conference P. F. Lomako, the Minister of Nonferrous Metallurgy, accepted the idea of multi-enterprise combines; but he made no mention of endowing associations or combines with greater rights, as party officials advocated. Moreover, he quoted Lenin's endorsement of "the greatest centralization of major production throughout the country" and Lenin's condemnation of "regional anarchosyndicalism" in economic administration as guides for his ministry's policies. Clearly Lomako opposed diffusing power to territorial associations.[339] Elsewhere another ministerial administrator stressed the technological virtues of the glavki and asserted that it was still not possible to apply "on a broad scale such a new form of administration as associations."[340] At round tables on reform and innovation during 1970, other ministerial spokesmen either ignored or attacked small-scale associations like those in electrical engineering, which were widely praised by party officials at the Leningrad conference.[341] The only concession by ministerial critics of small associations was an attempt to appropriate the term "association" for large ministerial glavki that had been formally transferred to commercial accounting.[342]

The ministries, in short, were fighting a bureaucratic holding action, and they achieved considerable success. The 24th Party Congress gave its imprimatur to "major associations" as the form of industrial organization toward which the regime should move, but the structure and scale of such associations remained undefined.[343] It was only in April 1973 that the regime finally decreed a general reorganization of industry to make the association the principal administrative unit, and the terms of the decree went a long way toward meeting the objections of ministerial officials.

The April decree instructed the ministries to submit general reorganization plans within the next six months; the approved plans were to be implemented during 1973–5. While it stated that most ministerial glavki were to be abolished, the associations slated to replace them were of several kinds, ranging from large all-union industrial associations to much smaller production associations. Many of the all-union associations were to have nationwide authority over their industrial sector. This raised the possibility that no real diffusion of power would occur, since the decree accented the prerogatives of all-union associations, which were to be the main agencies for integrating research and production, rather than the rights of production associations.[344]

Much bureaucratic infighting followed the decree. By the end of 1975 the number of science-production associations had grown to about 110, while the number of production associations had increased to roughly 2,200; together they produced about a quarter of total industrial output.[345] But many of these units had already existed before the promulgation of the 1973 decree, and further progress was slow. Proponents repeatedly charged that some ministries were simply relabeling their glavki as all-union associations but doing nothing to improve the quality of industrial administration.[346] Evidently some ministries also used the idea of a large "association" against the smaller industrial units to which that name had first been applied. As they formed all-union associations, they took the opportunity to dissolve existing production associations, or else to strip them of special administrative powers.[347] Finally, commentators repeatedly stated that the ministries were exercising detailed control over the constituent units of science-production and production associations, thereby disrupting the organizational unity which they theoretically possessed.[348] No doubt these problems contributed to the postponement of the deadline for completing the industrial reorganization from 1975 to 1980.[349]

That such measures generated conflict does not necessarily mean that they failed in substantive terms. The reform efforts after 1968 did achieve some successes, particularly where the science-production associations were concerned. Although measuring the length of the R & D cycle is very difficult, considerable evidence suggests that the science-production associations reduced this interval in several industrial branches. According to one Western assessment, the reduction averaged between 25 and 30 percent.[350] Other significant indicators, such as number of inventions and return on R & D expenditures, also climbed for these associations, and a fuller exchange of technological information reduced the degree of duplication of R & D.[351]

Although the science-production associations reduced the gestation period for the innovations on which they worked, innovation in the economy as a whole appeared not to increase. The national figure for new types of machines and equipment created each year declined from 3,007 in 1970 to 2,829 in 1976. Similarly, the number of new instruments and automation devices dropped from 1,032 in 1970 to 966 in 1976.[352] These figures were virtually identical with those for the second half of the 1960s, which some leaders had found disturbing. The creation of new types of indigenous technology thus appeared not to expand significantly after 1968.

Although more difficult to judge, the same seemed to be true of the introduction and diffusion of new technology at producing enterprises. Soviet statistical handbooks showed that the number of new industrial products assimilated into regular production grew from about 8,400 in

1966–70 to about 16,900 in 1971–5.[353] On its face this was an impressive figure. But such figures had been manipulated by industrial managers ever since the 1930s, and there was reason to doubt the significance of the increase in this case. Aside from other considerations,[354] the figure did not measure the change in the proportion of total production constituted by new output. The latter measure is a far more reliable gauge of the diffusion of innovations in industry, and it is less susceptible to manipulation in reporting. Comprehensive data of this kind were not published for the 1970s, but available figures gave a far less favorable picture of the rate of technological change. Statistics on the renewal of output in the machine-building branches in the Ukraine revealed that the proportion of new products had dropped between 1967 and 1974.[355] The same tendency to reduce the introduction of new engineering products was reported for Lithuania.[356] The proportion of new products in the output of the Ministry of Instrument Building, Means of Automation, and Control Systems likewise declined steadily between 1973 and 1976.[357] More encouragingly, the Ministry of Electrical Engineering reportedly increased the proportion of new products in its output from 6–7 percent in the late 1960s to 12–13 percent in 1975.[358] This was a noteworthy change, but it was by no means clear that it marked an improvement over the starting point of the reform effort in the mid-1960s, when the average proportion of new products for all the engineering branches was 13.8 percent.[359] Although we cannot be absolutely certain, the most plausible conclusion from these pieces of information is that the introduction and diffusion of new technology at regular enterprises did not increase significantly between 1968 and 1975.

This conclusion is consistent with the general performance of the economy between 1971 and 1975. According to one Western calculation, total GNP grew at an average annual rate of 3.7 to 3.8 percent, which was considerably lower than at any other time since World War II. The growth rate for industry also hit a postwar low, 0.8 percent under the average annual rate for 1966–70. Perhaps most strikingly, joint factor productivity actually decreased at an average of 0.2 percent per year.[360] A large part of the explanation for these low figures, of course, is the exceptionally poor harvests of 1972 and 1975. But part of the cause was the slow growth in the factor productivity of industry, which remained only slightly above the disappointing rate for 1966–70[361] and was strongly affected by the technological inertia of the bureaucratic structure.

Conclusion

During 1968–75, the debate over the USSR's comparative economic performance showed that the anxiety of many elite members on this score had risen dramatically. Brezhnev and others believed that the country faced unprecedented economic problems which required a major departure from past policies. Without such a departure, they feared that the Soviet Union might fall farther behind the West technologically. Moreover, they believed that moderating changes in the politics of Western regimes offered a new opportunity to make the necessary policy changes. This nontraditionalist strain of thought acquired growing support within the top leadership and the elite as a whole, and its spreading influence had fundamental political significance. At the same time, however, the traditionalist view of Soviet economic performance and foreign relations continued to enjoy firm support among many other leaders and elite members. Maintaining that future economic tasks were similar to those solved in the past, these persons felt that critical discussions of the economy needlessly weakened the regime's domestic legitimacy and foreign policy, and they discerned little change in the political and economic behavior of the West.

This shifting balance of elite opinion allowed some noteworthy departures in Soviet policy but also established limits on how far the new initiatives could be carried. It helped, for example, to bring about Soviet signature of the first SALT accord, and it played a role in the Soviet pursuit of a SALT II agreement. It did not, however, provide the political backing that the most assertive proponents of nontraditionalist ideas needed in order to shift R & D and investment priorities away from military programs. There is no evidence that any substantial change in these priorities took place, although it is significant that critics continued to raise the issue. The regime did, of course, sharply step up its pursuit of Western technology. But here, too, resistance from traditionalists kept the change of policy more modest than some nontraditionalists desired.

The changes in technological strategy after 1968 were shaped by disagreements not only among top leaders, but among less powerful elite groups guided by professional and bureaucratic interests. Social and natural scientists had a special responsibility to analyze the USSR's technological performance vis-à-vis the West, and the natural scientists belonged to the Soviet profession most noticeably associated with internationalism. Along with trade officials, many of these persons believed that an infusion of Western science and technology was needed to accelerate Soviet technological progress. Pressed by the party leadership to upgrade their efficiency, industrial officials in some lagging sectors likewise had a strong appetite for more foreign know-how. But bureaucratic interests and professional pride led some producer-goods manufacturers to oppose expanded technology transfers from abroad. Most military officials, skeptical of international cooperation and determined to protect their claim on domestic resources against erosion by the process of détente, also resisted greater reliance on Western know-how. Of all the groups, party ideologists and political police had the smallest professional reason to desire more foreign technology. Their duties focused on the preservation of nontechnical values that might be harmed by contacts with the West, and their lack of operational responsibility for technological progress probably kept them from recognizing the seriousness of domestic technological needs. Thus they acted from day to day in ways that impeded the acquisition of large technological benefits from such relations. Many impediments, of course, were also rooted in the USSR's internal economic organization and could be overcome only by improving the administration of the economy as a whole.

The leadership manifestly wanted to step up indigenous innovation but was anxious to avoid reforms that might undermine the political system. Hence it eschewed moves toward market and R & D competition. This concentration on a narrow range of reforms served only to provoke sharp conflicts among various bureaucratic elements eager to appropriate the slogan of economic efficiency for their own purposes. The reforms seemingly failed to generate significant improvement in the rate of domestic innovation. Despite the spreading anxiety about the economy's performance, the leaders were unable to achieve any substantial change in the operating style of the massive bureaucratic apparatus over which they presided. That apparatus was still a long way from the more flexible, organic style of administration they desired. Only a strong leader willing to attack the bureaucracy head-on, and probably committed to dismantling a major portion of it, could produce such a change of administrative conduct. In the Brezhnev years, no leader was willing or able to adopt such a course of action.

Politics and Technology: Past and Future

Official Soviet thought has changed markedly since the 1930s. Whereas Stalin and his lieutenants regarded the Western powers as a grave political menace that could be fended off only through the most extreme economic mobilization, later decision-makers, viewing the world in a less threatening light, eased Stalin's draconian domestic policies and entered into fuller relations with the West. The movement away from the traditionalist outlook, however, has been neither unilinear nor complete. Significant countertrends have occurred, as in the mid-1960s, and many elite members still hold a view of Western political behavior not much different from that prevailing under Stalin.

The balance of official opinion about the East-West economic race, on the other hand, has shifted more decisively toward nontraditionalist views. Although deeply apprehensive about the USSR's current technological means, the Stalinist elite was intensely confident of attaining technological supremacy over the West in the foreseeable future. The Varga controversy momentarily challenged but did not destroy that confidence. Today, however, official attitudes seem nearly the reverse of their Stalinist antecedents. While deriving deep satisfaction from the military and economic ground gained on the West since the 1930s, the regime appears increasingly uncertain of its ability to compete technologically with the West in the future. Sustained neither by dogged Stalinist conviction nor by heady Khrushchevian optimism, it has begun to grope for new policies to compensate for this possibility.

Soviet theoretical debates about politics and technology have played a vital part in this gradual reassessment of policy. Because public argument over high policy violates party norms, disputes over military spending and arms control, East-West economic ties, and even economic reorganization have frequently been cast in the form of conflicting descriptions of the objective situation facing the regime. Although phrased more subtly than in Khrushchev's time, such top-level theoretical disputes were extremely significant during the Brezhnev years, and the failure of many outside observers to notice them deserves comment. Even in an era seemingly characterized by colorless leadership and dry rhetoric, Western analysts should resist the temptation to dismiss leaders' pronouncements as nothing more than rehashed ideology. Careful collation and comparison of statements from a wide range of sources can shed much light on policymaking at the highest levels of the regime.

To be sure, the ideological justification of the domestic political order has always been one aim of official pronouncements about Soviet technological capacities and the international environment. Under Stalin the traditionalist expectation of surpassing the West bolstered the elite's sense of its own legitimacy and its determination to impose enormous sacrifices on ordinary citizens. Recent opponents of "blackening" the economic system still attach great importance to this ideological goal. But public discussions of Soviet technological capacities have often been intended to serve other functions as well. One function has been to aid in formulating domestic policies that can generate rapid economic growth. Another purpose has been to shape the country's foreign relations. Some leaders have tried to use the traditionalist image of the Soviet economy to discourage Western diplomatic pressure, whether in the Marshall Plan negotiations or the SALT talks. As a rule, traditionalists have aimed to block even those East-West agreements based on mutual concessions, but they have probably argued privately that the traditionalist stance is essential to negotiating any worthwhile agreement—a claim that becomes harder to counter when American officials treat Soviet economic weaknesses as an invitation to outstrip the USSR in an open-ended arms race. On the other hand, Soviet proponents of East-West accommodation and arms control have usually made the nontraditionalist interpretation of the USSR's technological capacities a central plank in their argument. Finally, top leaders have tried repeatedly to use public analyses of Soviet technological problems to undermine their domestic political rivals. The history of official pronouncements on politics and technology is, among other things, a history of covert disagreement about which of these functions is most important.

The record of such discussions reveals a long-term trend toward expanded participation in policymaking by middle-level officials and advisors. To a greater degree than under Khrushchev, the post-1964 debates about the political and technological competition with the West went beyond the confines of the Politburo and involved a widening circle of party officials, economic administrators, economists, and foreign policy specialists. The contrast with the style of policymaking under Stalin is even more dramatic, with the arguable exception of the short-lived Varga affair. Although the power to decide such issues remains unquestionably in the Politburo, secondary figures have taken on a larger role, expressing their views more forcefully than in the past. Their influence is substantial, particularly when serious disagreements exist at the top of the party.

Closely related is a trend toward greater rationality in the form of policymaking, if not in the results. The rise of think-tanks like IMEMO and the U.S.A. Institute has injected more sophistication and factual accuracy into discussions of Soviet economic performance and relations with the West. Needless to say, these new standards of discussion still strike many commentators and decision-makers as politically unacceptable. Nontraditionalists have frequently been reproached by opponents with violating the canons of Marxism-Leninism, and they in turn have accused traditionalists of unwittingly encouraging revisionism with hackneyed propaganda. The larger the circulation of the periodicals in which the debates occur, the greater the temptation to fall back on such ideological charges.[1] Nevertheless, ideological considerations figure much less prominently in many Soviet policy analyses today. More than at any time since the advent of Stalinism, some members of the elite are striving to think empirically and systematically about the USSR's current situation and future prospects.

Ever since Stalin, Soviet technological strategy has been an effort to reconcile the diverse imperatives postulated by official thought. The Stalinists initially used the rapidly expanding domestic R & D establishment primarily to copy and adapt Western know-how. Although this policy yielded large economic benefits during the early 1930s, choosing wisely between foreign and indigenous technologies became increasingly difficult as Stalin and his followers closed off more and more channels of interchange with the outside world. Of course, the country's deepening isolation served important political ends. Apart from assuring greater strategic independence from capitalist states, it aided the Stalinists' attempt to legitimize their tyranny with the argument that life was better at home than in the crisis-ridden West, and it helped project an idealized image of the USSR to foreign sympathizers. But these gains in internal political consolidation and public diplomacy were achieved at the expense of drastically reduced access to useful Western R & D.

Since Stalin's death the prevailing trend of technological strategy has been to seek greater access to Western technology. As the traditionalist siege mentality diminished, the party leadership became less fearful of entanglement with the outside world and more cognizant of the costs of foregoing access to foreign know-how. Nonetheless, serious tensions persist between the regime's political and technological requirements. Many elite members remain extremely wary of striking political bargains in exchange for Western technology. The recurring internal campaigns against "ideological coexistence" have reflected not simply this wariness, but also a fear of spontaneous ideological contamination not intentionally fomented by Western governments.

These tensions have been mirrored in the disparate attitudes of middle-level officials. Some officials, confident that the Soviet system has the requisite international power and domestic political stability, have favored entering into a much more active technological interchange with the West. Others have doubted the wisdom of more economic interdependence and have shown great concern about Soviet vulnerability to outside political influences. Putting Western technology to good use has thus been not only a problem of making impersonal institutions function effectively, but of making some of the officials who staff those institutions cooperate in an endeavor they think unwise. Thanks to the weakening performance of the domestic economy, this problem became much more serious under Brezhnev than in earlier decades.

For a long time, the mobilizational system built by Stalin showed an exceptional ability to marshal economic resources and construct industrial facilities based on proven technologies. From the beginning, however, it showed much less ability to create original technology. Most attempts at indigenous innovation were plagued by a shortage of slack resources, the skewed incentives of high-pressure economic plans, poor circulation of information, and the scapegoating of technological "wreckers." Through special measures the party leadership managed to overcome many of these obstacles in the weapons industries, which made rapid technological strides. But in most other sectors the Stalinist system barred the organic pattern of administration conducive to the introduction of indigenous technologies.

Many of the contemporary obstacles to indigenous innovation have a long history that sometimes makes Soviet accounts of the last decade sound uncannily like those of the 1930s. The political significance of these barriers, however, is far greater today than in Stalin's day. As long as the mobilization and redirection of underutilized land and labor could sustain the industrialization campaign, and as long as the political elite was ruthlessly willing to squeeze investment resources out of popular consumption, these technological defects were tolerable. But in the post-Stalin era the supply of underutilized resources has gradually declined, and the elite has curbed the tyrannical practices of Stalinism— largely out of a desire for self-preservation. Under these conditions the system's technological limitations have become more troublesome economically and more salient politically, triggering a long search for effective reforms of science and industry.

The history of this search is a story of marginal successes and large failures. The essence of the failures is that it has proven virtually impossible to make scientific and industrial institutions more innovative without injecting a large dose of market competition into the economy. Yet the political constituency for this sort of reform has been very small. Most economic planners and ministerial officials have flatly opposed such a step. Many regional party officials have belied their Western reputation as pillars of the status quo by expressing concern about slow innovation and urging administrative changes. But the changes they have favored would merely shift decision-making authority to the regional level without enlarging the role of the market. Nor has there been much real support for drastic decentralization at the base of the administrative hierarchy. The majority of managers and researchers may have felt at odds with the existing bureaucratic system, but their desire for professional security has also made them part of that system, and they have been unwilling to see it radically recast.

This constellation of bureaucratic forces has created powerful resistance to the efforts of top leaders to devise effective reforms. The greatest concern for the efficiency of the economic system appears to have been concentrated among liberal economists employed outside the major bureaucratic hierarchies, and among some administrators and leaders at the bureaucratic apex of the system. As members of an oligarchic ruling group whose stability was based on "trust in cadres," reform-minded Politburo members like Kosygin encountered extreme difficulty in seeking a major decentralization, because their Politburo rivals could muster extensive lower-level support in any showdown over the issue. Entrenched bureaucratic constituencies have thus limited the range of administrative change that is politically feasible.

Brezhnev, for one, was clearly mindful of this fact. In the past, leaders holding a nontraditionalist view of the economy usually favored decentralization as a solution. But even after accepting much of Kosygin's diagnosis of Soviet economic troubles, Brezhnev avoided this prescription. Instead he advocated a combination of improved central planning and regional industrial associations that would not fuse various segments of officialdom into a negative coalition which might enable other Politburo members to topple him. This tactic gave Brezhnev the political strength to tame the Politburo opponents of his foreign policy initiatives, including the quest for a new infusion of Western technology, but it did not solve the problem of improving indigenous innovation. Despite Brezhnev's rhetorical commitment to a radical improvement of domestic innovation, his operational policy increasingly amounted to seeking Western technology as a substitute for effective domestic reform.

Although this book has traced the interaction of Soviet politics and technology only to the mid-1970s, the analysis should shed some light on what the future may bring. Little has occurred since 1975 to alter our general picture of the political era that recently came to a close with Brezhnev's death. The events of the past eight years both confirmed the Brezhnev administration's unwillingness to root out the domestic obstacles to technological progress and highlighted the economic and political difficulty of meeting Soviet technological needs solely through wider interchange with the West. There is every reason to believe that technological progress will become an even more critical issue for Soviet foreign and domestic policymakers during the coming years.

The debate between the traditionalists and the nontraditionalists appears certain to continue through the 1980s.[2] The diminishing economic and technological vitality of the Soviet system will provide strong ammunition for nontraditionalist politicians and scholars. The Soviet growth rate is currently at a historic low, and Western projections suggest that it will continue to decline, making the acceleration of technological change a still more pressing matter. The falling growth rate will greatly intensify the conflicting claims of defense, investment, and consumption on the budget. These pressures will probably embolden nontraditionalists to assert more vigorously that the USSR must strike diplomatic bargains with the West in order to divert part of Soviet military research and industry to civilian needs and prevent the West from channeling its superior current resources into a steep new spiral of the arms race. Nontraditionalists will also urge that the country expand its access to Western technology and credits as an aid to coping with the new economic stringencies.[3]

None of these arguments is likely to persuade most traditionalists. Traditionalists will be able to point to a period of worsening Western economic performance since 1974, particularly in the United States, as evidence that their analysis of Western capitalism is still correct. Moreover, heightened Soviet-American tensions and a surge of American defense spending will serve to strengthen their conviction that their assessment of imperialist political intentions is accurate. Under these conditions, more avid pursuit of Western know-how, and especially decelerated development of Soviet military technology, will strike many traditionalists as an ill-advised invitation for Western (or at least U.S.) pressure on the USSR.

It is impossible to forecast which of these two broad orientations will prevail during the coming years, or whether either will prevail. During the period of wholesale leadership turnover now beginning, the regime's policies may conceivably oscillate sharply between the traditionalist and nontraditionalist outlooks as power shifts among various party factions. Western observers should not rule out the possibility, however, that after a decade of expansive rhetoric and little concrete action, a future contingent of party leaders may actually be prepared to reduce the emphasis on military technology in favor of civilian requirements if the international situation is conducive to such a move. Judging from recent official statements, the economic and technological cost of the military effort is already becoming a more pressing political issue.[4]

Western governments may not be able to exert a decisive impact on future Soviet choices of this kind, but they should recognize the various ways they may influence the outcome. More than is usually recognized, specific Soviet decisions about military competition with the West have been influenced by the level of Western economic performance at the time. Although only one of several determinants of Soviet technological priorities, this factor is important. The healthier the Western economies in the coming years, the less inclined the Soviet elite will be to engage in an all-out arms race. The lesson for the makers of Western foreign policy is that in addition to Western military power and diplomacy, the economic health of their own societies has an important effect on relations with the USSR. In particular, a massive U.S. military buildup that saps the U.S. economy may not serve America's foreign policy goals, let alone its domestic social needs.

As for East-West technology transfers, the structural weakness of the Soviet economy will continue to draw Soviet leaders toward Western know-how, even as they debate how high a political price to pay in return. This price may be denominated in many ways—as changes in the administration of foreign economic relations, as less arbitrary treatment of political dissidents, and as limitations on Soviet assertiveness in the world arena, to name only the most obvious. Although the need for foreign technology will probably increase Western opportunities to influence Soviet affairs, such trade-offs will undoubtedly provoke deep controversy in Soviet governing circles. Like the Soviet choice of domestic technological priorities, the outcome will depend partly on the capacity of the Western powers to follow a differentiated approach. An approach of this sort must combine a willingness to apply sanctions against genuinely dangerous Soviet actions with a willingness to offer substantial rewards for more temperate Soviet behavior. In order to strengthen nontraditionalist trends within the elite, it must minimize high-visibility public efforts to dictate Soviet policy. Above all, it must be based on a consensus among Western nations about the kind of Soviet conduct that warrants the reduction of Western technology transfers. The events of the past decade suggest that this task requires a level of political sophistication and cooperation within the West that will heavily tax the Western polities.

Apart from these issues, the question of economic reform will play a critical role in Soviet politics after Brezhnev. Khrushchev's successors took power on a platform promising gradual reform with a minimum of institutional disruption. This platform was appealing because it offered rank-and-file officials security from Khrushchev's bureaucratic shake-ups and because it yielded a marginal improvement in growth rates over the short term. It did not, however, speed up technological change, nor did it stem the long-term decline of Soviet growth. Thus, just as the Brezhnev-Kosygin oligarchy abandoned Khrushchev's reform approach, which it regarded as discredited, the new generation of leaders may well discard the measured gradualism of the Brezhnev years as nothing more than a recipe for economic decline. This seems all the more plausible because Brezhnev and his supporters, although extremely cautious about restructuring economic institutions, harped on the economy's shortcomings and urged a dramatic improvement in its technological dynamism. They thereby helped impress on the elite the need for a reform much more drastic than they themselves were willing to contemplate. After a period of quiescence, advocates of economic decentralization and widened technological competition are again speaking out,[5] and in the coming years the rapid turnover of the Brezhnev generation and the intensification of conflicts over economic resources will make the political process far more volatile. In these circumstances one or another leader—either a figure who attains unchallenged dominance or one who lacks the bureaucratic means to compete successfully in conventional Soviet terms—may champion a major decentralization of the economy.

This is not to say that such a market-oriented reform will actually occur. In the first place, the outcome of future reform debates will depend on the severity of the country's economic problems. The poorer the performance of the economy, the more willing the elite will be to make major institutional changes and to accept the further short-term decline in growth that transition to a new economic system would entail. We cannot predict the future performance of the economy with enough accuracy to know whether economic pressures will generate the strong sense of crisis that is one precondition for the successful introduction of such a reform. Nor can we know to what degree the future economic performance of the West will sharpen or assuage Soviet economic anxieties. To date, however, the West's mounting economic troubles have not materially diminished Soviet respect for Western technological dynamism.[6] It is thus conceivable that the combined pressure of domestic economic needs and international political competition may produce a mood favoring drastic reform.

A related determinant of future reform efforts will be the attitudes of the top party leaders. In theory the members of the Politburo might decide unanimously to accept any decline in economic performance as the necessary price of avoiding fundamental reform. They might, in other words, simply abandon the hope of overtaking and surpassing the West technologically. This sort of consensus, however, seems almost inconceivable. Surpassing the West has been a central Soviet goal for more than five decades, and it has played a crucial role in legitimizing the party elite's claim to rule. In a Marxist political culture with a fixation on progress and the future, "getting the country moving again" is likely to be a potent political theme—particularly in a period of slow growth when ambitious political aspirants are looking for a serviceable issue by which to rise. It will seem all the more appealing if the regime faces excruciating economic choices.

On the other hand, future Politburo members are even less likely to agree on the opposite course of action—pursuit of better economic performance through a sharp increase in market competition. The long-standing Soviet tradition of state centralism makes it inevitable that some leaders will tenaciously oppose such a step as destroying the essence of the socialist political order. In taking this position, they will have the backing of many lower officials who fear such a reform for careerist as well as philosophical reasons.

It thus appears that the likelihood of a successful market reform will hinge on the ability of leaders favoring this course to build strong support among lower officials. To judge from the historical interplay of bureaucratic interests, the creation of such a coalition seems highly improbable, and the safest guess is that it will not occur. However, a substantial segment of Soviet officialdom might be willing to countenance a market reform if it perceived a large threat to the whole political system from extremely low levels of economic growth. If grave enough, the danger of falling further behind the West technologically might convince more officials of the need for this step. Alternatively, the danger of failing to satisfy mounting popular economic demands might produce the same effect. In the past, Western observers have usually overestimated the strength and significance of such demands. But today the growth of consumption is slowing from a period in which popular welfare increased rapidly, whereas in the 1950s and 1960s consumption was rising quickly from the abysmal levels and depressed expectations of the Stalin era.[7] Some analysts have identified a recent wave of economic pessimism within the Soviet middle class.[8] This sort of pessimism is likely to increase and could spread to the lower classes. Some party officials already fear that the citizenry, for reasons ranging from economic hardship to official corruption, may be losing faith in the idea of socialism.[9] While the elite may conceivably respond to this fear with more repression rather than liberalization, the possibility that popular dissatisfaction will strengthen the case for a fundamental economic decentralization should not be excluded. At the least, an intense conflict over such a reform program will probably occur, even if ultimately the idea goes down to political defeat.

In a sense, then, a critical chapter in the history of Soviet politics and technology remains to be written. Since the 1920s the Soviet system has demonstrated many economic strengths, confounding the dire predictions of hostile critics. At the same time, however, it has fallen short of the ambitions of its leaders, and the gap between elite ambitions and technological performance seems destined to widen in the future. For this reason, the interplay of politics and technology will exert a powerful influence on both the domestic evolution and foreign relations of the regime. The complex interaction is a process that Western scholars and officials should follow closely because it will affect the fate not only of the Soviet people, but of our own.

Notes

Chapter 1

1. Peter G. Filene, *Americans and the Soviet Experiment, 1917–1933* (Cambridge, Mass., 1967), 71–5, 93–6, 104–9, 139–40.

2. Ibid., 196–8, 224–6, 255–9; Richard H. Pells, *Radical Visions and American Dreams: Culture and Social Thought in the Depression Years*, pbk. ed. (New York, 1974), 61–9.

3. Many Western scholars have urged the closer study of linkage politics in the USSR—a notion which I take to mean the way that domestic needs, institutions, ideas, and individual ambitions interact with foreign events to shape the regime's policies. But most writings on the subject have remained on a general plane. Moreover, the writings, partly because they have focused on different aspects of the Soviet domestic order, have reached disparate conclusions about the nature of such linkages. See, for example, Richard E. Pipes, "Domestic Politics and Foreign Affairs," in *Russian Foreign Policy: Essays in Historical Perspective*, ed. Ivo Lederer (New Haven, 1962), 145–69; Vernon Aspaturian, "Internal Politics and Foreign Policy in the Soviet System," in *Approaches to Comparative and International Politics*, ed. R. Barry Farrel, pbk. ed. (Evanston, Ill., 1966), 212–87; John Armstrong, "The Domestic Roots of Soviet Foreign Policy," *International Affairs* 41 (January 1965), 37–47; Alexander Dallin, "Domestic Factors Influencing Soviet Foreign Policy," in *The U.S.S.R. and the Middle East*, eds. Michael Confino and Shimon Shamir (Jerusalem, 1973), 31–53; William Zimmerman, "Elite Perspectives and the Explanation of Soviet Foreign Policy," *Journal of International Affairs* 24, No. 1 (1970), 84–98; and *The Conduct of Soviet Foreign Policy*, eds. Erik P. Hoffmann and Frederic J. Fleron, Jr., 2nd pbk. ed. (New York, 1980), 31–35.

4. The most important of these writings are by Ronald Amann, Joseph Berliner, Julian Cooper, R. W. Davies, Nicholas DeWitt, Loren R. Graham, John Hardt, George D. Holliday, Marvin Jackson, Jr., Alexander Korol, Robert A. Lewis, Linda L. Lubrano and Susan Gross Solomon, Louvan E. Nolting, Gertrude Schroeder, Peter H. Solomon, Jr., John R. Thomas and Ursula Kruse-Vaucienne, Antony Sutton, and Eugene Zaleski et al. For full citations, see the bibliography.

5. See the writings by Jeremy Azrael, Kendall E. Bailes, Loren R. Graham, David Joravsky, David Holloway, Rensselaer W. Lee, and Linda Lubrano cited in the bibliography.

6. See the writings by R. V. Burks, Paul Cocks, Erik P. Hoffmann, Robbin F. Laird, T. H. Rigby and Robert F. Miller cited in the bibliography.

7. With a few exceptions, I have omitted technology generated in the agricultural sector.

8. V. I. Lenin, *Imperializm, kak vysshaia stadiia kapitalizma* (Moscow, 1965), 84–94.

9. Ibid., 95–6; V. I. Lenin, *Polnoe sobranie sochinenii*, 5th ed. (Moscow, 1958–65), XXXIX, pp. 21–2.

10. I have taken this definition from David Joravsky's superb "Soviet Ideology as a Problem," *Soviet Studies* 18 (July 1966), 2–19; the article is reprinted in slightly altered form as chapter 1 of his *The Lysenko Affair*. See also Seweryn Bialer, "Ideology and Soviet Foreign Policy," in *Ideology and Foreign Policy*, ed. George Schwab (New York, 1978), 100–101.

11. Kenneth E. Boulding, "National Images and International Systems," in *International Politics and Foreign Policy*, ed. James N. Rosenau (New York, 1969), 423–6.

12. For these patterns in Soviet discussions of capitalism during the 1960s, see Franklyn Griffiths, "Images, Politics and Learning in Soviet Behavior Toward the United States," Ph.D. dissertation, Columbia University, 1972.

13. In a broad sense, of course, the concept of a national technological strategy may encompass additional issues such as the choice of the particular economic sectors in which research and investment are to be concentrated. Because of the scope of this book, however, I have been able to give these subjects only cursory attention.

14. For the range of technological strategies open to contemporary states, see Robert Gilpin, "Technological Strategies and National Purpose," *Science* 169 (1970), 441–8.

15. Alexander Gerschenkron, *Economic Backwardness in Historical Perspective* (Cambridge, Mass., 1962).

16. "Science in Underdeveloped Countries," *Minerva* 9, No. 1 (January 1971), 102; John Roberts, "Engineering Consultancy, Industrialization and Development," *Journal of Development Studies* 9 (October 1972), 45–9.

17. Charles Cooper, "Science, Technology and Production in Underdeveloped Countries: An Introduction," *The Journal of Development Studies* 9 (October 1972), 5; Amilcar Herrera, "Social Determinants of Science Policy in Latin America," ibid., 20–22, 34–7; Edward Shils, *The Intellectuals and the Powers* (Chicago, 1972), 429–30, 460.

18. Cooper, "Science, Technology and Production," 12–13; "Science in Underdeveloped Countries," 106–7; "U.S. International Firms and R, D and E in Developing Countries," *Development Digest* 12, No. 1 (January 1974), 102–11.

19. "Science in Underdeveloped Countries," 102; Richard S. Rosenbloom and Francis W. Wolek, *Technology and Information Transfer: A Survey of Practice in Industrial Organizations* (Boston, 1970), 69–71; Nathan Rosenberg, "Economic Development and the Transfer of Technology: Some Historical Perspectives," *Technology and Culture* 11 (1970), 552–5.

20. Amartya Sen, "Brain Drain: Causes and Effects," in *Science and Technology in Economic Growth*, ed. B. R. Williams (London, 1973), 393–403.

21. See Herrera, "Social Determinants."

22. Tony Longrigg, "Science and the Closed Society: Some Problems Facing the USSR in the 1970s," paper presented to the Soviet and East European Study Group, Royal Institute of International Affairs, London, January 1972, 5–6 (xerox); Norton E. Long, "Open and Closed Systems," in *Approaches to Comparative and International Politics*, 155–66.

23. Alexander Vucinich, *Science in Russian Culture* (Stanford, 1963 and 1970), II, 475 et passim; David Joravsky, *The Lysenko Affair* (Cambridge, Mass., 1970), 22; William L. Blackwell, *The Beginnings of Russian Industrialization, 1800–1860* (Princeton, 1968), 364–74.

24. Kendall E. Bailes, *Technology and Society under Lenin and Stalin: Origins of the Soviet Technical Intelligentsia, 1917–1941*, (Princeton, 1978), 38. There were some indigenous innovations—for example, in water turbines and the manufacture of high-grade steel. (Blackwell, *The Beginnings*, 398). However, government and industry failed to develop important Russian technical initiatives in telegraphy, locomotives, power transmission, and military weaponry. (Ibid., 397–400; Nicholas DeWitt, "Scholarship in the Natural Sciences," in *The Transformation of Russian Society*, ed. Cyril E. Black (Cambridge, Mass., 1960), 397.)

25. Vucinich, *Science in Russian Culture*, II, 481–2; Jerome Blum, "Russia," in *European Landed Elites in the Nineteenth Century*, ed. David Spring (Baltimore, 1977), 77, 83–5; Alexander Gerschenkron, *Continuity in History and Other Essays* (Cambridge, Mass., 1968), 136–7. Only in the 1890s did the regime change its educational policies so as to begin promoting the wide dissemination of technical knowledge; see Patrick Alston, *Education and the State in Tsarist Russia* (Stanford, 1969).

26. Peter I. Lyashchenko, *History of the National Economy of Russia to the 1917 Revolution*, translated by L. M. Herman (New York, 1949), 714; R. W. Davies, *Science and the Soviet Economy: Inaugural Lecture* (Birmingham, England, 1967), 2; Robert Lewis, "Industrial Research and Development in the USSR, 1924–1935," Ph.D. dissertation, University of Birmingham, 1975, 5–6; John P. McKay, *Pioneers for Profit* (Chicago, 1970), 95–106.

27. Bailes, *Technology and Society*, 53–4.

28. Of the roughly 2,200 offers received between 1922 and 1927, 93 percent were rejected, more than half on the grounds that the projects suggested should not be in foreign hands. (Calculated from Alexander Baykov, *The Development of the Soviet Economic System* (Cambridge, Mass., 1947), 125–6.) See also Richard B. Day, *Leon Trotsky and the Politics of Economic Isolation* (Cambridge, England, 1973); Sutton, *Western Technology*, I, esp. 16–44, 77–86, 136, 167–77, 217–21, 340; chapter 2 below, n. 40.

29. This paragraph is adapted from my "The Organizational Environment of Soviet Applied Research," in *The Social Context of Soviet Science*, eds. Linda L. Lubrano and Susan Gross Solomon (Boulder, Colo., 1980), 70–71.

30. J. Langrish et al., *Wealth from Knowledge: A Study of Innovation in Industry* (London, 1972), 19, 23, 74; E. Layton, "Conditions of Technological Development," in *Science, Technology and Society: A Cross-Disciplinary Perspective*, eds. Ina Spiegel-Rösing and Derek de Solla Price (London, 1977), 204–208.

31. Don Price, *The Scientific Estate*, pbk. ed. (New York, 1965), 26; Genadii Dobrov, *Nauka o nauke*, 2nd ed. (Moscow, 1970), 198–9; D. Pelz and F. Andrews, *Scientists in Organizations* (New York, 1966), 60–63; R. S. Rosenbloom and F. W. Wolek, *Technology and Information Transfer* (Boston, 1970), 35, 94; Edwin Mansfield, *The Economics of Technological Change* (New York, 1968), 126–7; Edwin Mansfield et al., *Research and Innovation in the Modern Corporation* (New York, 1971), 221; Langrish et al., *Wealth from Knowledge*, 77.

32. For a suggestion that this may be happening in Western science, see Jerome R. Ravetz, *Scientific Knowledge and Its Social Problems*, pbk. ed. (New York, 1973), 44–7, 58–62. Cf. Norman Storer, *The Social System of Science* (New York, 1966), 139–41, and N. D. Ellis, "The Occupation of Science," in *Sociology of Science*, ed. Barry Barnes, pbk. ed. (Baltimore, 1972), 195. For a clear assertion that it is occurring in Soviet science, see *The Scientific Intelligentsia of the USSR (Structure and Dynamics of Personnel)*, eds. D. M. Gvishiani, S. R. Mikulinsky, and S. A. Kugel (Moscow, 1976), 27–8.

33. Naum Jasny, *Soviet Industrialization, 1928–1952* (Chicago, 1961), 73–9; Holland Hunter, "The Overambitious First Soviet Five-Year Plan," *Slavic Review* 32 (June 1973), 237–57.

34. E. H. Carr and R. W. Davies, *Foundations of a Planned Economy*, pbk. ed. (Harmondsworth, England, 1974), 622–27; and especially Bailes, *Technology and Society*, 69–158; Joseph Berliner, *Factory and Manager in the U.S.S.R.* (Cambridge, Mass., 1957), 323–4.

35. Leon Herman, *Varieties of Economic Secrecy in the Soviet Union*, RAND Paper P-2840 (Santa Monica, 1963), esp. 5–9. See also Naum Jasny, *The Socialized Agriculture of the USSR* (Stanford, 1949), 728–36; cf. O. Edmund Clubb, *Twentieth Century China* (New York, 1964), 365–6.

36. For a systematic analysis of the role of government in various types of economic systems, see Charles Lindblom, *Politics and Markets* (New Haven, 1977).

37. See especially Victor Thompson, *Bureaucracy and Innovation* (University, Alabama, 1969), 35, 43–6. Studies of technological change in American industry show that individual corporations are most inclined to innovate when operating at about 75 percent of their production capacity. (Mansfield, *The Economics of Technological Change*, 107.) Equally striking is the fact that during depressions, some Western economies have shown a large-scale increase in industrial R & D activities. (Keith Pavitt, "Technology, International Competition and Economic Growth: Some Lessons and Perspectives," *World Politics* 25 (January 1973), 183; Michael Sanderson, "Research and the Firm in British Industry, 1919–1939," *Science Studies* 2 (1972), 119, 146.)

38. This long-term change has been very important. It should be added, however, that it seems questionable whether a further dramatic increase in the contribution of technological progress to the growth of developed countries has occurred in the last two or three decades, as some writers claim. See especially Simon Kuznets, *The Economic Growth of Nations* (Cambridge, Mass., 1971), 74, 307, 315; Mansfield, *The Economics of Technological Change*, 4–5, 36–7; R. C. O. Matthews, "The Contribuion of Science and Technology to Economic Development," in *Science and Technology in Economic Growth*, 1–31.

39. Joseph S. Berliner and Franklyn D. Holzman, "The Soviet Economy: Domestic and International Issues," in *The Soviet Empire: Expansion and Detente*, ed. William E. Griffith (Lexington, Mass., 1976), 85–144.

40. The distinction between mechanistic and organic institutions is an adaptation from Tom Burns and G. M. Stalker, *The Management of Innovation*, pbk. ed. (London, 1966).

41. In such organizations, leaders rely more on generalized rules, less on case-by-case rulings. Cf. Robert K. Merton, *Social Theory and Social Structure* (Glencoe, Ill., 1957), 195–6; Michel Crozier, *The Bureaucratic Phenomenon*, pbk. ed. (Chicago, 1967), 182.

42. For a general discussion of information in organizations, see Harold Wilensky, *Organizational Intelligence* (New York, 1967).

43. See especially *From Max Weber: Essays in Sociology*, eds. H. H. Gerth and C. Wright Mills, pbk. ed. (New York, 1958); Burns and Stalker, *The Management of Innovation*; and Anthony Downs, *Inside Bureaucracy* (New York, 1967), 59–61.

44. For example, two scholars studied the impact of organizational structure on economic effectiveness (defined in terms of profits, sales, and product development over the span of five years) in three American industries. They found that in industries with stable technologies, companies with a highly centralized form of administration, which corresponds roughly to the mechanistic pattern, performed better than did their decentralized competitors in those sectors. (Paul R. Lawrence and Jay W. Lorsch, *Organization and Environment: Managing Differentiation and Integration* (Boston, 1967), 115 and passim.) See also F. Glenn Boseman and Robert E. Jones, "Market Conditions, Decentralization, and Organization Effectiveness," *Human Relations* 27 (September 1974),

669; Alfred D. Chandler, Jr., *Strategy and Structure*, pbk. ed. (Cambridge, Mass., 1962), 326–62.

45. The mechanistic specialization of tasks and the organic specialization of persons are fundamentally distinct. In the first instance, an employee is assigned a narrow, simple task that he performs repeatedly; in the second, he performs tasks that, although they may be narrowly delimited, require complex skills and are not routine. (Victor Thompson, *Modern Organization* (New York, 1965), 25–6.)

46. Ibid., 6, 13, 24; William Kornhauser, *Scientists in Industry* (Berkeley, 1962); Dobrov, *Nauka o nauke*, 185.

47. Thompson, *Bureaucracy and Innovation*, 20–21 and passim. For evidence of the higher incidence of specialized professionals in the newer industrial sectors of the American economy, see Arthur Stinchcombe, "Social Structure and Organizations," in *Handbook of Organizations*, ed. James G. March (New York, 1965), 156–9. For the lower incidence of hierarchy, functional specialization of subdivisions, and rule specification in American manufacturing firms with high rates of product innovation, see Edward Harvey, "Technology and the Structure of Organizations," *American Sociological Review* 33 (April 1969), 247–59. For evidence that in industries undergoing rapid technological change, decentralized companies outperform their centralized competitors, see Lawrence and Lorsch, *Organization and Environment*, 70, 129, 143. For indications that industrial laboratories are more "organic" than manufacturing plants, see Paul R. Lawrence and Jay W. Lorsch, "Differentiation and Integration in Complex Organizations," *Administrative Science Quarterly* 12 (June 1967), 5.

48. One of the noteworthy features of Western studies of formal organizations is the frequency with which this context is neglected. This proclivity has been pointed out by Stinchcombe, "Social Structure and Organizations," 142–3; see also Sanford Lakoff, "Private Government and the Managed Society," in *Private Government*, ed. Sanford A. Lakoff (Glenview, Ill., 1973), 234; Carl Beck, "Bureaucracy and Political Development in Eastern Europe," in *Bureaucracy and Political Development*, ed. Joseph LaPalombara, pbk. ed. (Princeton, 1963), 290–91.

49. Joseph Litterer, "Systematic Management: The Search for Order and Integration," *Business History Review* 35 (Winter 1961), 463, and "Systematic Management: Design for Organizational Recoupling in American Manufacturing Firms," ibid. 37 (Winter 1963), 380–81; H. S. Person, "Scientific Management," *Encyclopedia of the Social Sciences*, 1st ed. (New York, 1930–34), V, 606; and Chandler, *Strategy and Structure*, 387–90.

50. On the emergence of technological competition, see Andrew Shonfield, *Modern Capitalism: The Changing Balance of Public and Private Power*, pbk. ed. (New York, 1965), 40–61. On the process of "innovation by invasion" of a sector by outside firms, see Donald Schon, *Technology and Change* (New York, 1967), 139–61, and also Mansfield, *The Economics of Technological Change*, 59–60. On the increasingly organic structure of corporate organization, partic-

ularly the trend toward a more decentralized form of autonomous divisions, see Chandler, *Strategy and Structure*, 363–78; Schon, *Technology and Change*, 121–2; Lawrence and Lorsch, *Organization and Environment*, 70, 129, 143; Alfred D. Chandler, "The Multi-Unit Enterprise: A Historical and International Summary," in *Evolution of International Management Structures*, ed. Harold F. Williamson (Newark, N.J., 1975), 243.

51. Chandler, *Strategy and Structure*, 377; Peter F. Drucker, *The Practice of Management* (New York, 1954), 212–13. On the criteria applied by central management and its powers vis-à-vis the divisions in General Motors, see Peter F. Drucker, *Concept of the Corporation*, rev. ed. (New York, 1972), 50, 56, 65–8.

52. On the general problem of "authority leakage" in centrally planned economies, see Downs, *Inside Bureaucracy*, 164–5. See also Kenneth Boulding, *The Organizational Revolution* (1953; reprinted Chicago, 1968), 20, 35, 63.

53. Leonid Vladimirov, "Soviet Science: A Native's Opinion," *New Scientist*, 28 November 1968, 490; Robert V. Daniels, "Soviet Politics Since Khrushchev," in *The Soviet Union Under Brezhnev and Kosygin*, ed. John Strong (New York, 1971), 24.

54. For a historical analysis of the way that many American engineers were incorporated into business corporations and their autonomy was subordinated to the quest for profit, see Edwin Layton, *The Revolt of the Engineers* (Cleveland, 1971).

Chapter 2

1. *Resheniia partii i pravitel'stva po khoziaistvennym voprosam* (Moscow, 1967–74), I, 345–6.

2. *KPSS v rezoliutsiiakh i resheniiakh s"ezdov, konferentsii i plenumov TsK*, 8th ed. (Moscow, 1970–72), III, 244, IV, 14.

3. Ibid., III, 247.

4. Ibid., 365.

5. Ibid., IV, 123.

6. *Direktivy KPSS i sovetskogo pravitel'stva po khoziaistvennym voprosam*, I (Moscow, 1957), 668; *KPSS v rezoliutsiiakh*, IV, 46–7, 111.

7. Joseph Stalin, *Works*, X, 295, as quoted by Paul Marantz, "Peaceful Coexistence: From Heresy to Orthodoxy," in *The Dynamics of Soviet Politics*, eds. Paul Cocks et al. (Cambridge, Mass., 1976), 296. Emphasis in the Soviet original. The Comintern statements are taken from *Blueprint for World Conquest*, ed. W. H. Chamberlin (Washington, D.C., 1946), 221, and *International Press Correspondence*, No. 84 (28 November 1928), 150, as quoted by Frederick S. Burin, "The Communist Doctrine of the Inevitability of War," *American Political Science Review* 57 (1963), 338–9.

8. *KPSS v rezoliutsiiakh*, IV, 407.

9. Stalin and his cohorts were not the only Soviet politicians who tried to turn the fear of war to their own political advantage; in 1927 Trotsky and his allies did so as well. See Alfred Meyer, "The War Scare of 1927," *Soviet Union* 5, part 1 (1978), 14–25; see also Sheila Fitzpatrick, "The Foreign Threat during the First Five-Year Plan," ibid., 31–2.

10. Stalin, *Sochineniia*, XIII, 38–9, as quoted in Jerry Hough and Merle Fainsod, *How the Soviet Union Is Governed* (Cambridge, Mass., 1979), 163–4.

11. *KPSS v rezoliutsiiakh*, IV, 409.

12. Burin, "The Communist Doctrine," 340; Raymond L. Garthoff, *Soviet Military Policy* (New York, 1966), 68–9.

13. In a famous passage Stalin explained that if the USSR reached the stage of full communism while capitalist encirclement still existed, the state would not wither away as previous Marxist thought had predicted. Rather, it would continue to perform these three vital functions. (*XVIII s"ezd Vsesoiuznoi kommunisticheskoi partii(b): stenograficheskii otchet* (Moscow, 1939), 15, 34–6.)

14. *KPSS v rezoliutsiiakh*, IV, 407.

15. Stephen F. Cohen, *Bukharin and the Bolshevik Revolution* (New York, 1973), · 254–6. The Soviet debate over capitalism was also closely intertwined with the question of the likelihood of Western socialist revolutions (which Bukharin thought improbable in the absence of war), and the political tactics that foreign Communist parties should adopt. (Ibid., 256–8.)

16. *XVI s"ezd Vsesoiuznoi Kommunisticheskoi Partii(b): stenograficheskii otchet* (Moscow, 1935), I, 41, 50.

17. *XVI s"ezd Vsesoiuznoi Kommunisticheskoi Partii(b): stenograficheskii otchet* (Moscow, 1930), 19–20.

18. *XVII s"ezd Vsesoiuznoi Kommunisticheskoi Partii(b): stenograficheskii otchet* (Moscow, 1934), 9.

19. *XVII konferentsiia Vsesoiuznoi Kommunisticheskoi Partii(b): stenograficheskii otchet* (Moscow, 1932), 157–8.

20. *Bol'shevik*, No. 16 (1931), 36–40, 42, 44.

21. *KPSS v rezoliutsiiakh*, V, 72.

22. Nikolai Bukharin, *Socialist Reconstruction and Struggle for Technique* [sic] (Moscow, 1932), 8–9. Emphasis in the original. The Russian-language version of this book was published in 1931. See also *XVII konferentsiia*, 80, and Cohen, *Bukharin*, 331, 350.

23. *KPSS v rezoliutsiiakh*, V, 23, 130.

24. Ibid., 41.

25. *Bol'shevik*, Nos. 11–12 (1930), 12.

26. Charles P. Kindleberger, *The World in Depression, 1929–1939*, pbk. ed. (Berkeley, 1973), 280.

27. M. Ioel'son, Review of *Krizis i zagnivanie kapitalisticheskoi promyshlennosti,* eds. L. Mendel'son and E. Khmel'nitskaia, *Bol'shevik,* No. 11 (1934), 86; ibid., No. 16 (1934), 18–20.

28. Ibid., Nos. 19–20 (1934), 53.

29. *Planovoe khoziaistvo,* No. 10 (1935), 33–4. For an analysis of discussions of this issue within the Commissariat of Heavy Industry in 1936, see Julian Cooper, "The Development of the Soviet Machine Tool Industry, 1928–1941," Ph.D. dissertation, University of Birmingham, 1975, 288–93.

30. *Izobretatel',* No. 1 (1936), 10. See also *Tekhnika i vooruzhenie,* No. 8 (1935), 21.

31. See the third section of this chapter.

32. *Bol'shevik,* No. 16 (1937), 24–5, 34; ibid., No. 1 (1938), 42, 47, 49, 56.

33. Ibid., No. 9 (1939), 6, 9.

34. Bailes, *Technology and Society,* 381–406.

35. *XVIII s"ezd,* 16.

36. Ibid., 17.

37. *Bol'shevik,* Nos. 3–4 (1941), 8.

38. Ibid., No. 1 (1941), 5.

39. Alexander Baykov, *Soviet Foreign Trade* (Princeton, 1946), 46, 102, 104.

40. For the fullest study of the concessions, see Antony Sutton, *Western Technology and Soviet Economic Development,* I (Stanford, 1968). In his statistical summary, Sutton overstates the number of concessions, though not necessarily their technological importance. On p. 9 he cites A. A. Santalov and L. Segal, *Soviet Union Yearbook,* 1930 (London, 1930), 206, as showing the following number of concessions concluded in each year: 1922, 18; 1923, 44; 1924, 55; 1925, 103; 1926, 110. Adding these together, he concludes that in total 330 concessions were granted through 1926. In fact, the cited source does not contain such statistics, and Santalov and Segal, *Soviet Union Yearbook,* 1926 (London, 1926), which does have such a series (p. 167), contains two figures that contradict Sutton's: 25 concessions for 1924 and 30 for 1925. Santalov and Segal, *Soviet Union Yearbook,* 1927 (London, 1927), 166, gives a figure of 28 for 1926. Hence the total number of concessions granted through 1926 should be fixed at 145, roughly half the number Sutton gives. In the fall of 1927, 99 of these agreements remained in effect. (Santalov and Segal, *Soviet Union Yearbook,* 1928 (London, 1928), 176.)

41. A. A. Santalov and L. Segal, *Soviet Union Yearbook,* 1930 (London, 1930), 210–11.

42. Alexander Baykov, *The Development of the Soviet Economic System* (Cambridge, Mass., 1947), 126. "Unembodied" technology comprises technical information and plans, as opposed to physical products.

43. Sutton, *Western Technology*, I, 16–44, 77–86, 136, 167–77, 217–21, 340.

44. *KPSS v rezoliutsiiakh*, III, 244, 247.

45. Robert A. Lewis, "Industrial Research and Development in the U.S.S.R., 1924–1935," Ph.D. dissertation, University of Birmingham, 1975, 24, Appendix 1.

46. Bailes, *Technology and Society*, 220.

47. *KPSS v rezoliutsiiakh*, IV, 47; *XVI s"ezd* (Moscow, 1935), I, 82.

48. Richard Day, *Leon Trotsky and the Politics of Economic Isolation* (Cambridge, England, 1973), 126–52 et passim; *Piatnadtsatyi s"ezd Vsesoiuznoi Kommunisticheskoi Partii(b): stenograficheskii otchet* (Moscow, 1962), II, 1431, 1443.

49. V. P. Miliutin in *Shestnadtsataia konferentsiia VKP(b): stenograficheskii otchet* (Moscow, 1962), 229–30. See also 247.

50. Quoted in V. D. Esakov, *Sovetskaia nauka v gody pervoi piatiletki* (Moscow, 1971), 94–5.

51. *Piatnadtsatyi s"ezd VKP(b)*, II, 1099–1100.

52. *Bol'shevik*, No. 10 (1928), 45–6. Emphasis in original.

53. A. F. Khavin, *Shagi industrii* (Moscow, 1957), 56. During the First Five-Year Plan Khavin was a prominent industrial journalist.

54. Calculated from Baykov, *Soviet Foreign Trade*, 104.

55. Alec Nove, *An Economic History of the U.S.S.R.* (London, 1968), 229; Herbert S. Levine, "An American View of Economic Relations with the U.S.S.R.," *Annals of the American Academy of Political and Social Science* 414 (July 1974), 11.

56. V. I. Kas'ianenko, *Zavoevanie ekonomicheskoi nezavisimosti SSSR* (Moscow, 1972), 179, 182; V. K. Furaev, *Sovetsko-amerikanskie otnosheniia (1917–1939)* (Moscow, 1964), 153; V. I. Kas'ianenko, *Kak byla zavoevana tekhniko-ekonomicheskaia samostoiatel'nost' SSSR* (Moscow, 1964), 205.

57. *Ekonomicheskaia zhizn' SSSR: khronika sobytii i faktov*, 2nd ed. (Moscow, 1967), I, 214, and Kas'ianenko, *Zavoevanie*, 186. Kas'ianenko states that in 1932 there were, in addition to foreign specialists, more than 10,000 foreign workers in the USSR, plus members of their families.

58. *Industrializatsiia SSSR, 1929–1932* (Moscow, 1970), 110. While the one-quarter figure pertains to projects designed entirely by foreign companies or jointly with Soviet organizations, there is evidence in the report that the role of the companies was critical even in the latter instances.

59. There were seventy such agreements in force in October 1929. (Sutton, *Western Technology*, II, 10–11.)

60. Ibid., 50.

61. Ibid., 55, 58.

62. Ibid., 47–8, 52–5.

63. In 1929, 400 of the 2,000 employees of Gipromez were foreigners. (Davies, *Science*, 7. See also *Bol'shevik*, No. 15 (1931), 59, 62, 64.)

64. Sutton, *Western Technology*, I, 75, and II, 62–3, 68, 77–8.

65. Ibid., II, 98, 100–101, 113–14.

66. Ibid., 193-4; *Planovoe khoziaistvo*, Nos. 6–7 (1932), 236–61.

67. Furaev, *Sovetsko-amerikanskie otnosheniia*, 153–4; Sutton, *Western Technology*, II, 81, 141, 146, 152–4.

68. Sutton, *Western Technology*, II, 81, 141, 146.

69. Bailes, *Technology and Society*, 376–9; Sutton, *Western Technology*, II, 124; *Planovoe khoziaistvo*, Nos. 6–7 (1932), 262–71.

70. *Bol'shevik*, No. 5 (1930), 14–15; *Marks, Engel's, Lenin, Stalin o tekhnike*, ed. V. F. Asmus (Moscow, 1934), 559.

71. *Nauchnyi rabotnik*, No. 4 (1929), 46, 57; ibid., No. 2 (1928), 97; ibid., Nos. 11–12 (1930), 155.

72. Cooper, "The Development of the Soviet Machine Tool Industry," 341.

73. *Resheniia partii i pravitel'stva*, I, 492.

74. Ibid., 708–9.

75. Loren R. Graham, *The Soviet Academy of Sciences and the Communist Party, 1927 to 1932* (Princeton, 1967), 121–30; Kendall E. Bailes, "The Politics of Technology: Stalin and Technocratic Thinking Among Soviet Engineers," *American Historical Review* 79 (1974), 445–69.

76. *XVI s"ezd VKP(b)* (Moscow, 1935), I, 73, 86; *Nauchnyi rabotnik*, Nos. 11–12 (1930), 23; *Bol'shevik*, No. 10 (1931), 37. Jeremy Azrael, *Managerial Power and Soviet Politics* (Cambridge, Mass., 1966), 56, 217n.

77. *XVI konferentsiia*, 554–5.

78. *XVI s"ezd* (Moscow, 1935), I, 560.

79. The year 1929 was not specified for this figure, but the context implied it. (Ibid., 560, 605, 658, 709.)

80. Zhores Medvedev, *The Medvedev Papers* (London, 1971), 93–4.

81. *Bol'shevik*, Nos. 23–24 (1929), 60.

82. According to an editorial note, Kaganovich's article was "an abbreviated and reworked stenogram" of his speech to the key November 1929 plenum of the Central Committee, where a major showdown occurred between Stalin and the Right Opposition (Bukharin, Rykov, Tomskii, and others).

83. *Bol'shevik*, Nos. 23–24 (1929), 67–8.

84. *XVI s"ezd* (Moscow, 1935), I, 551.

85. A. E. Ioffe, *Mezhdunarodnye sviazi sovetskoi nauki, tekhniki i kul'tury, 1917–1932* (Moscow, 1975), 284; Dana C. Dalrymple, "American Technology and Soviet Agricultural Development, 1924–1933," *Agricultural History* 40, No. 3 (July 1966), 191n.

86. Kas'ianenko, *Zavoevanie*, 190. Both figures were below the number of trips called for in the regime's economic plans. This work gives no figures for 1930.

87. *XVI s"ezd* (Moscow, 1935), I, 560–61, 658.

88. *Nauchnyi rabotnik*, No. 1 (1929), 113.

89. Ibid., and ibid., No. 11 (1929), 107–109.

90. Ibid.; Medvedev, *The Medvedev Papers*, 228. Cf. A. E. Ioffe, *Internatsional'nye nauchnye i kul'turnye sviazi Sovetskogo Soiuza (1928–1932 gg.)* (Moscow, 1969), 139.

91. It is possible that the crackdown in the early 1930s was followed by a partial relaxation and then by new restrictions. Cf. Andrew Swatkovsky, "United States-Soviet Scientific and Technical Exchanges and Contacts, 1945–Present: An Evaluation," Ph.D. dissertation, Columbia University, 1972, 55.

92. *Nauchnyi rabotnik*, No. 1 (1929), 114–15; Medvedev, *The Medvedev Papers*, 228.

93. *Vestnik Akademii nauk SSSR* (hereafter abbreviated *VAN*), Nos. 10–11 (1937), 299.

94. Philip Hanson, "International Technology Transfer from the West to the USSR," in U.S. Congress, Joint Economic Committee, *Soviet Economy in a New Perspective* (Washington, D.C., 1976), 788.

95. *XVII s"ezd*, 665, 669.

96. Calculated from Lewis, "Industrial Research," 24; Robert A. Lewis, "Some Aspects of the Research and Development Effort in the Soviet Union, 1924–1935," *Science Studies* 2 (1972), 162.

97. Lewis, "Some Aspects," 164; W. W. Leontieff, Sr., "Scientific and Technological Research in Russia," *American Slavic and East European Review* 4 (1945), 70–71.

98. *Vestnik metallopromyshlennosti*, No. 9 (1932), 5.

99. Ibid., No. 10 (1932), 8. See also *Zavodskaia laboratoriia*, No. 1 (1935), 125; *Bol'shevik*, No. 13 (1932), 44.

100. *Vestnik metallopromyshlennosti*, No. 8 (1933), 36.

101. *7-i s"ezd sovetov: stenograficheskii otchet. Biulleten' No. 6* (Moscow, 1935), 40. See also 11.

102. Levine, "An American View," 11; Nove, *An Economic History*, 229.

103. Kas'ianenko, *Zavoevanie*, 179, 182; Kas'ianenko, *Kak byla*, 205.

104. Kas'ianenko, *Kak byla*, 205. All seventy-eight agreements removed from effect between 1931 and 1933 are described by Kas'ianenko as having been "abolished." While some material compensation was paid by the Soviet side to Western companies, savings on dissolved and renegotiated agreements still came to 13.6 million rubles between May 1931 and May 1932. (Ibid.)

105. Iosif Stalin, *Sochineniia*, XIII, 70–72, as quoted by Bailes, *Technology and Society*, 155. For Stalin's other possible motives, see Bailes, 154–5.

106. Sutton, *Western Technology*, II, 271; Floyd J. Fithian, "Soviet-American Economic Relations, 1918–1933: American Business in Russia During the Period of Nonrecognition," Ph.D. dissertation, University of Nebraska, 1964, 229.

107. This conclusion was arrived at by reading Sutton, *Western Technology*, and categorizing all the datable agreements as being in effect during the First, Second, or Third Five-Year Plan.

108. Kas'ianenko, *Zavoevanie*, 300; *Vneshniaia torgovlia*, No. 1 (1938), 19.

109. Sutton, *Western Technology*, II, 82, 99, 101, 114, 128, 133, 152, 166, 178.

110. *Interavia*, No. 428 (1 May 1937), 7. No source was identified, but in view of the newsletter's function as a trade journal, it is likely that the information came from persons within the foreign companies concerned.

111. Michael R. Dohan, "The Economic Origins of Soviet Autarky, 1927–28–1934," *Slavic Review* 35 (1976), 619, 624, 628. The changing terms of international trade due to the Great Depression intensified the damage to export earnings done by the collapse of Soviet agriculture after collectivization. One Soviet author calculates that between 1929 and 1932 the change caused the USSR a net loss of more than 1 billion rubles in foreign trade transactions. (Kas'ianenko, *Zavoevanie*, 159.) In this period, a special Politburo commission on foreign currency under the chairmanship of Ia. E. Rudzutak was charged with generating exports. (*Voprosy istorii KPSS*, No. 11 (1964), 30–31.)

112. Dohan, "The Economic Origins," 620–22, 630.

113. Cf. ibid.

114. One such attempt is in *Bol'shevik*, No. 13 (1932), 34–5, 45. The author also found it necessary, however, to criticize those "opportunists" who wished to rely heavily on foreign imports for underestimating the USSR's domestic capacities. (Ibid., 43.)

115. *Bol'shevik*, Nos. 19–20 (1930), 61–2; ibid., Nos. 23–24 (1930), 74; Cooper, "The Development of the Soviet Machine Tool Industry," 343–4.

116. Baykov, *Soviet Foreign Trade*, 102; Sheldon Rabin, "Soviet-Owned Banks in Europe," Ph.D. dissertation, The Johns Hopkins School of Advanced International Studies, 1977, 109n. See also Molotov's comments in *7–i s"ezd sovetov: stenograficheskii otchet, Biulleten' No. 1* (Moscow, 1935), 22.

117. For example, *Bol'shevik*, Nos. 9–10 (1934), 26.

118. So far as foreign inventions are concerned, "a reform of the law on invention in 1931 lauded the rejection of the principle of novelty and encouraged the

widespread 'plagiarism' and application of inventions as already described in foreign literature." (Bailes, *Technology and Society*, 345.)

119. *Industrializatsiia SSSR, 1933–1937* (Moscow, 1971), 248.

120. *Sotsialisticheskaia rekonstruktsiia i nauka*, No. 8 (1936), 142 (hereafter cited as *Sorena*).

121. Ibid., 134.

122. Ibid., 138, 141.

123. *Bol'shevik*, No. 16 (1934), 27.

124. *Za industrializatsiiu*, 28 June 1936, 3. See also ibid., 22 October 1936, 2.

125. Ibid., 11 June 1935, 2.

126. Ibid., 14 November 1934, 1.

127. Ibid., 11 June 1935, 2. See ibid., 16 March 1936, 2, for similar complaints.

128. Ibid., 30 June 1935, 6; ibid., 10 July 1935, 3. (The first source is particularly critical of the lag in updating technology at the Gorkii Automobile Plant.) Cf. ibid., 2 July 1935, 2.

129. *XVII konferentsiia*, 154, 272, 275. See also A. Fediukin, *Velikii Oktiabr' i intelligentsiia* (Moscow, 1972), 391; Richard Lowenthal, "Development vs. Utopia in Communist Policy," in *Change in Communist Systems*, ed. Chalmers Johnson, pbk. ed. (Stanford, 1970), 56; Bailes, *Technology and Society*, 265–280.

130. Kendall E. Bailes, "Stalin and Revolution from Above: the Formation of the Soviet Technical Intelligentsia, 1928–1934," Ph.D. dissertation, Columbia University, 1971, 195.

131. Bailes, *Technology and Society*, 265–81.

132. Bailes, "Stalin and Revolution from Above," 205–206.

133. See, for example, *Pravda*, 27 July 1934, 6, and *Za industrializatsiiu*, 27 July 1934, 3. One article, by the party secretary of the plant, clearly implied that the German companies Demag and Krupp were behind the sabotage; see *Izvestiia*, 6 August 1934, 3.

134. *Za industrializatsiiu*, 30 July 1934, 3; ibid., 1 August 1934, 4; ibid., 2 August 1934, 1.

135. These trips were well publicized in the industrial press. For two of many possible examples, see *Za industrializatsiiu*, 6 July 1935, 3, and ibid., 11 July 1935, 3. For an unusually forceful criticism of this practice and a plea for fuller contacts with foreign R & D specialists, see the comments by P. A. Bogdanov in *Sovet pri narodnom komissare tiazheloi promyshlennosti SSSR: pervyi plenum, 10–12 maia 1935 g.* (Moscow and Leningrad, 1935), 220–27. A few months before this speech, Bogdanov had returned from five years in the United States as Soviet trade representative.

136. D. Danin, *Rezerford* (Moscow, 1966), 593, 596; C. L. Boltz, *Ernest Rutherford* (Berkeley, 1970), 183–4.

137. *VAN*, No. 5 (1934), 59. For other cases of this kind see Swatkowsky, "United States–Soviet Scientific and Technical Exchanges," 55.

138. In 1935 the Academy Presidium adopted a decision to join the International Union of Scientific Radiotechnology. (N. M. Mitriakova in *Istoriia SSSR*, No. 3 (1974), 48–9; *Materialy k istorii Akademii nauk za sovetskie gody (1917–1947)*, ed. S. I. Vavilov (Moscow and Leningrad, 1950), 145.) Notwithstanding Mitriakova's efforts to give the opposite impression, the USSR did not actually join this union until after Stalin's death. (E. D. Lebedkina, *Mezhdunarodnyi sovet nauchnykh soiuzov i Akademiia nauk SSSR* (Moscow, 1974), 112.) In 1936 one of the Academy's most prominent geophysicists endorsed an invitation from the International Geodesic and Geophysical Union (IGGU) for the Academy to join, and in May 1936 the Academy reportedly acted to do so. (*Materialy k istorii Akademii nauk*, 161, states that the Academy "entered" into the IGGU.) However, this decision was apparently abrogated or put in abeyance, since in 1955 a committee of the Academy formally joined the IGGU as the national Soviet representative. (Lebedkina, *Mezhdunarodnyi sovet*, 137.)

139. *Ezhegodnik germanskoi istorii*, 1970 (Moscow, 1971), 166n; B. N. Molas, "Mezhdunarodnye snosheniia Akademii," in *Akademiia nauk SSSR za 10 let: 1917–1927*, ed. A. E. Fersman (Leningrad, 1927), 202–203; Brigette Schroeder-Gudehus, "Challenge to Transnational Loyalties: International Scientific Organizations After the First World War," *Science Studies* 3 (1973), 110–13; Lebedkina, *Mezhdunarodnyi sovet*, 86, 110–41.

140. *Sorena*, No. 8 (1936), 136–7, 140–41.

141. *VAN*, No. 6 (1936), 3–4, 18.

142. *Bol'shevik*, No. 16 (1934), 24; *Za industrializatsiiu*, 16 March 1936, 2.

143. *Pravda*, 3 July 1936, 3. The occasion for the attack was an innocuous article by Luzin about Soviet secondary education in *Izvestiia*, 27 June 1936, 3. See also *Pravda*, 2 July 1936, 3.

144. *Pravda*, 9 July 1936, 3; ibid., 15 July 1936, 4.

145. Ibid., 10 July 1936, 3; ibid., 12 July 1936, 3; *Front nauki i tekhniki*, No. 7 (1936), 132.

146. *VAN*, Nos. 8–9 (1936), 7–10. A separate commission was appointed to investigate *Pravda*'s similar charges against a group of several scientists, but its findings were apparently never published. (Ibid., 93.)

147. In July *Pravda* listed fourteen members of the commission. In August, *Izvestiia* published a list from which three of the previous four members who did not belong to the Academy had been removed; the one new person on *Izvestiia*'s list was a member of the Academy. (*Pravda*, 14 July 1936, 3; *Izvestiia*, 6 August 1936, 4.) The changes apparently represented an effort to keep control of the issue in the hands of the Academy's membership.

148. *VAN*, Nos. 8–9 (1936), 7–10.

149. Ibid., 3–4; *Pravda*, 6 August 1936, 1. From internal evidence in the editorial and the dating of the Presidium's resolution on Luzin, it is clear that the editorial was written after the Presidium acted. Aside from the directness of the attack on the Academy, which the Presidium would hardly have printed on its own initiative, the vitriolic language of the editorial was completely uncharacteristic of the journal's regular articles. (Cf. *Bol'shevik*, No. 16 (1936), 13.) A few months later a journal on science and technology edited by pro-Stalin scientists remarked that the struggle against "Luzinism" had been begun not by scientific institutions but at the initiative of "the Stalinist Central Committee of the party and its central organ, *Pravda*." (*Front nauki i tekhniki*, No. 1 (1937), 10–11.)

150. *VAN*, No. 10 (1936), 22–3.

151. Both had been abroad since 1930. Ipat'ev, in particular, was conducting pioneering research on oil refining that the Soviet government was anxious to control directly. (Vladimir N. Ipatieff, *My Life in the United States: The Memoirs of a Chemist* (Evanston, Ill., 1959), 56–8, 74–6, 85–90, 103.)

152. They were deprived of their Soviet citizenship at the same time. (*VAN*, No. 1 (1937), 20–21.)

153. *Pravda*, 21 December 1936, 2.

154. *Front nauki i tekhniki*, No. 1 (1937), 122–3.

155. *VAN*, No. 1 (1937), 7–8.

156. Ibid.; ibid., Nos. 4–5 (1937), 17–18, 22–23.

157. Robert Conquest, *The Great Terror* (London, 1968), 151–2.

158. Bailes, *Technology and Society*, 101–109; *Bol'shevik*, No. 1 (1934), 156–8.

159. In early 1936 Bukharin made a trip to Western Europe as part of a Soviet delegation to purchase some of Marx's manuscripts. According to an emigré Menshevik with whom he held many private discussions, Bukharin was very interested in the idea of uniting the Soviet intelligentsia to take part in elections: "On this, Bukharin said: 'Some second party is necessary. . . .' His idea was that the second party, composed of the intelligentsia, would not be a force opposed in principle to the regime, but one making proposals for changes and remedies." (B. N. Nicolaevsky, *Power and the Soviet Elite*, ed. Janet D. Zagoria (New York, 1965), 15–16, 59–61.) Although the idea of meaningful elections in Stalin's Russia now sounds absurd, we must remember that in 1936 the regime was writing a new constitution that was formally highly democratic, and that Bukharin was closely involved in drafting it. For Bukharin's election to the Academy Presidium, see *Izvestiia*, 24 November 1935, 2. For his role in industrial research see the following section.

160. *VAN*, No. 1 (1937), 7–8. A major journal on science and technology edited by Bukharin, *Sotsialisticheskaia rekonstruktsiia i nauka*, was closed at the end of 1936, and in February 1937 *Pravda* charged that it had been "a captive of bourgeois ideology." (Cohen, *Bukharin*, 466n.) The attack on Bukharin focused especially on his influence on Soviet studies of the history of science and technology. See *VAN*, Nos. 4–5 (1937), 7–8, 17–23.

161. *Izvestiia*, 6 August 1936, 4; *Pravda*, 6 August 1936, 1; Cohen, *Bukharin*, 472n.

162. *VAN*, Nos. 4–5 (1937), 8–9, 12–13.

163. Conquest, *The Great Terror*, 191–5. At an extended meeting of the Academy's aktiv (core personnel) between March 27 and 29, the Academy was heavily criticized in connection with Stalin's speech to the plenum. Academy Vice President G. M. Krzhizhanovskii took the lead in identifying wrecking within the Academy and criticizing the Presidium. (*Materialy k istorii Akademii nauk*, 175–6.)

164. *Front nauki i tekhniki*, No. 7 (1937), 138–40. One of the charges made against Bukharin and other "enemies of the people" was that they had obstructed the "Soviet reconstruction" of the Academy for "a whole seven years," that is, since the earlier Stalinist purge of the Academy in 1929–30. (Ibid.)

165. *VAN*, No. 6 (1937), 68–9, 75–6.

166. *Front nauki i tekhniki*, No. 5 (1937), 10, 18; *Sovetskaia nauka*, No. 1 (1938), 39, 43, 47.

167. *VAN*, Nos. 8–9 (1936), 5.

168. Roy Medvedev, *Let History Judge* (New York, 1971), 525.

169. Frederick Barghoorn, *The Soviet Cultural Offensive* (Princeton, 1960), 51. For one such case, see Medvedev, *The Medvedev Papers*, 62.

170. *Kul'turnaia zhizn' v SSSR, 1928–1941: khronika* (Moscow, 1976), 47, 192, 394, 468, 471, 573.

171. E. A. Andreevich, "Structure and Functions of the Soviet Secret Police," in *The Soviet Secret Police*, eds. Robert Slusser and Simon Wolin (New York, 1957), 121. See also Medvedev, *The Medvedev Papers*, 351–66.

172. *VAN*, No. 6 (1936), 4.

173. Ibid., No. 4 (1941), 63.

174. Ibid., Nos. 10–11 (1937), 299. Roy Medvedev, *Let History Judge*, 525, confirms this trend.

175. *Resheniia partii i pravitel'stva*, II, 722, 724; *Stanki i instrument*, Nos. 10–11 (1939), 2.

176. *Resheniia partii i pravitel'stva*, II, 786.

177. Kas'ianenko, *Zavoevanie*, 198.

178. Calculated from Cooper, "The Development of the Soviet Machine Tool Industry," 558–63. Cf. *Stanki i instrument*, Nos. 10–11 (1939), 1.

179. *Planovoe khoziaistvo*, No. 9 (1938), 46.

180. *Stanki i instrument*, No. 2 (1941), 1–2; ibid., No. 4 (1941), 1–2.

181. *Bol'shevik*, No. 12 (1937), 41–2.

182. Ibid.

183. Ibid., 44. On the inadequacies of copying as a strategy, see also *Vestnik metallopromyshlennosti*, Nos. 8–9 (1938), 134.

184. During the early 1930s the Soviets bought and copied a number of foreign weapons such as tanks and airplanes. (David Holloway, "Innovation in the Defence Sector: Battle Tanks and ICBMs," in *Industrial Innovation in the Soviet Union*, eds. Ronald Amann and Julian Cooper (New Haven, 1982).) After 1935 they continued to seek foreign technical aid in the military fields where they still lagged far behind the West. Between 1936 and 1939, for example, Stalin sought unsuccessfully to have American shipyards construct Soviet warships that the USSR was still incapable of manufacturing. (John Lewis Gaddis, *Russia, The Soviet Union and the United States: An Interpretive History* (New York, 1978), 136.) In most military sectors, however, the regime, while still attempting to learn about Western designs, concentrated on building a domestic R & D establishment that could create superior original weapons. Historians differ on whether Soviet military technology had already drawn abreast of foreign weaponry before World War II began. By the end of the war, however, Soviet weapons were undoubtedly equal to or better than German weapons, which was possible only because the technological gap had been dramatically narrowed during the 1930s.

185. Sutton, *Western Technology*, II, 94–5; *Ocherki razvitiia tekhniki v SSSR* (Moscow, 1969), II, 228–9.

186. G. D. Komkov, B. V. Levshin, and L. K. Semenov, *Akademiia nauk SSSR: kratkii istoricheskii ocherk* (Moscow, 1974), 327.

187. *VAN*, No. 3 (1934), col. 12; Kas'ianenko, *Kak byla*, 122. The figures in this paragraph do not include the smaller republican academies in the Ukraine and Belorussia, which were loosely affiliated with the USSR Academy and also grew rapidly. A higher figure of 3.9 million rubles for the USSR Academy in 1928 is given by N. M. Mitriakova, "Struktura, nauchnye uchrezhdeniia i kadry AN SSSR (1917–1940 gg.)," in *Organizatsiia nauchnoi deiatel'nosti* (Moscow, 1968), 231.

188. Loren R. Graham, "The Formation of Soviet Research Institutes: A Combination of Revolutionary Innovation and International Borrowing," in *Russian and Slavic History*, eds. Don Karl Rowney and G. Edward Orchard (Columbus, Ohio, 1977), 65–6.

189. *Nauchnyi rabotnik*, No. 11 (1929), 35–6, and Lewis, "Industrial Research and Development," 41.

190. Lewis, "Some Aspects," 166. Between 1933 and 1937 total state budgetary allocations for these institutions rose from 780 to 1,982 million rubles. (*KPSS vo glave kul'turnoi revoliutsii v SSSR* (Moscow, 1972), 87.)

191. Graham, *The Soviet Academy*, chapters 3 and 4.

192. In 1918 only about 10.4 percent of the territory of the former Tsarist Empire had been covered by geological surveys; by 1937 more than 43.2 percent

had been surveyed. (Bailes, *Technology and Society*, 347.) A large part of this work was done by the Academy. (*VAN*, No. 5 (1934), 68.)

193. Graham, *The Soviet Academy*, 139; *220 let Akademii nauk SSSR* (Moscow, 1945), 17–24, 44–7; Esakov, *Sovetskaia nauka v gody pervoi piatiletki*, 214; Mitriakova, "Struktura," 231.

194. This idea was broached by at least two speakers at the 16th party conference (*Shestnadtsataia konferentsiia Vsesoiuzoi Kommunisticheskoi Partii(b): stenograficheskii otchet* (Moscow, 1962), 165, 249–50.) Shortly afterward, in May 1929, the Central Committee resolved to create "a center (or a permanent conference)" for this purpose, and ordered a special commission to submit a draft charter within two months. (*Izvestiia Tsentral'nogo Komiteta Vsesoiuznoi Kommunisticheskoi Partii(b)*, 29 July 1929, 20–21.) A month later the Central Committee resolved to speed up the formulation of the charter for an "all-union center planning scientific research work." (Ibid., 25 August 1929, 18.)

195. *VAN*, No. 8 (1932), cols. 8–15.

196. *VAN*, No. 12 (1932), col. 71.

197. *Sobranie zakonov i rasporiazhenii Raboche-krestianskogo Pravitel'stva* (1933), 542.

198. *VAN*, No. 4 (1935), col. 24.

199. *VAN*, No. 10 (1935), cols. 27–8, 31. An attempt in 1935 to write the formal power of coordination into the Academy's new charter, however, failed. (*Ustavy Akademii nauk SSSR, 1724–1974* (Moscow, 1974), 194.)

200. *Planovoe khoziaistvo*, No. 4 (1936), 9.

201. Komkov et al., *Akademiia nauk SSSR*, 323, 326.

202. Mitriakova, "Struktura," 231.

203. Komkov et al., *Akademiia nauk SSSR*, 316; *VAN*, No. 11 (1967), 58, 60, and No. 4 (1935), col. 17.

204. I. Dubinskii-Mukhadze, *Ordzhonikidze* (Moscow, 1963), 353.

205. *Front nauki i tekhniki*, Nos. 5–6 (1934), 144, 148–9; Robert A. Lewis, "Government and the Technological Sciences in the Soviet Union: The Rise of the Academy of Sciences," *Minerva* 15, No. 2 (Summer 1977), 196.

206. *VAN*, No. 12 (1937), 35; *Planovoe khoziaistvo*, No. 4 (1936), 5.

207. *VAN*, No. 4 (1935), col. 73; ibid., No. 6 (1935), col. 56. This draft was not published.

208. Ibid., No. 1 (1936), col. 102.

209. Cf. *VAN*, No. 6 (1936), 7.

210. *Izvestiia*, 24 November 1935, 2. V. P. Volgin, the Academy's permanent secretary, was replaced in November by N. P. Gorbunov. It is conceivable that his removal was related to the conflict over the new charter. (*VAN*, No. 6 (1935), col. 56; ibid., No. 1 (1936), cols. 102–103.)

211. Lewis, "Industrial Research and Development," 41.

212. *Sorena*, No. 4 (1936), 7–8. Emphasis in the original.

213. *VAN*, Nos. 10–11 (1937), 345–57; Mitriakova, "Struktura," 214–16.

214. *VAN*, No. 6 (1936), 10–11, 16–17, and Nos. 8–9 (1936), 97.

215. Ibid., Nos. 2–3 (1938), 56.

216. Ibid., No. 4 (1938), 90.

217. Ibid.

218. On this aspect of the criticism leveled against the Academy of Sciences, see Joravsky, *The Lysenko Affair*, 105–106. See also *VAN*, No. 5 (1938), 73.

219. In 1937, the Division's two existing institutes (for power engineering and mineral fuels) were working on industrial problems. A listing of the commissions in the same year showed none working in the agricultural area. (*VAN*, Nos. 10–11 (1937), 345–47.)

220. *Pravda*, 11 May 1938, 3; *Materialy k istorii Akademii nauk*, 190.

221. *Front nauki i tekhniki*, No. 6 (1938), 5.

222. In his speech Stalin offered an open invitation to use political techniques in scientific controversies, since he cited Lenin as the preeminent man of science. In 1917, he said, Lenin had waged a courageous struggle to disprove false scientific theories—in this case, theories suggesting the impossibility of revolution in Russia. (Ibid., 5–6.) See also *Sovetskaia nauka*, No. 1 (1938), 39, 49.

223. *VAN*, No. 5 (1938), 78.

224. The four were the Institute for the Study of Machines, the Institute of Mechanics, the Institute of Metallurgy, and the Institute of Mining. (*VAN*, Nos. 11–12 (1942), 122.) This source mistakenly includes in the list the Institute of Automation and Telemechanics, which was actually established in 1939. (*220 let*, 236.) See also Mitriakova, "Struktura," 216.

225. *VAN*, Nos. 9–10 (1938), 134–6, 147–8; ibid., Nos. 4–5 (1939), 172; *Materialy k istorii Akademii nauk*, 195.

226. *VAN*, Nos. 4–5 (1939), 178.

227. Ibid., Nos. 8–9 (1939), 178.

228. Ibid.; cf. Nos. 1–2 (1940), 30.

229. *VAN*, Nos. 9–10 (1938), 80.

230. Ibid., Nos. 1–2 (1940), 31; B. V. Levshin, *Akademiia nauk SSSR v gody Velikoi Otechestvennoi voiny* (Moscow, 1966), 10.

231. Komkov et al., *Akademiia nauk SSSR*, 341.

232. *VAN*, Nos. 8–9 (1939), 178.

233. Ibid.

234. Ibid., Nos. 1–2 (1940), 31.

235. Ibid., Nos. 4–5 (1939), 172, No. 4 (1941), 65–7, 109–10, and Nos. 5–6 (1941), 119.

236. *Planovoe khoziaistvo*, No. 11 (1940), 101.

237. *VAN*, No. 4 (1941), 67.

238. Ibid., 20.

239. Ibid., 65, 67.

240. *Resheniia partii i pravitel'stva*, I, 740–41; *Nauchnyi rabotnik*, No. 1 (1930), 16–17; *Bol'shevik*, No. 15 (1931), 68; *XVII konferentsiia*, 103; Cooper, "The Development of the Soviet Machine Tool Industry," 322; *VARNITSO*, Nos. 11–12 (1930), 41; *Vestnik metallopromyshlennosti*, No. 10 (1932), 10.

241. Lewis, "Industrial Research and Development," 287–8.

242. Esakov, *Sovetskaia nauka v gody pervoi piatiletki*, 115, 118–20.

243. Lewis, "Industrial Research and Development," 59–60. In fairness it should be added that Lewis emphasizes original research in these institutions somewhat more than I do.

244. Ibid., 76.

245. Ibid., 81–4; VSNKh was the central government body that supervised large-scale industry until 1932, when it was abolished.

246. *Nauchnyi rabotnik*, No. 11 (1929), 97, and Lewis, "Industrial Research and Development," 83–5.

247. For details, see Lewis, "Industrial Research and Development," 86–92. See also *Biulleten' finansovogo i khoziaistvennogo zakonodatel'stva* (hereafter cited as BFKhZ), No. 8 (1931), 24.

248. Bailes, "The Politics of Technology," 460.

249. *Bol'shevik*, Nos. 23–24 (1929), 70, and *XVI s"ezd* (Moscow, 1935), I, 147.

250. Bailes, "The Politics of Technology," especially 457–9, 462–3.

251. Ibid., 446–7.

252. A November 1930 VSNKh decree resolved to remove all but one institute from the control of NIS. In February 1931 VSNKh passed another decree indicating that only two institutes should remain under NIS control. (Lewis, "Industrial Research," 90–91; BFKhZ, No. 8 (1931), 24.)

253. Lewis, "Industrial Research," 92–3.

254. In August 1933 Bukharin called for a *"radical redistribution of qualified cadres,"* including the transfer of many more research institutions to the control of large plants, in order to promote a major decentralization of authority to enterprises. (*Pravda*, 4 August 1933, 2–3; emphasis in the original.) He may have hoped a comprehensive decentralization would make plants more receptive

to innovation and thereby bridge the institute-enterprise gap, which clearly worried him. Having become head of the Council of Factory Laboratories in January 1933, he may also have felt more confident of his ability to protect researchers from misuse by enterprise directors. Moreover, Bukharin lost control of the central Scientific Research Sector sometime between June 1933 and 1934 and so had less personal interest in maintaining its power over the institutes. (*Sorena*, No. 5 (1933), 170–71; *BFKhZ*, No. 15 (1934), 12–13; *Sorena*, No. 8 (1936), 143; Bailes, *Technology and Society*, 277n.)

255. Lewis, "Industrial Research," 34.

256. VSNKh was replaced by three commissariats, for heavy, light, and wood industry, at the beginning of 1932.

257. Lewis, "Industrial Research," 99.

258. Ibid., 93, 95–6.

259. *Sorena*, No. 3 (1934), 4–5.

260. Ibid., 7.

261. Ibid.; emphasis in the original.

262. From Bukharin's foreword to *Marks, Engel's, Lenin, Stalin o tekhnike*, ed. V. F. Asmus, 6. The book was set and printed in the first two months of 1934.

263. *BFKhZ*, No. 14 (1934), 2.

264. Ibid., No. 15 (1934), 12.

265. Those expenditures amounted to 123,704,000 rubles. Total expenditures came to 218,885,700 rubles. (*Sorena*, No. 3 (1934), 6.)

266. *BFKhZ*, No. 15 (1934), 12–13.

267. For the method used to make this estimate, see Bruce Parrott, "Technology and the Soviet Polity: The Problem of Industrial Innovation, 1928 to 1973," Ph.D. dissertation, Columbia University, 1976, 176–7n.

268. *Sorena*, No. 3 (1934), 6; *Nauchno-issledovatel'skie instituty tiazheloi promyshlennosti*, ed. A. A. Armand (Moscow-Leningrad, 1935), xviii.

269. The decree slashing research funds was signed by M. Kaganovich, a Deputy Commissar of Heavy Industry and the brother of Lazar Kaganovich, one of Stalin's most trusted lieutenants.

270. Calculated from the sources cited in note 268.

271. Calculated from *Socialist Construction in the U.S.S.R.* (Moscow, 1936), 467, and Nicholas DeWitt, *Education and Professional Employment in the U.S.S.R.* (Washington, D.C., 1961), 428.

272. *Industrializatsiia SSSR, 1933–1937* (Moscow, 1971), 363.

273. *VAN*, No. 6 (1936), 19.

274. *VAN*, Nos. 4–5 (1936), 67.

275. *Sorena*, No. 8 (1936), 134, 138.

276. Lewis, "Some Aspects," 174.

277. *Sorena*, No. 8 (1936), 135; *Front nauki i tekhniki*, No. 10 (1936), 102.

278. Calculated from *Industrializatsiia SSSR, 1933–1937*, 363.

279. Cooper, "The Development of the Soviet Machine Tool Industry," 304–305. Cooper indicates that this incident probably occurred in 1932. See also *Voprosy istorii KPSS*, No. 11 (1964), 32.

280. *Puti razvitiia tekhniki v SSSR* (Moscow, 1967), 227. In Russian, "ENIMS" stands for Experimental Scientific-Research Institute of Metal-Cutting Tools.

281. For an analysis of the debates among Soviet engineers over the proper choice of foreign models for copying, see Cooper, "The Development of the Soviet Machine Tool Industry."

282. *Pravda*, 5 July 1935.

283. See, for example, *Direktivy*, II, 426–8.

284. *BFKhZ*, Nos. 27–28 (1934), 28.

285. *Za industrializatsiiu*, 17 July 1935, 2; Lewis, "Industrial Research and Development," 289.

286. *BFKhZ*, No. 9 (1934), 4–5; *Resheniia*, II, 582–3; and *Finansovoe i khoziaistvennoe zakonodatel'stvo*, No. 10 (1936), 26. (Hereafter cited as *FKhZ*.)

287. *FKhZ*, No. 12 (1939), 23–5. This system based premiums for design organizations primarily on the degree to which they reduced the expense of building a plant below the cost anticipated before it was designed.

288. *Resheniia partii i pravitel'stva*, II, 636.

289. *Sorena*, No. 8 (1936), 134.

290. Marvin R. Jackson, Jr., "Soviet Project and Design Organizations: Technological Decision Making in a Command Economy," Ph.D. dissertation, University of California at Berkeley, 1967, 19.

291. Ibid., 57.

292. *Planovoe khoziaistvo*, No. 10 (1935), 36.

293. Ibid., 34–5.

294. *XVII konferentsiia*, 276; *Zavodskie laboratorii tiazheloi promyshlennosti*, ed. A. A. Armand (Moscow-Leningrad, 1935), xvii–xviii.

295. *Zavodskaia laboratoriia*, No. 1 (1935), 128, 113–14; *Sorena*, No. 1 (1935), 154–6, and No. 5 (1935), 144.

296. *Bol'shevik*, No. 16 (1934), 27.

297. *BFKhZ*, No. 22 (1935), 11–12; *Planovoe khoziaistvo*, No. 10 (1935), 37.

298. *Sorena*, No. 10 (1935), 6.

299. Ibid., 8.

300. *Za industrializatsiiu*, 2 February 1936, 3.

301. Kas'ianenko, *Zavoevanie*, 256, citing unpublished archival sources.

302. *Planovoe khoziaistvo*, No. 10 (1935), 37–8.

303. See, for example, *Za industrializatsiiu*, 9 August 1935, 3.

304. Ibid., 4 March 1936, 1; *Planovoe khoziaistvo*, No. 10 (1935), 37–8.

305. *Za industrializatsiiu*, 15 October 1935, 3; *Planovoe khoziaistvo*, Nos. 11–12 (1937), 76, and No. 9 (1938), 45–6.

306. *Sorena*, No. 8 (1936), 142.

307. *Front nauki i tekhniki*, No. 10 (1936), 107–8.

308. *Zavodskaia laboratoriia*, No. 1 (1935), 115, 126–8. For direct evidence of such hostility, see 124–5.

309. *Sorena*, No. 8 (1936), 138–9, 141.

310. Ibid., 143.

311. For Bukharin's views, see n. 254. Ordzhonikidze's position on the issue of decentralizing research in 1936 remains cloudy, despite the publication of the 1936 order in his name. He had, after all, endorsed an opposite course for machine-building design in 1935, and his speech to the August 1936 conference was not published.

312. Armand, for example, expressed this view. (*Sorena*, No. 10 (1935), 7–8.)

313. Joseph Berliner, *Factory and Manager in the U.S.S.R.* (Cambridge, Mass., 1957).

314. *Za industrializatsiiu*, 16 March 1936, 2; ibid., 9 August 1936, 2.

315. *Sorena*, No. 3 (1935), 60–78, and No. 7 (1936), 159–63.

316. *Stanki i instrument*, No. 8 (1937), 1–2.

317. David Granick, *Management of the Industrial Firm in the USSR* (New York, 1954), 115 and 161, citing *Za industrializatsiiu* for 14 March 1936 and 28 October 1936; *Vestnik metallopromyshlennosti*, No. 9 (1939), 88.

318. *Resheniia partii i pravitel'stva*, II, 722.

319. In a December 1940 resolution the Council of People's Commissars stressed the absolute requirement that technical designs for machines should be centrally approved; the resolution made departures from approved designs a crime. (*Resheniia partii i pravitel'stva*, II, 180–81.)

320. *Planovoe khoziaistvo*, No. 1 (1937), 20, 35–6; Joseph R. Barse, "The Planning of Technological Progress in the U.S.S.R.," Master's Essay, Columbia University Department of Political Science, 1954.

321. *Bol'shevik*, Nos. 3–4 (1941), 26.

322. *Planovoe khoziaistvo*, No. 12 (1940), 6.

323. *Tekhnika vozdushnogo flota*, No. 6 (1940), 5, as quoted in Institute for Research in Social Science, *The Soviet Aircraft Industry* (Chapel Hill, 1955), 133.

324. See, for example, *VARNITSO*, Nos. 9–10 (1930), 79–83. During the antibourgeois specialist campaign two leading officials of the Committee on Inventions were shot and fourteen were jailed. (*XVI s"ezd VKP(b)* (1931), 515.)

325. *Planovoe khoziaistvo*, No. 7 (1937), 45; *VAN*, Nos. 4–5 (1937), 7. One of the cases cited by Molotov in an unusually vitriolic speech on wrecking concerned an industrial plant that was said to have introduced its own rubber formula without adequate testing, thereby producing 200,000 defective tires. Molotov asserted that this was part of a broader wrecking campaign intended to frustrate the start-up of the manufacture of synthetic rubber. (*Bol'shevik*, No. 8 (1937), 18–19.) The purges, incidentally, damaged military as well as civilian R & D. See David Holloway, "Innovation in the Defence Sector," in *Industrial Innovation in the Soviet Union*, 337–8.

326. This trend dated from 1935, when the purge atmosphere began to intensify in the wake of Kirov's murder. (Bailes, *Technology and Society*, 341; *Industriia*, 1 April 1939.)

327. Group discussion (No. 403), B 2 Schedule, 4–5, Harvard Interview Project, as quoted by Bailes, *Technology and Society*, 354.

328. For an oblique reference, see *XVI s"ezd* (1931 ed.), 502. A foreign technical advisor reported that by 1931 one of these establishments was publishing its own limited-circulation technical bulletin. (Allan Monkhouse, *Moscow, 1911–1933* (Boston, 1934), 265).

329. Leonid Vladimirov, *The Russians* (London, 1968), 209; Bailes, *Technology and Society*, 357; A. S. Iakovlev, *Tsel' zhizni*, 3rd ed. (Moscow, 1972), 73; Nikolai Turov, "Zamestitel' A. N. Tupoleva: iz tiuremnykh vospominanii," *Novyi zhurnal*, 96 (1969), 123–4.

330. Bailes, *Technology and Society*, provides persuasive evidence of the existence of such a group of "moderates." The impact of their policies on the status of some imprisoned specialists is illustrated by an extraordinary article published by Ramzin in 1936. Ramzin recounted how he had worked in a prison design bureau under the control of the OGPU and had met Ordzhonikidze, who kept abreast of the bureau's work. On 8 July 1934 Ordzhonikidze visited the bureau and ordered that it be converted into a bureau of his own Commissariat of Heavy Industry—that is, that the prisoners be freed. This occurred two days before a general reorganization of the OGPU into the NKVD. At the time some commentators viewed the reorganization as a curb on police power (Merle Fainsod, *How Russia Is Ruled* (Cambridge, Mass., 1965), 433), and in the short run this was apparently true for Ramzin and some of his fellow specialists. In his article Ramzin heaped praise on Ordzhonikidze, giving the clear impression that he felt his rehabilitation was due to the latter's efforts. He also mentioned that this prison design bureau was designated as number eleven within the

OGPU's technical administration; this implies, but does not prove, that at least ten other such bureaus existed in early 1934. (*Za industrializatsiiu*, 6 February 1936, 3.)

331. A. Sharagin (G. S. Ozerov), *Tupolevskaia sharaga* (Frankfurt-am-Main, 1971), 19, 24–9.

332. This attitude is reported to have held up the introduction of Ramzin's boiler. (Bailes, *Technology and Society*, 357.)

333. Lewis, "Soviet Industrial Research," 67.

334. *BFKhZ*, No. 15 (1934), 12–13.

335. Lewis, "Soviet Industrial Research," 309–10.

336. Julian Cooper, *Defence Production and the Soviet Economy, 1929–41*, CREES Discussion Papers, Soviet Industrialization Project Series No. 3, Centre for Russian and East European Studies, University of Birmingham, 1976, 3n.

337. John Erickson, "Radio-location and the Air Defence Problem: the Design and Development of Soviet Radar," *Science Studies* 2 (1972), 247–8.

338. John Erickson, *The Soviet High Command* (London, 1962), 406–7.

339. G. K. Zhukov, *Vospominaniia i razmyshleniia*, 2nd ed. (Moscow, 1974), I, 213, 222.

340. *Voprosy istorii*, No. 10 (1968), 117–19. See also *7-i s"ezd sovetov: stenograficheskii otchet*, *Biulleten' No. 5*, 31, 33.

341. Cooper, *Defence Production*, 44. This move may also have been a Stalinist gambit designed to reduce Ordzhonikidze's operational control of the distribution of resources among industrial sectors. For evidence that Stalin and Ordzhonikidze disagreed about the pace of industrial investment, see Bailes, *Technology and Society*, ch. 11.

342. Iakovlev, *Tsel' zhizni*, 103–4.

343. See, for example, Bailes, *Technology and Society*, 399–406.

344. Zhukov, *Vospominaniia*, I, 215.

345. See, for example, *Voprosy istorii*, No. 10 (1968), 119.

346. David Holloway, "Soviet Military R & D: Managing the 'Research-Production Cycle,'" in *Soviet Science and Technology: Domestic and Foreign Perspectives*, eds. John R. Thomas and Ursula Kruse-Vaucienne (Washington, D.C., 1977), 205.

347. Lewis, "Soviet Industrial Research," 304–5.

348. *Sorena*, No. 3 (1936), 149.

349. Bailes, *Technology and Society*, 396–406; Robert A. Kilmarx, *A History of Soviet Airpower* (New York, 1962), 106–7; Iakovlev, *Tsel' zhizni*, 156, 159. It may be that the disappointments in Spain were one of Stalin's motives for expressing his dissatisfaction with Soviet science in May 1938.

350. A. N. Ponomarev, *Sovetskie aviatsionnye konstruktory* (Moscow, 1977), 14, 58. See also P. Astashenko, *Konstruktor legendarnykh ilov* (Moscow, 1972), 58.

351. *Voprosy istorii*, No. 10 (1968), 121; Erickson, "Radio-location," 247–8.

Chapter 3

1. From the standpoint of legitimation, of course, the war had a strong negative potential as well, since the early Nazi victories posed grave questions about the party's leadership. Stalin and his associates did not dwell on the causes of these early defeats, but instead accentuated the great Soviet victories beginning late in 1942. Stalin's successors, although they condemned some aspects of his rule, similarly sought to kindle heroic memories of the war as a means of political legitimation. (Seweryn Bialer, "Introduction," in *Stalin and His Generals*, ed. Seweryn Bialer (New York, 1969), especially 18–20.)

2. See, for example, the American Transcript of the Khrushchev Tapes, Part I, pp. 724–6 (available at the Russian Institute of Columbia University, New York City).

3. I. Stalin, *O Velikoi Otechestvennoi voine Sovetskogo Soiuza* (Moscow, 1943), 61–3; Iosif Stalin, *Sochineniia*, XV, pp. 69ff., as quoted by William O. McCagg, Jr., *Stalin Embattled, 1943–48* (Detroit, 1978), 64. In the present study, the volumes of the *Sochineniia* cited as XIV, XV, and XVI are those edited by Robert H. McNeal and published in book form by the Hoover Institution. They comprise the published works of Stalin not included in the thirteen official Soviet volumes of his works.

4. McCagg, *Stalin Embattled*, 64, citing Wolfgang Leonhard, *Die Revolution entlässt ihre Kinder*, 234.

5. Stalin, *Sochineniia*, XV, 163–4, as quoted by McCagg, *Stalin Embattled*, 73.

6. *Bol'shevik*, Nos. 23–4 (1942), 52–3.

7. Ibid.

8. Ibid., No. 9 (1942), 42.

9. See, for example, *Planovoe khoziaistvo*, No. 1 (1945), 82–6.

10. *VAN*, Nos. 2–3 (1942), 50.

11. *Bol'shevik*, No. 14 (1945), 13–14.

12. Ibid., 21.

13. In the later controversy over his book on capitalism, *Izmeneniia v ekonomike kapitalizma v itoge vtoroi mirovoi voiny* (Moscow, 1946), Academician Varga remarked that originally the intended audience for his book had been the aktiv, or core members, of the Soviet and foreign Communist parties. (*Soviet Views of the Post-War World Economy*, trans. Leo Gruliow (Washington, D.C., 1948), 5.) This intention was apparently never realized, but if it had been, the book would presumably have been disseminated to Soviet party members by Alek-

sandrov's Directorate. Aleksandrov, incidentally, was a philosopher by training and became a full member of the Academy in 1946.

14. In 1944 the three institutes of the Academy's Department of Economics and Law "fulfilled a large amount of research work on special assignments from the governmental organs." (*Izvestiia Akademii nauk SSSR: Otdelenie ekonomiki i prava*, No. 3 (1945), 41.) After 1942 a prominent scholar from one of the institutes, I. A. Trakhtenberg, belonged to the Council of Scientific-Technological Assessment attached to Gosplan. (*Bol'shaia Sovetskaia Entsiklopediia*, 1st ed., XLIV, 742.) By 1945 N. A. Voznesenskii—an Academy member, the chairman of Gosplan, and a candidate member of the Politburo—belonged to the Scholarly Council of another institute. (*Izvestiia Akademii nauk SSSR: Otdelenie ekonomiki i prava*, No. 6 (1945), 52.)

15. Molotov made the loan request on 3 January 1945. (Thomas G. Paterson, *Soviet-American Confrontation: Postwar Reconstruction and the Origins of the Cold War*, pbk. ed. (Baltimore, 1975), 37–8.) In the first half of 1945 the Academy's Division of Economics and Law discussed reports on technological changes in Western industry and the broader impact of the war on the capitalist system. According to a sketchy description of the meetings, the reports provoked "extremely lively discussion," that is, strong disagreements. (*VAN*, No. 4 (1945), 143–4. Cf. ibid., Nos. 3–4 (1943) and Nos. 4–5 (1943), 122; *Izvestiia Akademii nauk: Otdelenie ekonomiki i prava*, No. 3 (1948), 199–200.)

16. *Bol'shevik*, Nos. 23–4 (1944), 46–8, 50, 54; ibid., No. 16 (1945), 52–4.

17. Ibid., Nos. 17–18 (1945), 38–9.

18. A number of Western economists shared this expectation at the time. (W. S. Woytinsky, "What Was Wrong in Forecasts of Postwar Depression?", *Journal of Political Economy* 55 (1947), 143.)

19. *Bol'shevik*, Nos. 17–18 (1944), 45; ibid., Nos. 19–20 (1944), 27.

20. In late 1945 the Academy Presidium confirmed a new list of members of the Scholarly Council of this institute. The list included K. V. Ostrovitianov and I. I. Kuz'minov, whose views were cited in notes 16 and 17 above. It also included N. A. Voznesenskii, the candidate member of the Politburo who subsequently launched a frontal attack on Varga's unorthodox ideas about capitalism. (*Izvestiia Akademii nauk SSSR: Otdelenie ekonomiki i prava*, No. 6 (1945), 52.) According to a Western source, Voznesenskii began to involve the Institute in the work of Gosplan, which he headed, in 1943, when he became a member of the Academy. (Gerald Segal, "Automation, Cybernetics, and Party Control," *Problems of Communism* 25, No. 2 (March–April 1966), 4.)

21. For an outline of the research done by the Institute during the war years, see *Mirovoe khoziaistvo i mirovaia politika*, Nos. 2–3 (1945), 82. (Hereafter cited as *MKh i MP*.) See also ibid., No. 12 (1945), 82–6. Trakhtenberg, the author of the article on the American economy cited in note 10 above, was a full member of the Academy of Sciences and had worked in the Institute since 1931.

22. *MKh i MP*, No. 1 (1945), 11. Emphasis in the original. Cf. ibid., No. 9 (1945), 3–13.

23. *MKh i MP*, No. 1 (1945), 13. Speaking of the United States and Great Britain, Varga remarked that some groups there wanted to go back to prewar capitalism. "But these are exceptions. Basically everywhere they recognize that a deep reform of the capitalist system is necessary; everywhere one can find ideological currents such as the striving for a planned economy under capitalism, for the introduction of social insurance, for the strengthening of state capitalism, etc." The clear message was that major reforms under capitalism were not only possible, but quite likely. In view of the later controversy about the possibility of having a planned economy under capitalism, it is worth noting that in this article Varga neither put the phrase in the customary quotation marks nor disputed its feasibility. (*MKh i MP*, No. 6 (1946), 13.)

24. Varga, *Izmeneniia*, 86–90, 123.

25. Ibid., 12–13, 300–301.

26. *Planovoe khoziaistvo*, No. 1 (1945), 73–6, 86–9, and No. 3 (1945), 70–74, 77–9; *MKh i MP*, Nos. 2–3 (1945), 81.

27. William Zimmerman, "Choices in the Postwar World (1): Containment and the Soviet Union," in *Caging the Bear: Containment and the Cold War*, ed. Charles Gati, pbk. ed. (New York, 1974), 102–5.

28. Paterson, *Soviet-American Confrontation*, 24–9 and passim.

29. *MKh i MP*, No. 10 (1945), 51; see also ibid., Nos. 2–3 (1945), 80, 82.

30. *MKh i MP*, No. 3 (1946), 77–8. The figures published were taken from a U.S. government report. On the interruption, see George C. Herring, Jr., *Aid to Russia, 1941–1946: Strategy, Diplomacy, The Origins of the Cold War* (New York, 1973), ch. 8.

31. Varga, *Izmeneniia*, 319. In view of these statements, it is misleading to say that Varga's book suggested "that capitalism might live on, and that the trying interwar conditions of 'encirclement' would probably return." (McCagg, *Stalin Embattled*, 158.) Varga was indeed saying that capitalism would probably endure, but he was arguing that capitalist policies toward the USSR would be much more benign than in the past.

32. Varga also remarked that the newly established World Bank and International Monetary Fund (which were obviously going to be dominated by the United States) would facilitate the economic recovery of Europe. (Varga, *Izmeneniia*, 267.)

33. Varga, *Izmeneniia*, 258, 270.

34. *Bol'shevik*, Nos. 17–18 (1944), 43, 45, 47, and Nos. 19–20 (1944), 27.

35. *Bol'shevik*, Nos. 7–8 (1946), 21–2.

36. Ibid., Nos. 11–12 (1946), 2–4.

37. *Istoriia Kommunisticheskoi Partii Sovetskogo Soiuza*, V (Moscow, 1970), 678.

38. *Sovetskie Vooruzhennye Sily: istoriia stroitel'stva* (Moscow, 1978), 373, 389; McCagg, *Stalin Embattled*, 206.

39. Paterson, *Soviet-American Confrontation*, 53.

40. I am indebted to Werner Hahn for discussing this question with me and for allowing me to read the manuscript of his book, *Postwar Soviet Politics: The Fall of Zhdanov and the Defeat of Moderation, 1946–1953* (Ithaca, 1982), before its publication.

41. *Bol'shevik*, No. 1 (1946), 2–3, 6–8.

42. *Pravda*, 8 February 1946, 4.

43. *Pravda*, 6 February 1946, 2; ibid., 8 February 1946, 4.

44. Ibid., 8 February 1946, 4.

45. Ibid., 7 February 1946, 2; *Bol'shevik*, No. 21 (1945), 5–6, 8, 12–13.

46. *Bol'shevik*, No. 21 (1945), 12; *Pravda*, 7 February 1946, 2. See also *Pravda*, 7 February 1946, 3, and 8 February 1946, 2.

47. *Bol'shevik*, Nos. 17–18 (1945), 6. On p. 9 this editorial revealed that a recent resolution of the Central Committee had criticized the journal's board for ideological permissiveness in evaluating the theoretical content of articles. Between the appearance of Nos. 15 and 16 of the journal, there was a shake-up in the membership of the board.

48. Ibid., No. 1 (1946), 6–8.

49. *Pravda*, 7 February 1946, 4.

50. Ibid., 8 February 1946, 3.

51. *Bol'shevik*, Nos. 7–8 (1946), 14, 19–22; see also Michael C. Kaser, "Le débat sur la loi de la valeur en U.R.S.S.—Etude rétrospective 1941–1953," in *Annuaire de L'U.R.S.S.*, 1965, 555–70.

52. *Pravda*, 10 February 1946, 1–2.

53. Ibid.

54. *Bol'shevik*, No. 6 (1946), 71–2, 81.

55. Paterson, *Soviet-American Confrontation*, 151–5.

56. McCagg, *Stalin Embattled*, 166; *Bol'shevik*, No. 19 (1946), 3; Stalin, *Sochineniia*, XVI, 6.

57. Between 1945 and 1948 the regime reduced the total number of persons in the army and navy by some 8.5 million, to a floor of 2,874,000. (*Sovetskie Vooruzhennye Sily*, 374.) In 1946, the output of producer goods dipped to 73 percent of the 1944 level, whereas the production of consumer goods was 13 percent above the 1944 level. (*Narodnoe khoziaistvo SSSR v 1958*, 138, as cited by Hahn, *Postwar Soviet Politics*, ch. 1.)

58. The American offer was made by Secretary of State George Marshall on 5 June. The central idea had been adumbrated by Undersecretary Dean Acheson in a speech on 8 May. (Paterson, *Soviet-American Confrontation*, 28–9.)

59. These points are from an anonymously written account of a session of the Academy's Department of Economics and Law that heard a report by K. V. Ostrovitianov, a vigorous exponent of traditionalist views, on "The Economic Role of the Soviet State." The account chastised Ostrovitianov for failing to give adequate attention to the specific content of the "leading function" of the socialist state, which it labeled "the organizational-economic and cultural-educational function"—thereby notably omitting the defense function. Next it criticized him for failing to deal concretely with how Soviet postwar economic problems could be overcome. Finally, Ostrovitianov was censured for giving insufficient attention to "several timely and extremely interesting problems" of capitalism, particularly state regulation of the economy. (*VAN*, No. 2 (1947), 93–4.)

60. The gathering, chaired by Ostrovitianov, was jointly sponsored by the Academy's Institute of Economics and the Faculty of Economics of Moscow State University. One version of the proceedings was published as an appendix to the November 1947 issue of *MKh i MP*, which was signed to press in December 1947. For a translation, see *Soviet Views of the Post-War World Economy*, trans. Leo Gruliow (Washington, D.C., 1948). About the same time, attacks on the ideas of Varga's disciples also began to appear in print. See, for example, *Bol'shevik*, No. 13 (1947) and No. 23 (1947).

61. *Soviet Views*, 12, 17, 24, 34, 37–8, 85.

62. This speaker observed that even before World War II some Western political groups had been favorably disposed toward the USSR. His latent message was that the existence of such groups had not prevented the rise of Hitler and the Nazi invasion. (Ibid., 63.)

63. Ibid., 25–6, 39–41, 54, 81–2, 103.

64. One speaker asserted that Soviet economists "must help the Soviet people and our friends abroad to understand the contradictory phenomena of capitalist reality" and must "strengthen in Soviet people the confidence in our forces." Another commented that Varga's book had "disoriented" members of the Communist and non-Communist left in Eastern Europe. (Ibid., 5, 80, 113. See also Barrington Moore, Jr., *Soviet Politics—the Dilemma of Power*, pbk. ed. (New York, 1965), 387–8.)

65. *Soviet Views*, 5, 31–3.

66. Ibid., 116.

67. McCagg, *Stalin Embattled*, 263.

68. William O. McCagg, Jr., "Domestic Politics and Soviet Foreign Policy at the Cominform Conference in 1947," *Slavic and Soviet Series* (Russian and East European Research Center, Tel Aviv University) 2, No. 1 (1977), 4–5.

69. Adam Ulam, *Stalin: The Man and His Era*, pbk. ed. (New York, 1974), 658–9.

70. McCagg, *Stalin Embattled*, 263.

71. In 1945–6, leaving aside UNRRA relief, the USSR was the chief trading partner of Eastern Europe. But during 1946 and 1947, there was a marked tendency in several of the countries for the Soviet share of total exports and imports to decline, and for the Western share to increase sharply. This trend began to be reversed only in 1948. (Margaret Dewar, *Soviet Trade with Eastern Europe, 1945–1949* (London, 1951), 2–3, 96–8.)

72. Dewar, *Soviet Trade with Eastern Europe, 1945–1949*, 100–119; Ulam, *Stalin*, 659–60.

73. *Bol'shevik*, No. 20 (1947), 26.

74. Ibid.

75. Ibid., 24.

76. Ibid., 24–5.

77. Ibid., No. 19 (1947), 11–13.

78. V. Kolotov and G. Petrovichev, *N. A. Voznesenskii (biograficheskii ocherk)* (Moscow, 1963), 42.

79. *Narodnoe khoziaistvo SSSR v 1958*, 138, as cited in Hahn, *Postwar Soviet Politics*, ch. 1.

80. *Informatsionnoe soveshchanie predstavitelei nekotorykh kompartii v Pol'she v kontse sentiabria 1947 goda* (Moscow, 1948), 132–3, 151.

81. Ibid., 152.

82. Ibid., 136. I am grateful to Herbert Dinerstein for calling this aspect of Malenkov's speech to my attention.

83. McCagg, *Stalin Embattled*, 293.

84. Frederick G. Barghoorn, "The Varga Discussion and Its Significance," *American Slavic and East European Review* 7, No. 3 (Oct. 1948), 234n.

85. Nikolai Voznesensky, *The Economy of the USSR During World War II*, trans. Russian Translation Program, American Council of Learned Societies (Washington, D.C., 1948), 16, 99.

86. Ibid., 2, 16, 41, 92, 102.

87. *VAN*, No. 12 (1947), 103; V. V. Kolotov, *Nikolai Alekseevich Voznesenskii*, 2nd ed. (Moscow, 1976), 314. The traditionalist Ostrovitianov was named director of the institute. Although Varga and his associates occupied some administrative posts within the new establishment, they had little power to direct research or express their ideas in public. In 1948 Ostrovitianov announced plans to scrap eight monographs that had been scheduled for publication by Varga's institute, and to publish eight others only after they had been drastically rewritten. (*Voprosy ekonomiki*, No. 8 (1948), 75.)

88. In the last months of the war the journal of Gosplan (which Voznesenskii headed) printed articles by members of Varga's institute who were later attacked by the traditionalists for ideological errors. (See n. 26 above.)

89. McCagg, *Stalin Embattled*, 277, suggests that Stalin published an interview with Harold Stassen during the May 1947 meetings on Varga's book as a sign of his support for Varga. The timing was probably not coincidental, but Stalin's remarks about capitalism did not add up to an endorsement of Varga's position. His comments and questions to Stassen implied considerable skepticism that the bourgeois state could avoid an economic crisis through regulation. Moreover, he refused to disavow the doctrine of capitalist encirclement, and he deflected Stassen's efforts to draw an unequivocal distinction between Nazi Germany and the United States. (Stalin, *Sochineniia*, XVI, 78–80, 86–90.) Incidentally, it seems to me that McCagg seriously errs at this point by treating the classification of East European regimes as a minor part of the traditionalist-nontraditionalist dispute, when in fact it was central. (*Stalin Embattled*, 277.) See also Zbigniew Brzezinski, *The Soviet Bloc: Unity and Conflict*, rev. pbk. ed. (New York, 1961), 44–51, 71–7.

90. Kolotov, *Nikolai Alekseevich Voznesenskii*, 316–17.

91. Ibid.

92. *Planovoe khoziaistvo*, No. 2 (1948), 67; *Bol'shevik*, No. 1 (1948), 87–8; ibid., No. 24 (1948), 18–19, 22.

93. *Bol'shevik*, No. 19 (1947), 23–4, 26–7, No. 21 (1947), 31, 46–7, No. 7 (1948), 9–10, and No. 22 (1948), 41.

94. Ibid., No. 22 (1948), 41–2, No. 9 (1947), 40–41, and No. 5 (1948), 78.

95. Ibid., No. 11 (1948), 6, 8.

96. *Planovoe khoziaistvo*, No. 5 (1948), 88–9 (signed for printing on 1 November 1948); *Voprosy ekonomiki*, No. 9 (1948), 54–7, 66, 74 (signed for printing on 21 January 1949). Almost as surprisingly, other speakers revealed in passing that the 1949 research plan of the Institute of Economics called for Varga to write a book on "political changes in the capitalist world since the Second World War." (Ibid., 53, 60.) Varga had originally intended to publish such a book as a companion to his work on the capitalist economy, but the project had been shelved after the first book provoked a furor.

97. *Voprosy ekonomiki*, No. 9 (1948), 54–7, 66, 74.

98. *VAN*, No. 5 (1948), 85–7.

99. Gunnar Adler-Karlsson, *Western Economic Warfare, 1947–1967* (Stockholm, 1968), especially 22, 24–5.

100. Michael Kaser, *Comecon: Integration Problems of the Planned Economies*, 2nd ed. (New York, 1967), 9–12.

101. Marshall D. Shulman, *Stalin's Foreign Policy Reappraised* (Cambridge, 1963), 40, 45. A series of republican party congresses met in 1948–9. Such conferences

usually culminate in a national congress, but none occurred in 1949, suggesting that it had been canceled. (Jeremy Azrael, talk presented at the Kennan Institute of Advanced Russian Studies, Washington, D.C., 24 March 1977.)

102. *Pravda*, 15 March 1949, in *Current Digest of the Soviet Press* 1, No. 10 (1949), 45 (hereafter cited as *C.D.S.P.*). The policy aspects of Voznesenskii's fall remain cloudy. Reportedly the allocation of resources among ministries by Voznesenskii's Gosplan was one source of conflict between him and Beriia, who schemed with Malenkov to discredit him in Stalin's eyes. (*Literaturnaia gazeta*, 30 November 1963, 2; *Khrushchev Remembers*, ed. Strobe Talbott (Boston, 1970), 251.) In 1948, Voznesenskii seemed to be thinking very boldly about economic questions. In September, having formed a special commission to review preliminary drafts for the twenty-year plan, he revealed that he was also a member of the commission set up to draft a new Party Program and that he expected a new book he was writing on the political economy of communism to influence the Program's economic content. (Kolotov, *Nikolai Alekseevich Voznesenskii*, 236–7.)

Work on the twenty-year plan and on this book—especially if Voznesenskii took the high consumption targets of "full communism" seriously—could scarcely have failed to embroil him in hard economic questions about which not only Beriia but Stalin would be very sensitive. Voznesenskii had already shown an interest in using "money-commodity levers" to improve central direction of the economy, and his enemies apparently attacked his economic record as part of their offensive against him. (Kaser, "Le Débat"; Abraham Katz, *The Politics of Economic Reform in the Soviet Union* (New York, 1972), 27, 30, 34–6; Robert Conquest, *Power and Policy in the U.S.S.R.*, pbk. ed. (New York, 1961), 105; *Planovoe khoziaistvo*, No. 6 (1949), 25–7, 31–2; and the speech by Malenkov cited in note 107 below. See also Hahn, *Postwar Soviet Politics*, ch. 4.)

Seeking to square a political circle, Voznesenskii was apparently trying to demonstrate his orthodoxy by backing traditionalist views while also working to modify the regime's economic policies. In mid-1945 *Bol'shevik* had printed an obvious attempt to sanitize the ideological reputation of the Gosplan journal, which had run nontraditionalist analyses of capitalism during the war. Later that year the Central Committee, criticizing *Bol'shevik* for ideological laxity, had removed Voznesenskii and two other figures from the *Bol'shevik* editorial board. (*Bol'shevik*, No. 10 (1945), 70–75, and Nos. 17–18 (1945), 6–9.) Thus, whether or not Voznesenskii's traditionalist utterances after the war matched his private outlook, he had good reasons to champion them in public. They meshed with the views of Zhdanov (his closest political ally) and Stalin, who were pressing an ideological crackdown. In this atmosphere Voznesenskii may well have concluded that only by espousing the traditionalists' absolute contrast between socialism and capitalism could he introduce limited changes into Soviet economic plans without being condemned as a revisionist. (See also Kaser, *Comecon*, 25, and "Le Débat," 563.) But combining a "hard" foreign line and a "reformist" internal line was extremely difficult in the Stalinist system.

103. *Voprosy ekonomiki*, No. 3 (1949), 79–88, trans. in *C.D.S.P.* 1, No. 19 (1949), 4, 9.

104. Ibid., 4–5, 8–9.

105. For foreign policy differences within the elite during this period, see Shulman, *Stalin's Foreign Policy*, and Ronald L. Letteney, "Foreign Policy Factionalism Under Stalin, 1949–1950," Ph.D. dissertation, The Johns Hopkins School of Advanced International Studies, 1971, especially 35–7 and 51–4.

106. *Bol'shevik*, No. 24 (1949), 50.

107. Ibid., No. 21 (1949), 7, 14–15. Shortly afterward, Mikoian spoke at some length about economic planning, the devious efforts of Stalin's past enemies to subvert earlier plans, and the importance of high rates of growth. (*Bol'shevik*, No. 24 (1949), 51.) The most likely target of such remarks was Voznesenskii.

108. *Bol'shevik*, No. 23 (1949), 21–2, and No. 24 (1949), 17, 23–4.

109. Ibid., No. 21 (1949), 1–2. Apparently the first Western scholar to detect this aspect of the speech was Robert C. Tucker, *The Soviet Political Mind: Studies in Stalinism and Post-Stalin Change*, pbk. ed. (New York, 1963), 27–8. See also Shulman, *Stalin's Foreign Policy*, 122–3, and Letteney, "Foreign Policy Factionalism," 195.

110. *Bol'shevik*, No. 24 (1949), 48.

111. Ibid., No. 23 (1949), 21, and No. 24 (1949), 21; Letteney, "Foreign Policy Factionalism," 59, 197–9.

112. *Bol'shevik*, No. 24 (1949), 63, 67; Letteney, "Foreign Policy Factionalism," 195n. For later testimony that Stalin's theories were sometimes proclaimed "complete" in order to forestall further discussion of the underlying issues, see *Kommunist*, No. 10 (1956), 11.

113. *Bol'shevik*, No. 21 (1949), 8.

114. Ibid., No. 24 (1949), 49–50.

115. Ibid., 67; also Sidney Ploss, *Conflict and Decision-Making in Soviet Russia: A Case Study of Agricultural Policy, 1953–1963*, pbk. ed. (Princeton, 1965), 44.

116. Kaser, *Comecon*, 34.

117. Shulman, *Stalin's Foreign Policy*, 145.

118. Kaser, *Comecon*, 34, 49.

119. *Voprosy ekonomiki*, No. 10 (1950), 101–108. More than 200 people participated in the conference.

120. Precise measures of the increase are, of course, not available. But between 1947–8 and 1952, the size of the armed forces climbed from 2,874,000 to 4,600,000 (*Zasedaniia Verkhovnogo Soveta SSSR, piatogo sozyva (chetvertaia sessiia) 14–15 ian. 1960* (Moscow, 1960), 33; *The Military Balance 1971–1972* (London, 1971), 63); and between 1950 and 1951 the overt Soviet defense budget rose by 21 percent (calculated from Alec Nove, *An Economic History of the U.S.S.R.* (Baltimore, 1969), 319).

121. The increase in overt defense spending was 18 percent. (Calculated from Nove, *An Economic History*, 319.)

122. The discussions, which took place in November 1951, reportedly involved a broad circle of economists, philosophers, historians, and ideologists. Stalin wrote his *Economic Problems of Socialism in the U.S.S.R.*, analyzed below, as a follow-up to these discussions. (*Bol'shaia Sovetskaia Entsiklopediia*, 2nd ed., XXX, 204.)

123. *Bol'shevik*, No. 16 (1951), 58, 60–61.

124. Ibid., 61.

125. Ibid., 24–5, 32–3.

126. See the similar comments made by Ostrovitianov at about this time in *VAN*, No. 8 (1951), 45–6.

127. Varga raised the question of the inevitability of war in the November 1951 discussions of the text on political economy, but his suggestion that war might be avoidable was beaten down because Stalin disapproved. (E. Varga, *Ocherki po problemam politekonomii kapitalizma* (Moscow, 1964), 78, as cited by Zimmerman, "Choices in the Postwar World," 104.)

128. *Bol'shevik*, No. 21 (1951), 2, 8, 13.

129. Adler-Karlsson, *Western Economic Warfare*, 28.

130. *Bol'shevik*, No. 1 (1952), 4, 10–12.

131. Charles H. Fairbanks, Jr., has shown that in the early 1950s Beriia was directly at odds with Stalin over the treatment of Soviet minority nationalities, among other things, and that Stalin was preparing to purge Beriia on the grounds that he favored minority nationalism. (Charles H. Fairbanks, Jr., "National Cadres as a Force in the Soviet System: The Evidence of Beriia's Career, 1949–53," in *Soviet Nationality Policies and Practices*, ed. Jeremy R. Azrael (New York, 1978), 144–86.) The charge of collusion with foreign imperialists was a central part of Stalin's campaign against minority nationalism; it was also a key ingredient of the charges in the Doctors' Plot, which Stalin unveiled in 1953 and which was likewise intended to support the case for purging Beriia, among others. (Conquest, *Power and Policy*, 140, 164.) The divergence between Beriia and *Bol'shevik* over U.S.–Soviet relations occurred at the same time Stalin was moving against Beriia's political base in the secret police and elsewhere.

132. Capitalist encirclement would cease, said Stalin, only when socialism had triumphed in "a majority" of countries. (Stalin, *Sochineniia*, XVI, 165–66.)

133. Stalin, *Sochineniia*, XVI, 214–16, 226–31, 234, 240. Interpreting Stalin's discussion of the inevitability of war is difficult because he mixed together the two questions of war among imperialists and war between the capitalist and socialist camps. He rejected the belief of some Soviet commentators that two world wars had taught the imperialist states they must compose their differences with each other without resorting to war. He also criticized Soviet commentators who thought that the imperialist camp was now unified under American aegis

and that its internal contradictions were less significant than the contradiction between the capitalist and socialist blocs. This passage might be interpreted to mean that Stalin felt the international position of the USSR was qualitatively more secure than in the past. But given his stress on the persisting imperialist roots of war and on capitalist encirclement, I think it should be understood as a counterargument to those who wanted a relaxation of tensions with the West. Stalin was saying that while the long-term threat of war from imperialism remained, in the short term the West was not so united against the USSR that the Soviet leadership should try to defuse the situation through a change of foreign policy.

134. Ibid., 213, 223–6, 236–7.

135. *Khrushchev Remembers*, 71.

136. *Voprosy ekonomiki*, No. 12 (1952), 105, 109.

137. *Pravda*, 6 February 1953, as quoted by Garthoff, *Soviet Military Policy*, 71. It should be added that Stalin, even while he dwelt on the theme of imperialist aggressiveness, was not averse to using the lure of prospective trade deals as a diplomatic lever against the NATO allies, and he probably allowed some of his deputies to launch a campaign for expanded trade for that reason. (See Shulman, *Stalin's Foreign Policy*, 186; *Mezhdunarodnoe ekonomicheskoe soveshchanie v Moskve* (Moscow, 1952), especially 3–4, 54–5, 152.) On early tensions between the United States and Western Europe over East-West trade restrictions, see Adler-Karlsson, *Western Economic Warfare*, 15, 42–6.

138. Robert A. Lewis, *Science and Industrialisation in the USSR* (New York, 1979), 163.

139. A. Zverev in *Bol'shevik*, No. 2 (1944), 20.

140. Komkov et al., *Akademiia nauk*, 341, 383. Another source states that the number of junior scientists declined by 20 percent, though the number of senior researchers did not fall. (B. B. Levshin, *Akademiia nauk SSSR v gody Velikoi Otechestvennoi voiny* (Moscow, 1966), 158.)

141. *Istoriia Velikoi Otechestvennoi voiny Sovetskogo Soiuza* (Moscow, 1961–5), V, 409; *Bol'shevik*, No. 6 (1949), 20. Presumably these figures, like the one just quoted from Zverev, include R & D allocations from both the central state budget and other sources; if not, the jump from 1944 would be even larger. These figures are based on the Soviet concept of "science," which is roughly but not exactly equivalent to the standard Western definition of R & D. (Eugene Zaleski et al., *Science Policy in the USSR* (Paris, 1969), 95–6.) From 1945 to 1946 actual spending in the U.S.S.R. Academy increased 80.4 percent. (*VAN*, No. 9 (1947), 118.)

142. Figures for 1950 range from 8 to 10 billion. (Louvan E. Nolting, *Sources of Financing the Stages of the Research, Development, and Innovation Cycle in the U.S.S.R.*, Foreign Economic Reports, No. 3, U.S. Department of Commerce (Washington, D.C., 1973), 10.) These figures must be divided by ten to be commensurate with more recent figures, since the government revalued the ruble in 1961.

143. *Narodnoe khoziaistvo SSSR, 1958* (Moscow, 1959), 842–3; *Istoriia sotsialis-ticheskoi ekonomiki SSSR*, V (Moscow, 1978), 529. The 1941 and 1951 figures for institutes include institutes' affiliates; the 1945 figure includes affiliates and divisions (otdeleniia).

144. Herring, *Aid to Russia*, 21, 58.

145. Sutton, *Western Technology*, III, 5, 7–9. According to Soviet statistics, British and American production furnished roughly 11 percent of all Soviet tanks and aircraft during the war. (*Istoriia Velikoi Otechesvennoi voiny*, VI, 48, as cited by Alan S. Milward, *War, Economy and Society, 1939–1945* (Berkeley, 1977), 73.)

146. Thus, for example, the USSR received three times as many American trucks between 1942 and 1944 as it produced. Lend-Lease equipment provided one-fifth of the new power-generating capacity installed during the war. Lend-Lease also provided just under one-half of Soviet aluminum consumption, about three-eighths of Soviet copper consumption, and one-eighteenth of steel consumption. (Derived from Robert H. Jones, *The Roads to Russia: United States Lend-Lease to the Soviet Union* (Norman, Oklahoma, 1969), 220, 225, 229, and Milward, *War, Economy and Society*, 73.)

147. Milward, *War, Economy and Society*, 52; see also Jones, *The Roads to Russia*, 232.

148. For the period from 1941 through August 1945, the fragmentary evidence available suggests that there were probably about ten such agreements with U.S. firms. A tentative U.S. government survey in 1946 identified five agreements concluded during the war; the fields covered the manufacture of alcohol, the manufacture of neoprene, petroleum cracking, and telephone modernization. ("Technical Aid Contracts Between American Firms and the Soviet Government, 1940–1946," in File 1360-A, 1946–47, European Branch, Bureau of International Programs, Department of Commerce, Container 233, Record Group 285, U.S. National Archives, Washington, D.C.) A published source identifies two wartime contracts not included in the government survey; these were for assistance in the installation of rolling mills. (Stella K. Margold, *Let's Do Business with Russia* (New York, 1948), 100–104.)

149. For two such cases involving small numbers of Bell Aircraft and General Electric engineers, see State Department Decimal File 861.24/1566 and 861.24/1591, Record Group 59, U.S. Archives, Washington, D.C.

150. Jones, *The Roads to Russia*, 223; John R. Dean, *The Strange Alliance* (New York, 1947), 101–102.

151. See *Foreign Relations of the United States, 1944*, IV (Washington, D.C., 1966), 1101.

152. *VAN*, No. 6 (1943), 79. For a similar statement by a Vice-President of the Academy, see ibid., Nos. 11–12 (1943), 44.

153. Ibid., Nos. 4–5 (1944), 90.

154. P. L. Kapitsa, as quoted in Eric Ashby, *Scientist in Russia* (Harmondsworth, England, 1947), 137. Ashby gives an interesting firsthand account of the an-

niversary celebration. See also Swatkovsky, "United States–Soviet Scientific and Technical Exchanges," 75–6.

155. *Planovoe khoziaistvo*, No. 4 (1946), 30.

156. *VAN*, Nos. 7–8 (1945), 51; *VOKS Bulletin*, Nos. 3–4 (1945), 29–31, as cited in Swatkovsky, "United States–Soviet Scientific and Technical Exchanges," 67–8. See also *Bol'shevik*, No. 14 (1945), 73.

157. The impact of exposure to Europe on Soviet soldiers was obviously on the leaders' minds at this time. See Molotov's remarks in *Bol'shevik*, No. 21 (1945), 10. The leaders' suspicions are shown by the decision to imprison or execute a large number of Soviet prisoners of war repatriated from Germany. As for the social sciences, the jubilee meeting of the Academy's Department of Economics and Law pointed out that contacts with foreign social scientists had been absent for decades, and it urged a large increase in them. Varga chaired the meeting. (*VAN*, Nos. 7–8 (1945), 129.)

158. *Bol'shevik*, Nos. 7–8 (1944), 19, and No. 9 (1944), 3–7; *VAN*, Nos. 1–2 (1945), 8–9; and *Materialy k istorii Akademii nauk SSSR za sovetskie gody (1917–1947)*, ed. S. I. Vavilov (Moscow and Leningrad, 1950), 291.

159. Swatkovsky, "United States–Soviet Exchanges," 81.

160. S. Kaftanov in *Bol'shevik*, Nos. 13–14 (1946), 51.

161. "Technical Aid Contracts Between American Firms and the Soviet Government, 1940–1946." The twenty-five are in addition to the estimated ten agreements actually concluded during the war.

162. *Foreign Relations of the United States, 1944*, IV, 1079; "Technical Aid Contracts"; the latter source does not give a date for the contract tentatively negotiated for assistance with the Vega Machine Cannon. The existence of fifty contracts, most of them confidential, was revealed by E. C. Ropes, an official of the U.S. Department of Commerce, in "Russians Agree to Discuss Trade with Business Groups Here," *Journal of Commerce and Commercial* (New York), 18 December 1946, 1, 3.

163. Swatkovsky, "United States–Soviet Exchanges," 83–4; *Foreign Relations of the United States, 1947*, IV (Washington, D.C., 1972), 622–4.

164. Swatkovsky, "United States–Soviet Exchanges," 85. Presumably wartime U.S.–Soviet scientific ties were unusually close in this area. Although the United States transferred a very substantial amount of know-how about proven technology to the USSR during the war, the U.S. Office of Scientific Research and Development shared information about current R & D only in the field of medicine. (*Foreign Relations of the United States, 1944*, IV, 1101.)

165. Ulam, *Stalin*, 696.

166. V. L. Sokolov, *Soviet Use of German Science and Technology, 1945–1946*, Research Program on the USSR, Mimeographed Series No. 72 (New York, 1955), 1–8; Vladimir Alexandrov, "The Dismantling of German Industry," and Vladimir Rudolph, "The Execution of Policy" and "The Agencies of Control:

Their Organizations and Policies," in *Soviet Economic Policy in Postwar Germany*, ed. Robert Slusser (New York, 1953), 14–17, 23, 41.

167. Sokolov, *Soviet Use*, 26; Frederick I. Ordway III and Mitchell R. Sharpe, *The Rocket Team* (New York, 1979), 324; Gerald W. Schroder, "How Russian Engineering Looked to a Captured German Scientist," *Aviation Week* 62, No. 19 (9 May 1955), 27–34; Clarence Lasby, *Project Paperclip: German Scientists and the Cold War* (New York, 1971).

168. Sokolov, *Soviet Use*, 5.

169. G. A. Tokaev, *Stalin Means War* (London, 1951), 141–4.

170. Swatkovsky, "United States–Soviet Exchanges," 83–4.

171. *Krasnaia zvezda*, 28 March 1946, 4.

172. M. M. Lobanov, *Nachalo sovetskoi radiolokatsii* (Moscow, 1975), 258–63.

173. *Planovoe khoziaistvo*, No. 6 (1946), especially 68–74; Lobanov, *Nachalo*, 258–63. See also *MKh i MP*, No. 6 (1946), 117–20.

174. Iakovlev, *Tsel' zhizni*, 444–5. The Commissar of Aircraft Industry, A. I. Shakhurin, was replaced by M. V. Khrunichev.

175. Ibid.

176. Ibid., 455.

177. Raymond L. Garthoff, *Soviet Strategy in the Nuclear Age* (New York, 1958), 177; A. Fedoseev, *Zapadnia* (Frankfurt/Main, 1976), 115. A. N. Ponomarev, *Sovetskie aviatsionnye konstruktory* (Moscow, 1977), 47, observes uncharacteristically that the testing of the Tu-4 was accompanied by "extensive and lengthy" discussions between R & D personnel and the commissioning authority.

178. Garthoff, *Soviet Strategy*, 177; Jean Alexander, *Russian Aircraft Since 1940* (London, 1975), 357–9.

179. Institute for Research in Social Science, *The Soviet Aircraft Industry*, 80; Asher Lee, "Strategic Air Defense," in *The Soviet Air and Rocket Forces*, ed. Asher Lee (New York, 1959), 120–21; M. M. Lobanov, *Nachalo*, 260; I. Radunskaia, *Aksel' Berg* (Moscow, 1971), 226.

180. *Bol'shevik*, No. 21 (1947), 16.

181. Apart from the trade restrictions, Secretary of Commerce Harriman recommended in 1946 that captured German technical documents be withheld from the USSR as a bargaining chip. He also instructed Commerce's Office of Technical Services to stop filling Soviet requests for technical publications. (Paterson, *Soviet-American Confrontation*, 65–6.) For the Academy jubilee in June 1945, the United States and Great Britain restricted the travel of Western nuclear physicists to the USSR (Arnold Kramish, *Atomic Energy in the Soviet Union* (Stanford, 1959), 83–6). No doubt the motive was fear of Soviet atomic espionage—an account of which may be found in David J. Dallin, *Soviet Espionage* (New Haven, 1955), 453–72.

182. *VAN*, No. 2 (1948), 97–9.

183. Ibid., No. 6 (1948), 77.

184. For a detailed account of the whole postwar philosophical debate and Kedrov's role in it, see Hahn, *Postwar Soviet Politics.*

185. *Pravda*, 21 January 1949, 1.

186. *VAN*, No. 1 (1949), 35. Emphasis in the original.

187. For the criticism, see ibid., No. 3 (1948), 101–5, and *Voprosy filosofii*, No. 2 (1947), 379–81. The editorial against cosmopolitanism appeared in a number of the journal which still listed Kedrov as chief editor, but before the next number appeared, *Bol'shevik* announced his removal. (*Voprosy filosofii*, No. 2 (1948), especially 19–24; *Bol'shevik*, No. 5 (1949), 8.)

188. *Bol'shevik*, No. 5 (1949), 8, 31–2.

189. *Pravda*, 7 September 1949, 2.

190. Swatkovsky, "United States–Soviet Exchanges," 93; *Voprosy filosofii*, No. 2 (1948), 19–20; *Bol'shevik*, No. 21 (1951), 41.

191. *Bol'shevik*, No. 4 (1949), 32; *VAN*, No. 9 (1948), 23, and No. 7 (1949), 81; *Bol'shevik*, No. 21 (1951), 41; Maxim Mikulak, "Postwar Soviet Attacks on 'Bourgeois' Science, 1945–51," Certificate Essay, Russian Institute of Columbia University, 1952.

192. *Pravda*, 6 January 1949, trans. in *C.D.S.P.* 1, No. 2 (1949), 52.

193. Nicholas DeWitt, "Scholarship in the Natural Sciences," in *The Transformation of Russian Society*, ed. Cyril E. Black (Cambridge, Mass., 1960), 391.

194. *Pravda*, 17 December 1949, 1.

195. *VAN*, No. 4 (1949), 7, 13–14.

196. *Literaturnaia gazeta*, 5 March 1949, as summarized in *C.D.S.P.* 1, No. 12 (1949), 53; *Pravda*, 29 August 1949, in *C.D.S.P.* 1, No. 35 (1949), 57; *Literaturnaia gazeta*, 12 April 1950, in *C.D.S.P.* 2, No. 16 (1950), 42.

197. *Vestnik inzhenerov i tekhnikov*, No. 3 (1948), 84; see also *Stal'*, No. 11 (1948), 965–6.

198. *Pravda*, 3 August 1949, 2.

199. Kaser, *Comecon*, 17–18; Raymond Hutchings, *Soviet Economic Development* (New York, 1971), 238.

200. I know of no published figures on travel to the West by Soviet scientists and engineers during this period, but there are good reasons to believe the number remained minuscule. In 1941–53 the USSR joined no further international scientific unions. (Lebedkina, *Mezhdunarodnyi sovet*, 86, 110–41.) In one case, the Soviet chemist who was vice-president of a union to which the USSR already belonged failed to attend one of its major meetings and did not reply to the letters the organization addressed to him. (Swatkovsky, "United States–Soviet Exchanges," 491.)

201. *Biulleten' Gosudarstvennogo komiteta SSSR po koordinatsii nauchno-issledovatel'skikh rabot*, No. 3 (1962), 23.

202. In 1959, 1,272 Soviet specialists went to other socialist countries, and 3,113 specialists from these countries came to the Soviet Union. (Ibid.)

203. The acquisition of such literature within the Academy system was centralized in the early 1930s, during an extreme hard-currency shortage. For evidence of the serious impact of currency considerations on the acquisition of foreign scientific periodicals in the 1960s and 1970s, see my *Information Transfer in Soviet Science and Engineering: A Study of Documentary Channels* (Santa Monica: Rand Corporation, 1981), Section IV.

204. According to *VAN*, Nos. 10–11 (1937), 299, the Library of the Academy of Sciences received 68,149 foreign books and journals in 1936. Ibid., No. 4 (1941), 63, gives the comparable figure for 1940 as "more than 25,000." Ibid., No. 11 (1950), 105–6, states that between 1 January 1949 and 1 August 1950 the Library of the Academy received 24,158 books and journals, or slightly more than 15,000 per year.

205. In 1951 the Academy's international exchange of books was reportedly twice as large as in 1950. (*VAN*, No. 2 (1952), 24.) In July 1952 the Academy set up the All-Union Institute of Scientific Information to index and abstract foreign scientific publications and to produce Soviet-printed (and censored) reproductions of many important foreign scientific journals. (Parrott, *Information Transfer*, Section IV.) The officials of the Academy were eager for such a change. See *Literaturnaia gazeta*, 28 April 1951, 3, in *C.D.S.P.* 3, No. 18, 8–10; *Izvestiia*, 30 April 1953, 3; and Tony Longrigg, "Soviet Science and Foreign Policy," *Survey* 17, No. 4 (1971), 46.

206. J. Bernard Hutton, *The Traitor Trade* (New York, 1963), 67–74; Adler-Karlsson, *Western Economic Warfare*, 82.

207. E. A. Andreevich, "Structure and Functions of the Soviet Secret Police," in *The Soviet Secret Police*, 117.

208. Ibid.

209. Loren R. Graham, "The Development of Science Policy in the Soviet Union," in *Science Policies of Industrial Nations*, eds. T. Dixon Long and Christopher Wright (New York, 1975), 30–31.

210. Jerry Hough and Merle Fainsod, *How the Soviet Union Is Governed* (Cambridge, Mass., 1979), 178; Hahn, *Postwar Soviet Politics*, ch. 1; *Istoriia sotsialisticheskoi ekonomiki SSSR*, V (Moscow, 1978), 181–3.

211. *Voenno-istoricheskii zhurnal*, No. 6 (1961), 66; D. M. Kukin, "Partiinoe i gosudarstvennoe rukovodstvo narodnym khoziaistvom SSSR nakanune i v period voiny," in *Sovetskii tyl v Velikoi Otechestvennoi voine* (Moscow, 1974), I, 18.

212. *Istoriia Velikoi Otechestvennoi voiny Sovetskogo Soiuza, 1941–1945* (Moscow, 1961–5), II, 540–41; *Voprosy istorii estestvoznaniia i tekhniki*, vyp. 2 (1975), 25.

These sources name some Council members. Several belonged to the Academy of Sciences.

213. *Istoriia Velikoi Otechestvennoi voiny*, II, 541; *VAN*, Nos. 9–10 (1941), 73.

214. S. V. Kaftanov, "Organizatsiia nauchnykh issledovanii v gody voiny," in *Sovetskaia kul'tura v gody Velikoi Otechestvennoi voiny* (Moscow, 1976), 57.

215. Ibid., 57.

216. *Voprosy istorii estestvoznaniia i tekhniki*, vyp. 2 (1975), 27–8. While Kaftanov was not the only person with scientific credentials to serve as a plenipotentiary of the State Committee, the available evidence suggests that he had the broadest administrative authority over R & D at this time. See *Istoricheskie zapiski*, vyp. 60 (1957), 26–7; Vladimir Keller, *Sergei Vavilov* (Moscow, 1975), 220; S. I. Vavilov, *Sobranie sochinenii*, I (Moscow, 1954), 20; *Bol'shaia Sovetskaia Entsiklopediia*, 3rd ed., XIX, 157.

217. Levshin, *Akademiia nauk*, 34–5. Levshin describes the approving body as the State Committee's Technological Council, rather than the Scientific-Technological Council, but the context suggests it was the body headed by Kaftanov.

218. Ibid., 37; Komkov et al., *Akademiia nauk SSSR*, 347–9; *VAN*, Nos. 9–10 (1941), 68.

219. *VAN*, Nos. 9–10 (1941).

220. Ibid.; ibid., 9–10.

221. Ibid., No. 2 (1942), 91; ibid., No. 5 (1942), 77–9, 81–4; ibid., Nos. 4–5 (1943), 122.

222. One of the disputants, E. A. Chudakov, was an original member of the Projects Commission appointed in October 1941. Three others—P. L. Kapitsa, A. M. Terpigorev, and L. A. Orbeli—became members sometime between then and early 1942. (*VAN*, Nos. 9–10 (1941), 74; ibid., Nos. 2–3 (1942), 119.)

223. *Istoricheskie zapiski*, vyp. 60 (1957), 25. This source does not describe the particulars of the agreement.

224. *VAN*, No. 5 (1942), 13.

225. Stalin's telegram was dated 24 March 1942. The changes in the Academy's plan were confirmed by him in a second telegram of 12 April. (*Istoricheskie zapiski*, vyp. 60 (1957), 25; this source, incidentally, misdates the initial telegram from Stalin.) The new commission was appointed by the Academy Presidium on 3 April. (Komkov et al., *Akademiia nauk*, 347.) N. G. Bruevich, a specialist in precision mechanics who had been elected a corresponding member of the Academy in 1939, held the rank of major general. On 12 April he was appointed as the Academy's representative on another commission set up to study captured enemy equipment. On 8 May he became a full member of the Academy, and on 10 May he became Academician-Secretary. (*Materialy k istorii*, 259; *Akademiia nauk SSSR: personal'nyi sostav* (Moscow, 1974), II, 46.) For his role in planning secret R & D, see *VAN*, Nos. 1–2 (1944), 87.

226. *Istoricheskie zapiski*, vyp. 60 (1957), 25.

227. Komkov et al., *Akademiia nauk SSSR*, 353–64; Levshin, *Akademiia nauk*, 44–9, 79–81, 120–24.

228. See S. Kaftanov, *Sovetskaia intelligentsiia v Velikoi Otechestvennoi voine* (Moscow, 1945).

229. David Holloway, "Entering the Nuclear Arms Race: The Soviet Decision to Build the Atomic Bomb, 1939–1945," Working Paper Number 9, International Security Studies Program, The Wilson Center, Washington, D.C., 1979, 20–29.

230. A. S. Iakovlev, *Rasskazy aviakonstruktora* (Moscow, 1964), 285; *Voprosy istorii*, No. 3 (1975), 143–4; A. N. Ponomarev, *Sovetskie aviatsionnye konstruktory* (Moscow, 1977), 207–9, 248; Alexander, *Russian Aircraft*, 4.

231. Holloway, "Battle Tanks and ICBMs," 377.

232. Holloway, "Innovation in the Defence Sector," 282–3.

233. Ibid.

234. *Istoriia Vtoroi mirovoi voiny* (Moscow, 1973–80), XI, 297.

235. *VAN*, Nos. 9–10 (1943), 80–81; ibid., Nos. 11–12 (1943), 47.

236. Levshin, *Akademiia nauk*, 150; Komkov et al., *Akademiia nauk*, 379.

237. B. V. Levshin, "Razvitie sovetskoi nauki v gody voiny," in *Sovetskaia kul'tura*, 41–2.

238. *Planovoe khoziaistvo*, No. 2 (1945), 48; M. I. Khlusov, *Razvitie sovetskoi industrii, 1946–1958* (Moscow, 1977), 77.

239. Lobanov, *Nachalo*, 251–3; Radunskaia, *Aksel' Berg*, 203.

240. Lobanov, *Nachalo*, 251–3.

241. Ibid., 253–6; M. M. Lobanov, *Iz proshlogo radiolokatsii* (Moscow, 1969), 168–9.

242. Lobanov, *Nachalo*, 254; Radunskaia, *Aksel' Berg*, 204.

243. Radunskaia, *Aksel' Berg*, 212. At another point in the diary, Berg noted that charges had been raised in high government meetings that the Council was mismanaging the radar program. The context suggests that this probably occurred in 1944. (Ibid., 226.)

244. Ibid., 221, 224.

245. Lobanov, *Nachalo*, 255.

246. *Sovetskaia aviatsionnaia tekhnika* (Moscow, 1970), 51–2; A. N. Ponomarev, *Sovetskie aviatsionnye konstruktory* (Moscow, 1977), 248.

247. *Sovetskaia aviatsionnaia tekhnika*, 51–2. *Istoriia Velikoi Otechestvennoi voiny*, V, 412, misleadingly suggests that the plane designed by Artem Mikoian and tested in March 1945 was a pure jet; in fact it was based on a power plant that combined piston and jet engines.

248. *Voenno-istoricheskii zhurnal*, No. 4 (1973), 89. A. Iakovlev and Artem Mikoian (not to be confused with his brother Anastas) designed the first Soviet jet planes. The party-state authorization for Mikoian to work on such a fighter came in February 1945. (A. Arzumanian, *General'nyi konstruktor A. I. Mikoian* (Moscow, 1961), 34.)

249. Holloway, "Entering," 29–30.

250. Ibid., 31–2.

251. *Planovoe khoziaistvo*, No. 1 (1946), 34–5, No. 5 (1946), 24, and No. 6 (1951), 64.

252. *Resheniia partii i pravitel'stva*, III, 467–8.

253. *Planovoe khoziaistvo*, No. 3 (1949), 64.

254. *Sbornik postanovlenii, prikazov i instruktsii po finansovo-khoziaistvennym voprosam*, No. 12 (1946), 31. See also *Bol'shevik*, No. 6 (1946), 77, 84.

255. Zhores Medvedev, *Soviet Science* (New York, 1978), 44. For confirmation of this trend, see *VAN*, No. 3 (1947), 37, and No. 7 (1947), 6–7.

256. *Pravda*, 23 October 1952, 2; also *VAN*, Nos. 8–9 (1946), 26.

257. Joseph Berliner, *The Innovation Decision in Soviet Industry* (Cambridge, Mass., 1976), 452.

258. Roy A. Medvedev and Zhores A. Medvedev, *Khrushchev: The Years in Power*, trans. Andrew Durkin (New York, 1976), 38, as cited by Holloway, "Innovation in the Defence Sector," 336.

259. The only reports of postwar arrests of specialists that I have discovered concern the missile and A-bomb programs. Both reports imply that the number of arrests was not large. (Kramish, *Atomic Energy*, 109–10; G. A. Tokaty, "Soviet Rocket Technology," *Technology and Culture* 4 (1963), 525ff.)

260. Calculations based on 1926–7 prices show that during the war heavy industrial output grew by about an eighth, whereas consumer industry shrank by more than two-fifths. In 1945–50 heavy industrial output grew by 83 percent, while light industrial production climbed by 108 percent. Overall, the 1950 figure for heavy industry was 105 percent above the 1940 level; the figure for consumer industry was only 23 percent above 1940. (Calculated from Nove, *An Economic History of the USSR*, 291; see Nove's caveat about the effects of using indices based on 1926–7 prices.) In 1945–50, 87.9 percent of industrial investments went to the producer goods branches, while 12.1 percent went to the light and food industries. (Ibid., 290.)

261. *Bol'shevik*, Nos. 17–18 (1945), 43, and No. 9 (1946), 45; *Planovoe khoziaistvo*, No. 4 (1946), 30, and No. 1 (1947), 33.

262. *Istoriia Vtoroi mirovoi voiny, 1939–1945* (Moscow, 1973–80), VIII, 358, and X, 409; *Bol'shevik*, Nos. 3–4 (1945), 31, and Nos. 17–18 (1945), 43.

263. *Sovetskie Vooruzhennye sily: istoriia stroitel'stva*, 373, 389. See also Radunskaia, *Aksel' Berg*, 227.

264. *Bol'shevik*, No. 6 (1946), 73, 77. According to the memoirs of the then Minister of Finance, the worsening international situation meant that during the writing of the new five-year plan "defense expenditure was not reduced to the extent that we had calculated. Besides, the rapid progress of military technology required significant resources." (A. G. Zverev, *Zapiski ministra* (Moscow, 1973), 227.) I am grateful to David Holloway for bringing this passage to my attention.

265. *Bol'shevik*, No. 9 (1946), 41.

266. *Planovoe khoziaistvo*, No. 5 (1945), 35. (Signed for printing on 20 October 1945).

267. Holloway, "Entering," 40–46.

268. Iakovlev, *Tsel' zhizni*, 444–5; *Sovetskaia voennaia entsiklopediia* (Moscow, 1976), I, 52; *Voenno-istoricheskii zhurnal*, No. 4 (1973), 89.

269. Holloway, "Battle Tanks and ICBMs," 391.

270. The head was A. A. Blagonravov. (*Sovetskaia voennaia entsiklopediia*, I, 130–131.)

271. Lobanov, *Nachalo*, 260–261. In September 1945, when the State Defense Committee was abolished, the Council on Radar was attached to the Council of People's Commissars.

272. *VAN*, No. 9 (1944), 102, and No. 1 (1946), 87; *Materialy k istorii*, 321.

273. Medvedev, *Soviet Science*, 46, 51n, 53; Iakovlev, *Tsel' zhizni*, 461–2; Lobanov, *Nachalo*, 261–2; Fedoseev, *Zapadnia*, 113–14.

274. This decision, which probably revived the issue of how to utilize the ideas of captured German rocket specialists, may have been linked to publication of Stalin's letter condemning the overestimation of German military theory. (*Bol'shevik*, No. 3 (1947), 4–8.) The letter, dated February 1946, when the Politburo was winding up a major military review, was published in March 1947, the month the decision on rockets was made.

275. Tokaty, "Soviet Rocket Technology," 524–5 and passim; Holloway, "Battle Tanks and ICBMs," 391.

276. *Aviatsiia i kosmonavtika*, No. 10 (1968), 35; Ponomarev, *Sovetskie aviatsionnye konstruktory*, 63, 101–2.

277. *VAN*, Nos. 5–6 (1946), 29.

278. *VAN*, No. 2 (1946), 103.

279. Arnold Kramish, *Atomic Energy in the Soviet Union* (Stanford, 1959), 86.

280. Quoted from George Ginsburgs and Armino Rusis, "Soviet Criminal Law and the Protection of State Secrets," *Law in Eastern Europe* 7 (1963), 27–8. New categories of economic information were added to the classified list as well. (*Sovetskoe gosudarstvo i pravo*, No. 8 (1947), 18.)

281. *Sovetskoe gosudarstvo i pravo*, No. 8 (1947), 15, 17. Under the new decrees it was no longer necessary to prove the accused person's intent to reveal the information. Transmission of information for purposes of espionage remained punishable under separate statutes.

282. *Bol'shevik*, No. 9 (1946), 46; *Planovoe khoziaistvo*, No. 1 (1947), 33.

283. *Istoriia Velikoi Otechestvennoi voiny*, V, 526.

284. *VAN*, No. 4 (1946), 10.

285. *Pravda*, 12 August 1946, 1.

286. *Stal'*, Nos. 11–12 (1946), 625–6, and No. 11 (1948), 964–5.

287. *Pravda*, 2 April 1949, 1; *Izvestiia*, 13 April 1949, 1; *Ekonomicheskaia zhizn' SSSR*, I, 423–5.

288. *VAN*, No. 7 (1949), 54.

289. Ibid., 52–3.

290. The Council of Ministers had previously ordered the Academy to make changes in its research plans for 1946–50 and 1947, but I know of no evidence that these changes were prompted by military considerations. (*VAN*, No. 7 (1947), 5–7.)

291. *VAN*, No. 7 (1949), 50, 54. In 1946 the Presidium had established a commission to revise the Academy Charter, but nothing had come of it. (*Materialy k istorii*, 336.)

292. *Izvestiia*, 15 April 1949, 2; *VAN*, No. 4 (1949), 78–9. The precise date of Bruevich's removal is not clear. One source gives it as 17 March 1949. But the source gives the same date for A. V. Topchiev's appointment as Chief Scholarly Secretary, a position formally established by the Council of Ministers only on 7 April. (*Akademiia nauk SSSR: personal'nyi sostav* (Moscow, 1975), II, 348; *Ustavy Akademii nauk SSSR*, 196.)

Bruevich had been Academician-Secretary of the Academy since May 1942. In both 1945 and 1947 he was publicly identified as head of the Presidium's Department of Special Projects and its Secret Department, which presumably administered military-related projects within the Academy. (*VAN*, No. 9 (1945), 82–3, and No. 8 (1947), 125. Cf. *VAN*, No. 7 (1946), 81.)

293. *VAN*, No. 4 (1949), 79. Iu. A. Zhdanov should not be confused with his father, A. A. Zhdanov, who died in mid-1948.

294. A. V. Topchiev in *VAN*, No. 3 (1950), 32.

295. Ibid., 30; ibid., No. 8 (1949), 58–9.

296. *Resheniia partii i pravitel'stva*, III, 564.

297. Cf. the more sanguine picture of the development of such equipment given by the Bureau of the Academy's Department of Technological Sciences: *Izvestiia Akademii nauk SSSR: Otdelenie tekhnicheskikh nauk*, No. 6 (1949), 960. For the Presidium membership in 1949, see *Soviet Science 1917–1970*, part I,

54–5. For the members' specialties, see *Akademiia nauk SSSR: personal'nyi sostav*, II, and the entries in the *Bol'shaia Sovetskaia Entsiklopediia*, 3rd ed.

298. *Resheniia partii i pravitel'stva*, III, 565–9. This resolution has not been published in its entirety.

299. *VAN*, No. 6 (1949), 19–22.

300. *Izvestiia*, 13 April 1949, 1.

301. *Izvestiia*, 22 June 1949, 3, 17 August 1949, 1, and 27 November 1949, 2.

302. *Izvestiia*, 24 January 1950, 2.

303. *Pravda*, 17 September 1951, 1, 6 October 1951, 1, and 3 October 1952, 2; *VAN*, No. 2 (1950), 25.

304. *Bol'shevik*, No. 21 (1951), 28–41.

305. Ibid., No. 3 (1952), 8; ibid., No. 4 (1952), 21.

306. See in particular ibid., No. 12 (1952), 3.

307. Ibid., No. 4 (1952), 21, refers explicitly to Stalin's 1938 speech stressing "young forces" as a means of combating scientific stagnation.

308. For a discussion of this theme in relation to the development of Soviet cybernetics, see David Holloway, "Innovation in Science—The Case of Cybernetics in the Soviet Union," *Science Studies* 4 (1974), 299–337.

309. For a list of organized technological debates supposedly inspired by the call for more discussion in science, see *Izvestiia Akademii nauk SSSR: Otdelenie tekhnicheskikh nauk*, No. 4 (1951), 490–96.

310. For an especially vivid example of its antiscientific uses, see *Bol'shevik*, No. 21 (1951), 42 and passim.

Chapter 4

1. Wolfgang Leonhard, *The Kremlin Since Stalin*, pbk. ed. (New York, 1962); Herbert S. Dinerstein, *War and the Soviet Union*, rev. pbk. ed. (New York, 1962).

2. The USSR was not at war in Korea, but it was supplying materiel to Communist China and North Korea, which were. In April 1953, after Stalin's death but before the signature of the armistice, an article in *Kommunist* remarked defensively that the USSR believed in peaceful coexistence not on the fallacious grounds that the capitalists would voluntarily give up their hostile designs, but because international coexistence and cooperation had been made possible by a change of the worldwide correlation of forces in favor of socialism. (As shown below, the impact of the USSR's growing power on capitalist intentions was a central point of argument among the Soviet leaders at this time.) A few pages later, the article added: "It would be incorrect, however, to close one's eyes to the fact that even now there are not only overt but also covert opponents of the peaceful regulation of the Korean question. Precisely because peace in

Korea does not suit them, they are making efforts in all sorts of ways to complicate the Korean question, in order to drag out the war in Korea." Strikingly, the passage did not identify these hidden opponents as Westerners and instead placed this remark closest to its discussion of the Chinese Communist, North Korean, and Soviet positions on the war. The statement may therefore have been directed at Soviet politicians who opposed steps to terminate the conflict. (*Kommunist*, No. 7 (1953), 29.)

3. *Kommunist*, No. 12 (1953), 23, 30.

4. Ibid., 29–31, 34.

5. *Kommunist* (Erevan), 12 March 1954, cited by Dinerstein, *War and the Soviet Union*, 71–2; Donald Zagoria, *The Sino-Soviet Conflict, 1956–61*, pbk. ed. (New York, 1964), 156.

6. *Izvestiia*, 11 March 1954, quoted by Dinerstein, *War and the Soviet Union*, 73.

7. Dinerstein, *War and the Soviet Union*, 15, 73, 103, 141.

8. Thus, advocates of nontraditionalist views excised from the *Pravda* version of a speech by Khrushchev his quotation of an alleged statement from Lenin that "as long as there is capitalist encirclement any understanding [with the West] will be very difficult and complicated." (Leonhard, *The Kremlin*, 88–9, quoting the version of the speech as delivered in Czechoslovakia.)

9. *Kommunist*, No. 7 (1954), 87, 89. Cf. ibid., No. 4 (1955), 51.

10. Ibid., No. 1 (1954), 21–2. Cf. *VAN*, No. 3 (1954), 11–12.

11. On the substantial changes introduced into the Fifth Five-Year Plan during the fall of 1953, see Leonhard, *The Kremlin Since Stalin*, 84–5.

12. Dinerstein, *War and the Soviet Union*, 100.

13. *Kommunist*, No. 12 (1953), 14–15.

14. Thus, for instance, a *Kommunist* editorial condemned the "fetishization" of the laws of socialism, asserting that such objective laws change as historical conditions change, and should be understood in connection with the party's concrete economic and political tasks. The failure to draw such connections, said the editorial, does "enormous harm" and keeps cadres from understanding party policy. (*Kommunist*, No. 14 (1953), 58–61; see also ibid., 114.)

15. Ibid., No. 13 (1953), 77–8.

16. Dinerstein, *War and the Soviet Union*, 100.

17. Ibid., 139–41.

18. *Kommunist*, No. 2 (1955), 22.

19. *Izvestiia*, 11 March 1954, quoted in Dinerstein, *War and the Soviet Union*, 104.

20. *Pravda*, 24 January 1955, quoted in Dinerstein, *War and the Soviet Union*, 142.

21. The Soviet debate over nuclear deterrence is skillfully analyzed in Dinerstein, *War and the Soviet Union*. In 1954–5, the USSR was preparing to put its first two models of intercontinental bomber into regular production, and it was still developing its first ICBMs. (Asher Lee, *The Soviet Air Force* (London, 1961), 134–6.) We do not know that Malenkov opposed these programs, but we do know that he was more inclined than his critics to think that simple possession of the hydrogen bomb would curb the imperialists. (Cf. Dinerstein, 95, 143.)

22. *Kommunist*, No. 7 (1953), 39; ibid., No. 6 (1953), 20–21; ibid., No. 3 (1954), 120; ibid., No. 15 (1955), 5; *Voprosy ekonomiki*, No. 11 (1953), 78–9. To judge by a censorious aside in the third of these articles, "some economists" in the USSR thought expanded Western military spending might improve Western growth rates, but such views apparently failed to find their way into print.

23. *Pravda*, 20 January 1955, 1, in *C.D.S.P.* 7, No. 3 (1955–6), 32–3.

24. *Pravda*, 17 July 1955, 1–6, in *C.D.S.P.* 7, No. 28 (1955–6), 7.

25. According to one Soviet report, industrial labor productivity grew at 12 percent in 1950, 10 percent in 1951, 7 percent in 1952, 6 percent in 1953, and 7 percent in 1954. (*VAN*, No. 9 (1955), 8.)

26. *Kommunist*, No. 3 (1955), 26, 31–5; ibid., No. 10 (1955), 5. The Central Committee plenum that presumably removed Malenkov from the chairmanship of the Council of Ministers met from 25 to 31 January 1955; Malenkov tendered his formal resignation to the Supreme Soviet in early February. (Conquest, *Power and Policy*, 255.) The editorial against technologically complacent administrators who harmed the state appeared on 20 January. Shortly after the plenum, one economics journal editorialized against a large number of scholarly proponents of light industry. (*Voprosy ekonomiki*, No. 1 (1955), 15–25.) At an academic meeting that ended on the day before the plenum began, some economists were still trying to deflect the criticism equating priority for heavy industry with technological progress and growth. (*VAN*, No. 5 (1955), 93–4, 96; *Voprosy ekonomiki*, No. 5 (1955), 138, 147.)

27. *Kommunist*, No. 7 (1955), 6–7, 9; ibid., No. 10 (1955), 4–5; ibid., No. 11 (1955), 60.

28. Ibid., No. 5 (1954), 18–19, 28–9; Roger Pethybridge, *A Key to Soviet Politics: The Crisis of the Anti-Party Group* (New York, 1962), 49–53, 72–3.

29. Dinerstein, *War and the Soviet Union*, 113.

30. Pethybridge, *A Key*, 54–5.

31. Dinerstein, *War and the Soviet Union*, 144–5. Lincoln P. Bloomfield, Walter C. Clemens, Jr., and Franklyn Griffiths, *Khrushchev and the Arms Race: Soviet Interests in Arms Control and Disarmament, 1954–1964* (Cambridge, Mass., 1966), 51, indicates that Soviet military manpower was cut by 2 million in 1955–6. See also *Sovetskie Vooruzhennye Sily: istoriia stroitel'stva* (Moscow, 1978), 413.

32. Pethybridge, *A Key*, 89n; Bloomfield et al., *Khrushchev and the Arms Race*, 41–2, 51–3. The latter source's suggestion concerning the acceleration of the

missile program seems highly plausible, although the published Soviet data on R & D manpower used to document it are less than conclusive.

33. *XX s"ezd Kommunisticheskoi Partii Sovetskogo Soiuza: stenograficheskii otchet* (Moscow, 1956), I, 36–8.

34. Paul Marantz, "Peaceful Coexistence: From Heresy to Orthodoxy," in *The Dynamics of Soviet Politics*, ed. Paul Cocks et al. (Cambridge, Mass., 1976), 300–301.

35. *XX s"ezd*, II, 417; Dinerstein, *War and the Soviet Union*, 95.

36. *XX s"ezd*, I, 456–60; Dinerstein, *War and the Soviet Union*, 81.

37. *Kommunist*, No. 6 (1956), 32.

38. *Krasnaia zvezda*, 22 March 1957, quoted in Dinerstein, *War and the Soviet Union*, 83–4.

39. *XX s"ezd*, I, 322–4.

40. Ibid., 15.

41. *Kommunist*, No. 6 (1956), 22; *Mirovaia ekonomika i mezhdunarodnye otnosheniia*, No. 1 (1957), 158; hereafter cited as *MEMO*.

42. *Voprosy ekonomiki*, No. 12 (1955), 150–53. Despite formal approval from the Academy Presidium, none of the sources I have consulted shows that this institute was set up, and in February 1956 Mikoian identified the Institute of Economics as the only Academy institution working on economic questions. (*XX s"ezd*, I, 323–4.)

43. *VAN*, No. 6 (1956), 118. There is a hint that an earlier effort to set up a new Academy institute on international relations was made just before Stalin's death. See *VAN*, No. 3 (1953), 34–5.

44. William Zimmerman, *Soviet Perspectives on International Relations, 1956–1967* (Princeton, 1969), 34, 64; *MEMO*, No. 3 (1957), 7, 11.

45. *XX s"ezd*, II, 417; *Kommunist*, No. 6 (1956), 15, 25.

46. Among the new factors that Khrushchev expected would slow down Western growth were high arms spending, increasing international economic competition among the United States, Western Europe, and Japan, and a shrinking market for capital goods. (*XX s"ezd*, I, 15–16.) See also *Kommunist*, No. 1 (1956), 103–4, and ibid., No. 9 (1956), 29–32, 35–6.

47. Ibid., No. 4 (1956), 14–15, 31–3. This view represented a limited modification of the two-tiered prognosis for the West that Varga had offered in 1946.

48. *Kommunist*, No. 2 (1956), 8.

49. Pethybridge, *A Key*, 72–6, 81n, 132–3; Conquest, *Power and Policy*, 459.

50. Nancy Nimitz, "The Lean Years," *Problems of Communism* 14, No. 3 (May–June 1965), 12, 14; Pethybridge, *A Key*, 79–80; Carl A. Linden, *Khrushchev and the Soviet Leadership, 1957–1964*, pbk. ed. (Baltimore, 1966), 49–50. In view

of his past record, Malenkov probably joined in the advocacy of heavy industry for reasons of political expediency.

51. *Kommunist*, No. 10 (1957), 4–5.

52. Pethybridge, *A Key*, 99.

53. Marantz, "Peaceful Coexistence," 304–5.

54. Ibid.; *Kommunist*, No. 11 (1957), 11; Conquest, *Power and Policy*, 461; Pethybridge, *A Key*, 62.

55. Late in June 1957 the Central Committee voted to remove Molotov, Kaganovich, Malenkov, and Shepilov from the Presidium. After reciting the foreign policy issues on which Molotov had resisted a more moderate approach, it added that Kaganovich had sided with Molotov in "many" of these instances, and that Malenkov had gone along "in a number of cases." (Quoted in Conquest, *Power and Policy*, 461.) The wording suggested that of the three, Malenkov had taken the least assertive stance on foreign policy matters. See also Pethybridge, *A Key*, 105–6, 122; Linden, *Khrushchev*, 52–3.

56. *Pravda*, 7 November 1957, 2–6, in *C.D.S.P.* 9, No. 45 (1957–8), 6, 10–12, 18.

57. Ibid., 18; Dinerstein, *War and the Soviet Union*, 79.

58. L. Ilichev in *International Affairs* (Moscow), No. 3 (1958), 7. Shortly after this article appeared, Khrushchev elevated Ilichev to a key ideological post in the party secretariat. (Linden, *Khrushchev*, 66n; Michel Tatu, *Power in the Kremlin* (New York, 1969), 200–201.)

59. Major-General N. Talensky in *International Affairs* (Moscow), No. 3 (1958), 26–30, hinted more broadly at this possibility.

60. Ibid., No. 3 (1958), 11; emphasis in the original.

61. Marshall Zhukov was purged from the Presidium and the Ministry of Defense in November 1957.

62. Arnold Horelick and Myron Rush, *Strategic Power and Soviet Foreign Policy* (Chicago, 1965), 47.

63. Linden, *Khrushchev*, 68. However, no new steps to reduce the military were taken until 1960.

64. Linden, *Khrushchev*, 69.

65. *International Affairs* (Moscow), No. 1 (1958), 71–2; emphasis in the original.

66. Both Communist China and Yugoslavia come to mind as possible sources of this skeptical view. However, the Chinese Communists argued at the time that Sputnik marked a shift in the world correlation of forces which was even larger than that claimed by the Soviets, and that the deterrent power of Soviet nuclear weapons was so great that the USSR could foment local revolution without fear of a nuclear war with the West. (Zagoria, *The Sino-Soviet Conflict*, 152–71.) The article's foreign policy aim indicates that it was not directed at

the Yugoslavs, who by Communist standards viewed the intentions of the West in benign terms fully compatible with Khrushchev's desire to put more emphasis on peaceful coexistence.

67. Linden, *Khrushchev*, 50–52, 69. For Mikoian's support for Khrushchev, see *Kommunist* (Erevan), 12 March 1958, 2.

68. *Kommunist* (Erevan), 12 March 1958, 2.

69. *Pravda*, 27 March 1958, quoted in Horelick and Rush, *Strategic Power and Soviet Foreign Policy*, 54.

70. Cf. *Kommunist*, No. 3 (1957), 6.

71. *Kommunist* (Erevan), 12 March 1958, 2. For Mikoian's application of this principle to dealings with West Germany, see Angela Stent, *From Embargo to Ostpolitik: The Political Economy of West German–Soviet Relations, 1955–1980* (New York, 1981), 82.

72. *Vneshniaia torgovlia*, No. 4 (1958), 2–3. The Western review of the strategic embargo ran from late March until mid-July and produced a substantial relaxation of export controls. (Adler-Karlsson, *Western Economic Warfare*, 97.)

73. *Pravda*, 19 May 1958, 1–5.

74. *Keesing's Contemporary Archives*, September 6–13, 1958, p. 16385. As possible Soviet exports, Khrushchev mentioned manganese and chromium ore, ferrous metals, and platinum.

75. *MEMO*, No. 9 (1958), 3–5.

76. *Kommunist*, No. 17 (1958), 36; A. Arzumanian in ibid., No. 2 (1959), 35; ibid., No. 13 (1959), 87–8, 93–5.

77. *Kommunist* (Erevan), 12 March 1958, 3.

78. *Pravda*, 19 May 1958, 1–5; Linden, *Khrushchev*, 69–70.

79. I. Lemin in *MEMO*, No. 5 (1958), 21–4, 33; signed for printing on 15 May 1958. It is worth adding that while Soviet journals such as *MEMO* often show a distinct tendency in the political slant of their articles, exceptions occur in all journals. In this case, the article was at odds with the tenor of most others printed by *MEMO*.

80. *Vneocherednoi XXI s"ezd Kommunisticheskoi Partii Sovetskogo Soiuza: stenograficheskii otchet* (Moscow, 1959), I, 19, 72–3, 106–7.

81. Ibid., 17–18, 64–5, 72–3.

82. Ibid., 82–4.

83. Ibid., 553–4, 556. In November 1958 the USSR had established a six-month deadline for the satisfaction of its demands for a drastic change in the political status of Berlin; it retreated from this deadline in March 1959. Mikoian's reference to the Far East concerned the confrontation of Communist China with Nationalist China and the United States over the islands of Quemoy and Matsu. For other

statements by Mikoian favoring wider East-West economic ties, see *MEMO*, No. 4 (1960), 55; ibid., No. 6 (1960), 7.

84. *Vneocherednoi XXI s"ezd*, I, 155–6.

85. Ibid., 494–8; *Plenum Tsentral'nogo Komiteta Kommunisticheskoi Partii Sovetkogo Soiuza, 24–29 iiunia 1959 g.* (Moscow, 1959), 101–21, 129–31, 391–402. These divergent priorities were reflected in the post-Congress writings of planners and economists. See *MEMO*, No. 6 (1959), 10; *Kommunist*, No. 5 (1960), 61–2; *Voprosy ekonomiki*, No. 2 (1961), 151; Naum Jasny, "Plan and Superplan," *Survey*, No. 38 (October 1961), 29–43.

86. *Vneocherednoi XXI s"ezd*, I, 501–2.

87. Interestingly, Suslov signaled that the main Congress report represented only Khrushchev's personal opinions, not the views of the leadership as a whole. (Linden, *Khrushchev*, 88–9.) It is not clear, however, whether Suslov took this position because of differences over economic policy.

88. *XXII s"ezd Kommunisticheskoi Partii Sovetskogo Soiuza: stenograficheskii otchet* (Moscow, 1961), III, 229–335.

89. This was the Scientific Council on the Complex Problem of the Economic Competition of the Two Systems and the Underdeveloped Countries. Chartered in February 1962, it was attached to the Academy Presidium and chaired by A. Arzumanian, the head of IMEMO. The Council's general research plans required the approval of Gosplan, the State Committee on Foreign Economic Relations, the Ministry of Foreign Trade, the Ministry of Foreign Affairs, and others. Arzumanian's description emphasized both the Council's relation to government policy and its freedom to initiate new studies when "life" demanded them. (*VAN*, No. 8 (1962), 14–22.)

90. See, for example, *Kommunist*, No. 1 (1960), 82, 85; *MEMO*, No. 2 (1960), 152. A reviewer of one Soviet East-West comparison called for a change to more productive sectors like chemistry (*MEMO*, No. 4 (1963), 151–5). Khrushchev adduced this argument in his December 1963 plea for chemicalization, and it was reiterated by A. Arzumanian in *Pravda*, 24 and 25 February 1964, 2–3.

91. *MEMO*, No. 11 (1962), 66–71; ibid., No. 7 (1964), 37–8; ibid., No. 9 (1964), 138–9; *VAN*, No. 3 (1964), 65–6.

92. The regime had been engaged in a dialogue with the West on disarmament questions since 1954. For an analysis of the Soviet proposals, see Bloomfield et al., *Khrushchev and the Arms Race*.

93. *Pravda*, 15 January 1960, 1–5, in *C.D.S.P.* 12, No. 2 (1960–61), 7–8, 11–12.

94. Ibid., 6.

95. Ibid., 10, 14.

96. Ibid., 14. Around the time of the speech, some Soviet commentators stressed that arms control would mean that "first-class scientific laboratories presently

occupied in creating and improving means of destruction could be shifted to . . . putting out new, improved models of [nonmilitary] production." These benefits were discussed in a way which suggested that they would accrue not only to the West, but to the USSR as well. (*MEMO*, No. 11 (1959), 35; ibid., No. 1 (1960), 5–6; *Kommunist*, No. 13 (1959), 89.)

97. Leonhard, *The Kremlin Since Stalin*, 370–71.

98. *Pravda*, 16 January 1960, 9, emphasis in the original; ibid., 2 February 1960, 3; Tatu, *Power*, 79–80.

99. Tatu, *Power*, 70, 80–84.

100. *XXII s"ezd*, II, 186–7.

101. During the U-2 crisis, Khrushchev declared that the affair "must not compel us to revise our plans by increasing appropriations for weapons and for the army, must not compel us to halt the process of reducing the army." (*Pravda*, 8 May 1960, quoted in Horelick and Rush, *Strategic Power*, 71.) For the impact of the incident on Khrushchev's position, see Tatu, *Power*, and Linden, *Khrushchev*.

102. Thomas W. Wolfe, *Soviet Strategy at the Crossroads* (Cambridge, Mass., 1964), 141.

103. Linden, *Khrushchev*, 114–15, 187–91.

104. *Keesing's Contemporary Archives*, June 6–13, 1959, pp. 16851–2; "Credit is King in Red Chemical Plant Deals," *Chemical Week* 94, No. 12 (21 March 1964), 27–8.

105. Harry B. Ellis, "Moscow Timing: Bonn Takes Notice," *Christian Science Monitor*, 5 September 1964, 1. In 1962 the United States and West Germany also embargoed exports of large-diameter pipe to the Soviet oil industry. For an analysis of the pipe embargo and of Western wheat sales to the USSR in this same period, see Stent, *From Embargo to Ostpolitik*, 93–126.

106. Sheldon Rabin, "Soviet-Owned Banks in Europe," Ph.D. dissertation, The Johns Hopkins School of Advanced International Studies, Washington, D.C., 1977, 230–31.

107. One of the opponents was Presidium member V. V. Grishin. On 25 October 1963 Khrushchev remarked that the harvest had been inadequate, forcing the USSR to buy grain from Canada and other countries; negotiations with the United States were still under way. Three days later Grishin said the harvest was "fully adequate to meet the population's needs." The next month, following Khrushchev's criticism of unnamed persons who were willing to countenance starvation rather than import grain, *Kommunist* stated that foreign purchases were necessary because the domestic grain harvest "at present is inadequate." (Linden, *Khrushchev*, 188; *Pravda*, 27 October 1963, 1–3; *Trud*, 29 October 1963, 1–3; *Kommunist*, No. 18 (1963), 5, 11.)

108. *Plenum Tsentral'nogo Komiteta Kommunisticheskoi Partii Sovetskogo Soiuza, 9–13 dekabria 1963 goda* (Moscow, 1964), 73.

109. Ibid., 73–4, 395.

110. Linden, *Khrushchev*, 190–96.

111. Thomas W. Wolfe, *Soviet Power and Europe, 1945–1970*, pbk. ed. (Baltimore, 1970), 121–2.

112. "Germany's Erhard Talks about Trade, Tariffs, the U.S. Dollar," *U.S. News and World Report* 56, No. 23 (8 June 1964), 63–4. Cf. Stent, *From Embargo to Ostpolitik*, 149.

113. In April the East German press put the onus for the continuing Sino-Soviet polemics on the CPSU, and the Soviet press published an esoteric counterattack on the East German party for undermining the CPSU's international authority. (Carola Stern, "East Germany," in *Communism in Europe: Continuity, Change, and the Sino-Soviet Dispute*, ed. William Griffith (Cambridge, Mass., 1964–6), II, 130n; *Kommunist*, No. 6 (1964), 14–15.)

114. Wolfe, *Soviet Power*, 122.

115. Ibid., 123.

116. Arthur J. Olsen, "Khrushchev States Readiness to Visit Bonn on Invitation," *New York Times*, 4 September 1964, 1, 6; Wolfe, *Soviet Power*, 123; Stent, *From Embargo to Ostpolitik*, 118–19.

117. Wolfe, *Soviet Power*, 125; *Pravda*, 25 September 1964, 1–2.

118. *Pravda*, 2 October 1964, 1.

119. *Izvestiia*, 4 October 1964, 2. *Pravda*, over which Khrushchev presumably had less control, printed a more typical, favorable account of East German technical achievements. (*Pravda*, 3 October 1964, 4.)

120. The October 2 account of the meeting said that it had occurred "in recent days" (na dniakh).

121. Among other things, the leadership could not agree on the proper length for the planning period. (Tatu, *Power*, 391–3.)

122. *Pravda*, 6 October 1964, 2.

123. *Pravda*, 7 October 1964, 3.

124. *Pravda*, 17 July 1955, 1–6, in *C.D.S.P.* 7, No. 28 (1955–6), 8.

125. *Pravda*, 8 July 1955, 1, in *C.D.S.P.* 7, No. 27 (1955–6), 30.

126. *Pravda*, 22 February 1956, 1–6, in *C.D.S.P.* 8, No. 12 (1956–7), 19–20.

127. *Pravda*, 20 January 1955, 1, in *C.D.S.P.* 7, No. 3 (1955–6), 32–3; see also *Pravda*, 8 July 1955, 1, in *C.D.S.P.* 7, No. 27 (1955–6), 30.

128. *Pravda*, 17 July 1955, 1–6, in *C.D.S.P.* 7, No. 28 (1955–6), 8; *Kommunist*, No. 13 (1955), 112; ibid., No. 15 (1955), 101–3.

129. George Ginsburgs and Armino Rusis, "Soviet Criminal Law and the Protection of State Secrets," *Law in Eastern Europe* 7 (1963), 27–8.

130. *XX s"ezd*, I, 116; *Kommunist*, No. 13 (1957), 81.

131. *VAN*, No. 3 (1954), 102; ibid., No. 11 (1954), 88–9.

132. *Kommunist*, No. 13 (1956), 82–3; see also ibid., No. 15 (1958), 42–3.

133. *VAN*, No. 6 (1956), 48–9; ibid., No. 9 (1956), 86.

134. *Kommunist*, No. 13 (1956), 82–3.

135. *Sovetskaia pechat'*, No. 9 (1958), 29, in *C.D.S.P.* 10, No. 39 (1958–9), 25–6.

136. *Znamia*, No. 10 (1960), 127–38, in *C.D.S.P.* 12, No. 47 (1960–61), 8–10.

137. *Pravda*, 21 December 1957, 6, in *C.D.S.P.* 9, No. 51 (1957–8), 18–20.

138. *Ekonomicheskaia gazeta*, 24 August 1960, 3–4, in *C.D.S.P.* 12, No. 36 (1960–61), 25.

139. John Lewis Gaddis, *Russia, the Soviet Union and the United States: An Interpretive History* (New York, 1978), 229.

140. *Kommunist*, No. 11 (1960), especially 40–48; see also 110–11. In 1963 the KGB set up a new directorate to steal Western technology. (John Barron, *KGB: The Secret Work of Soviet Secret Agents*, pbk. ed. (New York, 1974), 107.) One motive for this step may have been to prove that espionage was an effective alternative to aboveboard contacts as a means of acquiring Western know-how.

141. Soviet statements identified Penkovskii as a civilian employee of the Committee and a colonel in the military reserve. An American book published with the cooperation of the CIA states that he was a regular colonel of Soviet military intelligence on assignment to the State Committee. (Frank Gibney, "General Introduction," in Oleg Penkovskiy, *The Penkovskiy Papers* (New York, 1965), 6–7. See also Thomas Powers, *The Man Who Kept the Secrets: Richard Helms and the CIA*, pbk. ed. (New York, 1981), 127, 362.)

142. *Pravda*, 17 May 1963, 1.

143. *Izvestiia*, 30 May 1963, 4, in *C.D.S.P.* 15, No. 20 (1963–4), 11.

144. Tatu, *Power*, and Linden, *Khrushchev*. By May 1963 Khrushchev had recovered from the political damage he suffered from the Cuban Missile Crisis—thanks partly to a heart attack suffered by Kozlov, one of his main conservative opponents, in April. (Tatu, *Power*, 341–3.)

145. This may be one reason the KGB lost some prerogatives to the Ministry of Internal Affairs in April 1963. (Tatu, *Power*, 327.) Two noteworthy elements of the *Izvestiia* interview were the effort of the military prosecutor to establish that the KGB had hindered Penkovskii's espionage well before arresting him, and the prosecutor's exasperated attempt to explain to the persistent *Izvestiia* correspondents why Penkovskii had not been immediately arrested when he came under suspicion. (*Izvestiia*, 30 May 1963, 4, in *C.D.S.P.* 15, No. 20 (1963–4), 10.)

146. In 1955–65 total imports from CEMA and the West grew by about 250 and 350 percent respectively. (Leon Herman, "The Promise of Self-Sufficiency

under Soviet Socialism," in *The Development of the Soviet Economy: Plan and Performance*, ed. Vladimir G. Treml (New York, 1968), 236–7.)

147. According to one estimate, the ratio between imports of Western equipment and domestic equipment investment in the following year was 1.0 percent in 1955; 0.8 percent in 1958; 1.8 percent in 1962; and 1.6 percent in 1964. A different calculation of relative prices might conceivably double the percentage for each year, but would not affect the direction of the trend. Between 1961 and 1966 the Western share of equipment investment in the chemical industry was about 30 percent. Another important sector was shipbuilding, where the Western share was about 10 percent. (Philip Hanson, "The Import of Western Technology," in *The Soviet Union Since the Fall of Khrushchev*, eds. Archie Brown and Michael Kaser (London, 1975), 19, 31–6, 38–40.)

148. *VAN*, No. 8 (1955), 72, and No. 3 (1964), 83; Tony Longrigg, "Soviet Science and Foreign Policy," *Survey* 17, No. 4 (Autumn 1971), 46.

149. Bruce Parrott, *Information Transfer in Soviet Science and Engineering: A Study of Documentary Channels*, Rand Report R–2667–ARPA (Santa Monica, 1981), Section IV.

150. Longrigg, "Soviet Science," 47.

151. Lloyd Jordan, "Scientific and Technical Relations among East European Communist Countries," *Minerva* 8 (1970), 376–95; *Kommunist*, No. 13 (1959), 81.

152. Jordan, "Scientific and Technical Relations," 379; *VAN*, No. 12 (1958), 56, No. 8 (1959), 47–8, and No. 3 (1960), 97–8.

153. On travel by scientists, see Longrigg, "Soviet Science," 42–3, who discusses the uncertainties of these data and the possible causes of their short-term fluctuations. Soviet sources also give the number of "specialists" from Communist countries who visited the USSR. This figure rose from 1,645 for all of 1948–52 to 3,150 in 1961. (Jordan, "Scientific and Technical Relations," 377.)

154. Parrott, *Information Transfer in Soviet Science and Engineering*, Section IV.

155. Longrigg, "Soviet Science," 43.

156. On the security barriers to scientific correspondence, see Zhores Medvedev, *The Medvedev Papers* (London, 1970). In 1962 and 1964 the USSR Council of Ministers adopted two resolutions on scientific and technical information. While both described domestic information networks very negatively, they did not refer to the problem of secrecy. (*Resheniia partii i pravitel'stva*, V, 75–8 and 500–504.) But see Leonid Vladimirov, *The Russian Space Bluff* (London, 1971), 178–85.

157. Calculated from Nolting, *Sources of Financing*, 10.

158. In mid-1961 A. N. Kosygin, the head of Gosplan, addressed a nationwide meeting of scientific researchers. After detailing the problems of Soviet R & D, he remarked that the authorities often were told that science allocations were insufficient. But increasing these allocations meant cutting other parts of the

budget, which was permissible only if the research yielded economic results quickly. Future increases in research spending, he added pointedly, were thus in the hands of the scholars themselves; it depended on how effectively they used the resources they had already received. (*VAN*, No. 7 (1961), 103–4.)

159. Nicholas DeWitt, "The Polity of Russian and Soviet Science: A Century of Continuity and Change," in *The Social Reality of Scientific Myth: Science and Social Change*, ed. Kalman Silvert (New York, 1969), 174–5; *Bol'shevik*, No. 17 (1951), 29–30. No doubt the advocates of such ideas were encouraged to speak out by Stalin's statement in 1950 that science did not belong to the social superstructure—although he suggested that it did not belong to the economic base either. See Loren R. Graham, "Reorganization of the USSR Academy of Sciences," in *Soviet Policy-Making*, eds. Peter H. Juviler and Henry W. Morton (New York, 1967), 152.

160. *Kommunist*, No. 6 (1954), 60–61.

161. Ibid., No. 18 (1954), 54.

162. *Pravda*, 31 December 1955, 2, in *C.D.S.P.* 7, No. 52 (1955–6), 20–22, and *Kommunist*, No. 2 (1956), 34–5.

163. *Pravda*, 6 February 1956, 1.

164. *XX s"ezd*, I, 378.

165. *Kommunist*, No. 13 (1957), 70–84, in *C.D.S.P.* 9, No. 41 (1957–8), 3.

166. Graham, "Reorganization," 143.

167. *Izvestiia*, 6 September 1959, quoted in ibid., 144.

168. Quoted in Graham, "Reorganization," 153.

169. *VAN*, No. 11 (1953), 4–7. There are no quantitative measures of the shift that occurred, but the Academy President later remarked that the Academy had "substantially widened" its attention to the needs of these sectors. (Ibid., No. 3 (1954), 20; see also ibid., No. 8 (1954), 39–40.)

170. Nesmeianov and Topchiev in *Kommunist*, No. 8 (1953), 49; *VAN*, No. 5 (1954), 24.

171. *VAN*, No. 11 (1956), 4–5.

172. *VAN*, No. 3 (1955), 12–18, in *C.D.S.P.* 7, No. 25 (1955–6), 9.

173. *VAN*, No. 3 (1955), 19–38, in *C.D.S.P.* 7, No. 29 (1955–6), 9–10.

174. *VAN*, No. 6 (1956), 11.

175. *VAN*, No. 12 (1954), 67–8, No. 1 (1956), 39–40, No. 6 (1956), 28–30, 49, and No. 9 (1956), 86; *Kommunist*, No. 2 (1956), 40.

176. See, for example, *VAN*, No. 6 (1956), 46–9. The strength of this sentiment showed up during 1956 in the election of Academy leaders, a process usually characterized by outward unanimity. Seven of eight divisions supported the nomination of Nesmeianov for a new term as president. But the Division of

Physical-Mathematical Sciences recommended that the election be postponed for several months, until Nesmeianov could report on actual progress in democratizing the Academy, renewing scholarly collegiality, adopting a correct science policy, and eliminating other shortcomings. This suggestion was turned aside by a majority vote of the members. (*VAN*, No. 11 (1956), 7–9.)

177. *VAN*, No. 6 (1956), 11; *VAN*, No. 3 (1955), 12–18, in *C.D.S.P.* 7, No. 25 (1955–6), 9–10; *VAN*, No. 3 (1955), 19–38, in *C.D.S.P.* 7, No. 29 (1955–6), 6.

178. For the Central Committee resolution on Gostekhnika, see *Resheniia partii i pravitel'stva*, IV, 223. There were some discrepancies in the public descriptions of Gostekhnika's powers. Cf. *Pravda*, 19 May 1955, 1–2, in *C.D.S.P.* 7, No. 20 (1955–6), 4, and *Pravda*, 2 July 1955, 1, in *C.D.S.P.* 7, No. 25 (1955–6), 7.

179. *Pravda*, 31 December 1955, 2, in *C.D.S.P.* 7, No. 52 (1955–6), 20–22.

180. *Pravda*, 15 February 1956, 1–11, in *C.D.S.P.* 8, No. 5 (1956–7), 12; *Pravda*, 22 February 1956, 1–6, in *C.D.S.P.* 8, No. 12 (1956–7), 20.

181. In December 1956 Nesmeianov remarked that the Academy had prepared a series of memoranda on important future scientific and technological problems, but that because of "external circumstances" these memos had not been reviewed by the "directive organs" and had not been made binding on the ministries. (*VAN*, No. 2 (1957), 40.) It is reasonable to assume that Gostekhnika was one of the directive organs in question. Moreover, in 1956 the Academy Presidium set up a commission to revise the Academy Charter. The instructions to the commission put considerable emphasis on spelling out the Academy's nation-wide authority over research planning. The final version drawn up two years later, however, gave the Academy much less authority than the Presidium had earlier seemed to claim. (Ibid., No. 11 (1956), 105; *Ustavy Akademii nauk SSSR, 1724–1974* (Moscow, 1974), 151–2.)

182. *Pravda*, 19 February 1956, 3–4, in *C.D.S.P.* 8, No. 9 (1956–7), 24.

183. *VAN*, No. 3 (1955), 12–18, in *C.D.S.P.* 7, No. 25 (1955–6), 9.

184. *Pravda*, 19 February 1956, 3–4, in *C.D.S.P.* 8, No. 9 (1956–7), 22–3.

185. *VAN*, No. 2 (1957), 40. This is a shortened text of Nesmeianov's address to the December 1956 General Meeting.

186. *Pravda*, 23 February 1956, 5–6, in *C.D.S.P.* 8, No. 13 (1956–7), 15.

187. Calculated from *VAN*, No. 3 (1956), 8–9. The statement is based on combined totals for full and corresponding members.

188. Ibid.; *VAN*, No. 11 (1956), 6.

189. *VAN*, No. 3 (1956), 13–14.

190. Ibid., No. 6 (1956), 19, 23, and No. 2 (1957), 13, 17–18.

191. Ibid., No. 5 (1956), 4, and No. 6 (1956), 42–3, 48.

192. Ibid., 35–7.

193. Ibid., No. 3 (1957), 14. The speaker, B. N. Stechkin, was a cofounder of the Central Aero- and Hydrodynamic Institute (TsAGI) and an organizer of the Zhukovskii Air Engineering Academy. In 1954–62 he headed the Academy's Laboratory of Engines (renamed the Institute of Engines in 1961). (*Sovetskaia Voennaia entsiklopediia*, VII (Moscow, 1979), 541–2.)

194. *VAN*, No. 9 (1956), 85–6; *Pravda*, 24 February 1957, in *C.D.S.P.* 9, No. 8 (1957–8), 50; Nicholas DeWitt, "Reorganization of Science and Research in the U.S.S.R.," *Science* 133 (1961), p. 1985; *VAN*, No. 4 (1959), 7–8.

195. *VAN*, No. 5 (1958), 9–28, in *C.D.S.P.* 10, No. 27 (1958–9), 9–10. See also *VAN*, No. 4 (1959), 64–72, in *C.D.S.P.* 11, No. 25 (1959–60), 10.

196. Linda Lubrano Greenberg, "Policy-Making in the U.S.S.R. Academy of Sciences," *Journal of Contemporary History* 8, No. 4 (October 1973), 75–6.

197. *VAN*, No. 4 (1959), 70–71.

198. *Plenum Tsentral'nogo Komiteta Kommunisticheskoi Partii Sovetskogo Soiuza, 24–29 iiunia 1959 g.* (Moscow, 1959), 371, 465–6; Lubrano Greenberg, "Policy-Making in the U.S.S.R. Academy," 75.

199. *Izvestiia*, 9 August 1959, 3.

200. *Pravda*, 10 August 1959, 1.

201. This phase of the debate is well described by Graham, "Reorganization."

202. *Izvestiia*, 21 October 1959, 3.

203. *Izvestiia*, 28 August 1959, 3; *VAN*, No. 4 (1960), 93.

204. *Izvestiia*, 28 August 1959, 3. Bardin began this article by citing the speech in which Khrushchev had said that metallurgical research should be done outside the Academy; it was in this speech that Khrushchev mentioned Bardin as an example of the type of specialist who was really out of place as an Academy member.

205. See this chapter above, note 85.

206. See, for example, Zaleski et al., *Science Policy in the USSR*, 227.

207. *VAN*, No. 4 (1960), 97. The speaker was A. A. Blagonravov.

208. *Khrushchev Remembers: The Last Testament*, ed. Strobe Talbott (Boston, 1974), 48.

209. Ibid., 48–9. Presumably the panel submitted its conclusions before 7 January 1960, the date of Bardin's death. Thus the incident—either the commission's report, or both the report and news of the American decision—occurred in time to sharpen the infighting over the Academy reform.

210. *Izvestiia*, 12 June 1960, 3, in *C.D.S.P.* 12, No. 24 (1960–61), 14–15. For an example of the equivocation, see Nesmeianov's statement in *Izvestiia*, 25 February 1960, 2, in *C.D.S.P.* 12, No. 8 (1960–61), 30. Cf. *VAN*, No. 4 (1960), 68.

211. *Pravda*, 7 May 1957, 3, in *C.D.S.P.* 9, No. 18 (1957–8), 19.

212. *Pravda*, 11 May 1957, 1–2, in *C.D.S.P.* 9, No. 20 (1957–8), 6–7.

213. *Pravda*, 16 March 1959, 1, in *C.D.S.P.* 11, No. 20 (1959–60), 6–7; *VAN*, No. 4 (1959), 64–72.

214. *Pravda*, 29 January 1959, 2–3, in *C.D.S.P.* 11, No. 5 (1959–60), 22; *Pravda*, 30 January 1959, 3–4, in *C.D.S.P.* 11, Nos. 6–7 (1959–60), 6.

215. *Pravda*, 13 March 1959, cited in Graham, "Reorganization," 145.

216. *Pravda*, 12 April 1961, 1, in *C.D.S.P.* 13, No. 15 (1961–2), 13–14.

217. Calculated from Zaleski et al., *Science Policy in the USSR*, 202, 217.

218. *Pravda*, 12 April 1961, 1, in *C.D.S.P.* 13, No. 15 (1961–2), 13–14.

219. Ibid.

220. At the end of 1955 the Academy Presidium, having heard a report on the Geneva conference on the peaceful uses of atomic energy, said that it was necessary "substantially to increase the scale of the Academy's projects on the peaceful use of atomic energy and to improve the direction and coordination of them." (*VAN*, No. 12 (1955), 74.) Soon after, Nesmeianov remarked that "the time has arrived to put the scientific side of nuclear research under the supervision of the [Academy's] Division of Physical-Mathematical Sciences, to a much greater degree than has been the case up till now." (Ibid., No. 3 (1956), 13.) Although he did not indicate who currently controlled the research, Western sources generally agree that in 1953 authority over nuclear weapons research was vested in the new Ministry of Medium Machine Building. The post-Geneva emphasis on the peaceful uses of atomic energy, which implicitly raised the question of who should administer nuclear research, may have encouraged Nesmeianov to bid for a greater Academy role in the nuclear field. However, in April 1956 *Pravda* announced the creation of a Main Administration for the Utilization of Atomic Energy, which took over some of the Academy's responsibilities for work on power reactors and power stations. (George A. Modelski, *Atomic Energy in the Communist Bloc* (London, 1959), 9–13.)

221. Like some Western atomic scientists, Sakharov was concerned about the dangers of radioactive fallout from the tests. In 1958 he reportedly gained the support of Igor Kurchatov, the academician who headed the program of scientific research on atomic weapons, for his appeal to Khrushchev. (*Sakharov Speaks*, ed. Harrison E. Salisbury, pbk. ed. (New York, 1974), 10–12, 32–4.) See also *Khrushchev Remembers: The Last Testament*, 68–71.

222. *VAN*, No. 4 (1959), 48–50. The speaker was E. K. Fedorov, a geophysicist who was a corresponding member at the time and became a full member in 1960.

223. *VAN*, No. 12 (1959), 3, and No. 2 (1959), 51–2. Semenov, for example, told a Paris UNESCO meeting that atomic testing should be permanently terminated as a step toward disarmament. He added that scientists are not "creative machines" who exist only to do scientific research, but people concerned about

the humanistic goals of science. "It is painful to us when our scientific results are utilized to do evil to people. . . . We must feel our responsibility sharply and conduct ourselves in a corresponding manner."

224. *Khrushchev Remembers: The Last Testament,* 60–61.

225. *Plenum Tsentral'nogo Komiteta Kommunisticheskoi Partii Sovetskogo Soiuza 19–23 noiabria 1962 g.* (Moscow, 1963), 36; Zaleski et al., *Science Policy in the U.S.S.R.,* 227. In addition, some R & D financing seems to have been shifted from the "defense" to the "science" category of the budget in the late 1950s or early 1960s. (Holloway, "Innovation in the Defence Sector," 357.)

226. DeWitt, "Reorganization," p. 1988; Holloway, "Battle Tanks and ICBMs," 401–402.

227. This should not be construed to mean that Nesmeianov and Keldysh differed in principle on the general questions of the importance of basic research and the Academy's national role. On these matters Keldysh adhered to Nesmeianov's position. (*VAN,* No. 3 (1963), 10, and No. 11 (1963), 57–8.)

228. *Izvestiia,* 5 April 1962, 2.

229. *Planovoe khoziaistvo,* No. 10 (1962), 53.

230. *Biulleten' Gosudarstvennogo komiteta po koordinatsii nauchno-issledovatel'skikh rabot,* No. 7 (1963), 18.

231. The 1961 resolution called for a transfer of some applied-research institutes from the U.S.S.R. Academy, but not from the republican academies. In 1963, 50 applied-research institutes and 52 other applied-science institutions were transferred out of the republican academies. (*VAN,* No. 3 (1964), 20.)

232. Although one member of the State Committee was removed for his role in the affair, there reportedly was no purge of the organization. (Tatu, *Power,* 325–6.)

233. *Resheniia partii i pravitel'stva,* V, 306–7.

234. See, for example, *Pravda,* 23 October 1954, 1, in *C.D.S.P.* 6, No. 43 (1954–5), 26; *Pravda,* 20 March 1955, 2, in *C.D.S.P.* 7, No. 12 (1955–6), 3.

235. *Resheniia partii i pravitel'stva,* IV, 222. This resolution was adopted jointly with the Council of Ministers.

236. Ibid., 223.

237. Ibid., 227–8, 237.

238. "Report of the Central Committee," in *C.D.S.P.* 8, No. 4 (1956–7), 13, and No. 5 (1956), 14. See also *XX s"ezd,* II, 274.

239. *Kommunist,* No. 17 (1956), 3–4; Leonhard, *The Kremlin Since Stalin,* 233–5.

240. *Pravda,* 30 March 1957, 1–4, in *C.D.S.P.* 9, No. 13 (1957–8), 13. See also *Kommunist,* No. 8 (1957), 7.

241. *Pravda,* 30 March 1957, 1–4, in *C.D.S.P.* 9, No. 13 (1957–8), 7–9.

242. Leonhard, *The Kremlin Since Stalin*, 237.

243. *Pravda*, 30 March 1957, 1–4, in *C.D.S.P.* 9, No. 13 (1957–8), 7–8. In *Kommunist*, No. 15 (1957), 116, L. Leont'ev parried an article which leveled this charge shortly after the publication of Khrushchev's speech.

244. Leonhard, *The Kremlin Since Stalin*, 244–5.

245. *Pravda*, 11 May 1957, 1–2, in *C.D.S.P.* 9, No. 20 (1957–8), 15; Alec Nove, *The Soviet Economy*, 2nd rev. pbk. ed. (New York, 1968), 76.

246. *Pravda*, 11 May 1957, 3, in *C.D.S.P.* 9, No. 20 (1957–8), 13.

247. *Pravda*, 30 March 1957, 1–4, in *C.D.S.P.* 9, No. 13 (1957–8), 13.

248. By 1958 Gosplan directly controlled 323 industrial research institutes employing 19,000 researchers. (DeWitt, "Reorganization," p. 1985.) In 1956 there were 496 research institutions serving construction and industry—presumably both military-related and civilian. They employed 45,000 research workers. (*Pravda*, 30 March 1957, 1–4, in *C.D.S.P.* 9, No. 13 (1957–8), 13.)

249. *Pravda*, 11 May 1957, 1–2, in *C.D.S.P.* 9, No. 20 (1957–8), 15. It is worth noting that in 1957 the Council of Ministers was headed by Bulganin. In 1958 Bulganin was removed from this post, belatedly labeled a member of the anti-party group and expelled from the party Presidium.

250. Linden, *Khrushchev*, 52–3; *Kommunist*, No. 2 (1958), 49. The three ministries that were not abolished were for medium machine building (generally agreed in the West to be the manufacturer of atomic weapons), chemical industry, and power stations.

251. *Narodnoe khoziaistvo SSSR v 1959 godu* (Moscow, 1960), 135.

252. A substantial part of Moscow's remaining share of industry was administered through the three industrial ministries that still existed at the start of 1958. Conceivably the three may have accounted for all of Moscow's 6 percent share. On the other hand, the transformation of the Ministry of Chemical Industry into a state committee in mid-1958 did not produce a drop in Moscow's share in 1959, even though chemical and petrochemical output accounted for nearly 4 percent of total industrial production. (*Narodnoe khoziaistvo SSSR v 1972 g.* (Moscow, 1973), 166.) This suggests either that the State Committee for Chemistry kept control of its enterprises or that the series of figures on Moscow's share of industry is inaccurate.

253. *Ezhegodnik Bol'shoi Sovetskoi Entsiklopedii*, 1960, 9; ibid., 1961, 8; ibid., 1962, 13; *Sobranie postanovlenii i rasporiazhenii Soveta Ministrov Soiuza Sovetskikh Respublik*, 1963, 108–110.

254. *Zaria vostoka*, 25 November 1962, 2.

255. For an example of the criticism, see *Resheniia partii i pravitel'stva*, IV, 573.

256. Alexander G. Korol, *Soviet Research and Development: Its Organization, Personnel, and Funds* (Cambridge, Mass., 1964), 39.

257. *Pravda*, 26 April 1963, 1–7, in *C.D.S.P.* 15, No. 17 (1963–4), 5.

258. *Plenum Tsentral'nogo Komiteta Kommunisticheskoi Partii Sovetskogo Soiuza, 13–16 iiulia, 1960* (Moscow, 1960), 268.

259. *Kommunist,* No. 1 (1958), 50; *Pravda,* 30 January 1959, 3–4, in *C.D.S.P.* 11, Nos. 6–7 (1959–60), 6; *Plenum Tsentral'nogo Komiteta Kommunisticheskoi Partii Sovetskogo Soiuza, 24–29 iiunia, 1959 g.* (Moscow, 1959), 36–7.

260. Joseph Berliner, *The Innovation Decision in Soviet Industry* (Cambridge, Mass., 1976), 113.

261. This is an example of the ubiquitous Soviet practice of "overinsurance," in which an organization seeks minimum targets and maximum resources, in case the goals handed down from above should turn out to be unexpectedly demanding.

262. *Plenum . . . 13–16 iiulia, 1960 g.,* 43.

263. Ibid., 269.

264. Ibid.

265. *Kommunist,* No. 10 (1956), 75–6.

266. Berliner, *The Innovation Decision,* 277–89; A. N. Komin, *Problemy planovogo tsenoobrazovaniia* (Moscow, 1971), 150.

267. *Sobranie postanovlenii i rasporiazhenii Soveta Ministrov Soiuza Sovetskikh Sotsialisticheskikh Respublik,* 1957, 209–11; ibid., 1960, 334–45; ibid., 1964, 340–52; A. Basistov, "Effektivnost' novoi tekhniki i material'naia zainteresovannost' v tekhnicheskom progresse," in *Problemy ekonomicheskogo stimulirovaniia nauchno-tekhnicheskogo progressa,* ed. L. A. Gatovskii (Moscow, 1967), 281; *Planovoe khoziaistvo,* No. 10 (1969), 65.

268. *Resheniia partii i pravitel'stva,* IV, 572–3, 689–91.

269. Ibid., 678. In the first quarter of 1963 the plan for research and introduction of new technology was fulfilled in major industries as follows: ferrous metallurgy, 40 percent; energy and electrotechnical industry, 62 percent; chemistry, 70 percent; instruments, electronics and communications, 78 percent; nonferrous metallurgy, 82 percent; and machine building, 89 percent. (*Biulleten' Gosudarstvennogo komiteta po koordinatsii nauchno-issledovatel'skikh rabot,* No. 7 (1963), 15–16.)

270. *Resheniia partii i pravitel'stva,* IV, 571–2, 679, 689. This problem stemmed partly from the lack of a standard methodology for calculating the economic effects of new technology. It was worsened by the weak incentive for industrial officials to make such calculations in developing new products which they had been ordered to introduce. Finally, the situation was complicated by a shortage of trained economists. In 1962 almost 70 percent of the economic staff members of Leningrad plants lacked a secondary specialized education, and only 14 percent had a higher education—proportions much worse than those for the other staff offices of the plants. (*Kommunist,* No. 16 (1964), 30–31.)

271. *Resheniia partii i pravitel'stva,* IV, 680–82; Zaleski et al., *Science Policy in the USSR,* 434.

272. For a discussion of the extent of scientific secrecy by a Soviet science journalist who defected in the mid-1960s, see L. Vladimirov, "Soviet Science: A Native's Opinion," *New Scientist*, 28 November 1968, 490. See also *Resheniia partii i pravitel'stva*, V, 75–8, 500–504; *Ekonomicheskaia gazeta*, 21 March 1961, 3; Parrott, *Information Transfer*, passim; and Richard W. Judy, "Information, Control, and Soviet Economic Management," in *Mathematics and Computers in Soviet Economic Planning*, eds. John P. Hardt et al. (New Haven, 1967), 25–6.

Chapter 5

1. T. H. Rigby, "The Soviet Leadership: Toward a Self-Stabilizing Oligarchy?", *Soviet Studies* 22 (1970–71), 168–91; George W. Breslauer, "On the Adaptability of Soviet Welfare-State Authoritarianism," in *Soviet Society and the Communist Party*, ed. Karl W. Ryavec (Amherst, Mass., 1978), 6–9.

2. For the sake of convenience, throughout this chapter I have called the party's highest decision-making body the Politburo, although the party Presidium was not actually renamed the Politburo until April 1966.

3. Thomas B. Larson, *Disarmament and Soviet Policy, 1964–1968*, pbk. ed. (Englewood Cliffs, N.J., 1969), 55–6; Marantz, "Peaceful Coexistence," 307; Zimmerman, *Soviet Perspectives*, 232–6.

4. Percentages calculated from Larson, *Disarmament*, 82.

5. Leonid Brezhnev, *Leninskim kursom* (Moscow, 1973–6), I, 161–2.

6. Ibid., 159—63, 180.

7. M. A. Suslov, *Izbrannoe* (Moscow, 1972), 416. See also P. E. Shelest, *Idei Lenina pobezhdaiut* (Kiev, 1971), 245.

8. *Pravda*, 25 July 1965, quoted in Thomas W. Wolfe, *The Soviet Military Scene: Institutional and Defense Policy Considerations*, Rand Memorandum RM–4913–PR (Santa Monica, 1966), 65; *Pravda*, 18 August 1965, 4.

9. Wolfe, *The Soviet Military Scene*, 62–3, 86; Wolfe, *Soviet Power and Europe*, 451–6.

10. A. N. Kosygin, *Izbrannye rechi i stat'i* (Moscow, 1974), 194–6.

11. Kosygin, *Izbrannye rechi*, 201, 204.

12. Ibid., 221.

13. *Pravda*, 16 May 1965, 2.

14. Kosygin, *Izbrannye rechi*, 221, 231, 253–4. The reorganization occurred before Brezhnev made the speeches quoted above.

15. Ibid., 209–19. Two weeks before Kosygin's Gosplan speech, central ministries were reestablished for the defense industries. In this speech Kosygin signaled his opposition to reestablishing ministries to administer industry, including military industry. It is a plausible guess that he felt the organizational issue

and the level of defense spending were linked, since central ministries have customarily pressed for increased allocations to the sectors they represent.

16. *Pravda*, 22 May 1965, 2; *XXIII s"ezd Kommunisticheskoi Partii Sovetskogo Soiuza: stenograficheskii otchet* (Moscow, 1966), I, 245–6; Wolfe, *The Soviet Military Scene*, 65. A few months after the 23rd Party Congress endorsed heavier defense spending, Podgornyi fell in with this position. (*Pravda*, 10 June 1966, 2.)

17. *XXIII s"ezd*, I, 93, II, 305; Brezhnev, *Leninskim kursom*, I, 437.

18. *Pravda*, 3 June 1966, 2–3; ibid., 29 October 1966, 2.

19. Ibid., 8 June 1966, 3.

20. *XXIII s"ezd*, II, 8, 10, 17, 65.

21. Ibid., 17, 65.

22. Brezhnev, *Leninskim kursom*, I, 14.

23. *XXIII s"ezd*, 24–5, 53–4, 70.

24. A. P. Kirilenko, *Izbrannye rechi i stat'i* (Moscow, 1976), 8–9, 43, 55, 70–71; Suslov, *Izbrannoe*, 447–8; *XXIII s"ezd*, I, 132, 134.

25. Kosygin, *Izbrannye rechi*, 262, 266.

26. *XXIII s"ezd*, II, 13.

27. Those prognoses came back to haunt A. A. Arzumanian, the academic specialist who had played the leading advisory role in formulating them. Arzumanian was director of IMEMO. According to an emigré researcher who worked at IMEMO in the early 1960s and sometimes acted as a consultant to the Central Committee Secretariat, IMEMO drafted the 1961 Party Program's predictions concerning the East-West economic competition. Later when it became clear that the predictions could not be met, Arzumanian received an official reprimand from the Central Committee, suffered a stroke, and died soon afterward. (I. Glagolev, "Dom na Staroi ploshchadi," *Posev*, No. 9 (1978), 31.) Since Arzumanian, who died in July 1965, was still listed early in that year as heading the Academy's council on East-West economic competition, the reprimand apparently came after Khrushchev's fall. If so, the rebuke probably came from Kosygin or his supporters inside the party apparatus.

28. *XXIII s"ezd*, I, 406–7; *Kommunist*, No. 17 (1966), 32.

29. D. Gvishiani, *Sotsial'naia rol' nauki i nauchnaia politika* (Moscow, 1968), 6.

30. *XXIII s"ezd*, I, 44.

31. Ibid., 22.

32. For further evidence, see Brezhnev, *Leninskim kursom*, II, 124.

33. *XXIII s"ezd*, II, 60–63.

34. Kosygin, *Izbrannye rechi*, 381, 386.

35. *Pravda*, 27 June 1967, 2.

36. Kosygin said that recently the Politburo and Council of Ministers had considered "basic questions" connected with the plan and decided that it should be confirmed "approximately in 1969." (*Sovetskaia Belorussiia*, 15 February 1968, 1–2.) On 1 March *Kommunist* noted that in accordance with a party-state resolution, preparatory work on the plan was beginning. First, the planners would identify the main tendencies of scientific-technological progress. Then they would write a detailed plan. (*Kommunist*, No. 4 (1968), 10.)

37. *Sovetskaia Belorussiia*, 15 February 1968, 1–2; see also Kirilenko, *Izbrannye rechi i stat'i*, 90–91. A later version of Kosygin's speech consists of "excerpts" badly distorted by unmarked deletions, rearrangement, and inclusion of new material. (A. N. Kosygin, *K velikoi tseli: Izbrannye rechi i stat'i* (Moscow, 1979), I, 523–39.)

38. *Sovetskaia Belorussiia*, 15 February 1968, 1–2.

39. Ibid.

40. Ibid. I owe my awareness of this speech and its relation to Brezhnev's March 1968 pronouncement to Peter Wiles, "On the Prevention of Technology Transfer," in NATO Directorate of Economic Affairs, *East-West Technological Cooperation: Main Findings of a Colloquium Held 17th–19th March, 1976 in Brussels* (Brussels, 1976), 26–7.

41. *Pravda*, 30 March 1968, 1; Wiles, "On the Prevention of Technology Transfer," 26–7.

42. Brezhnev, *Leninskim kursom*, II, 312.

43. Ibid., 2. For a speech by a party ideological secretary seconding these points, see P. N. Demichev in *Pravda*, 20 June 1968, 1.

44. In his February speech, Kosygin issued a vigorous warning against ideological permissiveness and Western propaganda intended to undermine the cohesion of Soviet society.

45. Without itself recognizing East Germany, the new West German coalition formed in December 1966 abandoned the Hallstein doctrine and indicated its readiness to establish diplomatic relations with other states that recognized East Germany. Rumania established diplomatic relations with West Germany in January 1967. Czechoslovakia and Hungary likewise showed an interest in diplomatic ties, until Soviet intercession squelched the idea. (N. Edwina Moreton, *East Germany and the Warsaw Alliance: The Politics of Detente* (Boulder, Colo., 1978), 57–60, 67, 71; see also Wolfe, *Soviet Power*, 285, 299–301, 348–51.)

46. Moreton, *East Germany*, 58–9; Brezhnev, *Leninskim kursom*, I, 518–21, and II, 57, 60, 124.

47. Wolfe, *Soviet Power*, 351. The increasing East European pressure for greater access to Western technology was clearly expressed in May 1968 at a large Moscow symposium on R & D held under CEMA auspices. Two Polish participants outspokenly remarked that "in many branches of science and tech-

nology the socialist countries are still lagging." Asserting that the development of science and technology solely within CEMA had "no social or economic foundation" and threatened a "permanent lag" for the socialist countries, they called for wider inclusion of those countries in world technological relations. Another Pole complained that joint CEMA R & D plans did not "coincide . . . with the main dimensions of the so-called 'technological gap' " and that the actual exchange of findings from joint projects was unreliable, thanks partly to "border restrictions" on personal contacts among researchers from different socialist countries. (Iu. Paestka and Z. Madei, "Vliianie mirovoi obshchestvenno-ekonomicheskoi sistemy na perspektivy razvitiia tekhniki i ekonomiki v Pol'she," and Z. Ziulkovski, "Aspekty effektivnosti mezhdunarodnogo sotrudnichestva v oblasti nauchnykh issledovanii," in *Upravlenie, planirovanie i organizatsiia nauchnykh i tekhnicheskikh issledovanii* (Moscow, 1970–71), V, 141–7, 192–4.)

48. Brezhnev, *Leninskim kursom*, II, 187–8.

49. Moreton, *East Germany*, 76; Wolfe, *Soviet Power*, 362.

50. Brezhnev, *Leninskim kursom*, II, 204, 212–3.

51. Given Kosygin's criticism of Soviet computers and Brezhnev's praise of socialist achievements in electronics, it is possible that one specific technological question dividing them was whether to pattern the first computer jointly developed by CEMA members (the *Riad*) on the design of the IBM S/360. This issue was a source of dispute within the USSR in the late 1960s. (N. C. Davis and S. E. Goodman, "The Soviet Bloc's Unified System of Computers," *Computing Surveys* 10, No. 2 (June 1978), 101.) The IBM design was finally adopted. As for the diplomatic benefits of commerce, veiled differences surfaced in discussions of Lenin's policy toward Western concessions in the USSR during the 1920s. On the day that Brezhnev attacked Kosygin's views, the main journal of party history printed an article which emphasized Lenin's desire to attain rapid economic independence for Soviet Russia and commented that during the 1920s the party had had to conduct "a persistent struggle against capitulationist inclinations to receive 'any sort of concessions,' 'any sort of credits,' 'at any price.' " The discussion of specific Western concession offers focused exclusively on bids that Lenin had rejected. (*Voprosy istorii KPSS*, No. 4 (1968), 3–15.) In late May, however, a second article in the same journal emphasized Lenin's confidence that the Soviet system could resist capitalist inroads in such dealings and cited his dissatisfaction with the slow implementation of the concessions program. Lenin had especially underscored the diplomatic value of the program. Thanks to the positive effect of the concessions policy on Western bourgeois circles, the article concluded, "in significant measure . . . the Leninist idea of peaceful coexistence of states with different systems of property became a reality in the world arena." (Ibid., No. 6 (1968), 3–13; see also Wolfe, *Soviet Power in Europe*, 318.)

52. Thomas W. Wolfe, *The SALT Experience* (Cambridge, Mass., 1978), 2; Lyndon Johnson, *The Vantage Point* (New York, 1971), 480.

53. Wolfe, *The SALT Experience*, 2–3; John Newhouse, *Cold Dawn: The Story of SALT* (New York, 1973), 92, 96–7.

54. Wolfe, *The SALT Experience*, 2–3; Newhouse, *Cold Dawn*, 101–2, 106–7; Ted Greenwood, *Making the MIRV: A Study of Defense Decision Making* (Cambridge, Mass., 1975), 10.

55. *Pravda*, 7 March 1967, 1–2.

56. Ibid., 11 May 1967, 1.

57. Ibid., 27 June 1967, 2. Cf. Johnson, *The Vantage Point*, 484.

58. *Sovetskaia Moldaviia*, 1 October 1967, 1–2; ibid., 3 October 1967, 2.

59. *Pravda*, 7 November 1967, 1–2.

60. In his memoirs, Johnson suggests this explanation, supporting it with a reference to unspecified "evidence"—presumably government intelligence. (Johnson, *The Vantage Point*, 480.)

61. Brezhnev, *Leninskim kursom*, II, 46, 48–9, 61.

62. *Trud*, 1 November 1967, 2.

63. Suslov, *Izbrannoe*, 513.

64. Brezhnev, *Leninskim kursom*, II, 101; *Trud*, 21 October 1967, 2.

65. Suslov, *Izbrannoe*, 512–3.

66. On 14 August 1967 the Central Committee adopted a long resolution on the role of the social sciences. The resolution stressed that social scientists must expound the party's leading role and develop Marxist theory in order to combat revisionism. But it also complained that the Academy's institutes sometimes failed to give "a deep and objective analysis of the real processes of social life," which "hinders the correct evaluation of our society's historical experience and prospects of development." The resolution called for social science institutions to play a greater role "in preparing scientifically grounded recommendations necessary for developing the policy of the Communist party and the Soviet state." Among important research topics, the decree listed means of ensuring rapid Soviet economic development and effective use of science and technology. It also urged the study of "new phenomena in the economy of contemporary capitalism," such as state economic regulation, and analysis of "the socioeconomic contradictions of imperialism." (*KPSS v rezoliutsiiakh i resheniiakh*, IX, 342–57.)

67. *MEMO*, No. 1 (1968), 121–2. Previously a deputy editor of *Pravda* and deputy director of IMEMO, Inozemtsev had assumed the IMEMO directorship in mid–1966.

68. Ibid., 125.

69. Ibid., 122.

70. Ibid.

71. Ibid. Inozemtsev had earlier highlighted the special significance of such prognoses in *Pravda*, 19 May 1967, 5.

72. For Suslov's role in 1952, see Conquest, *Power*, 102–3.

73. *Sovetskaia Belorussiia*, 15 February 1968, 1–3.

74. The increase in overt defense spending is calculated from Larson, *Disarmament*, 82.

75. Ibid.

76. Ibid.

77. Brezhnev, *Leninskim kursom*, II, 46, 48–9, 61. Brezhnev's statement dated from July 1967, but as we shall see presently, he repeated this view in mid-1968.

78. *Pravda*, 30 March 1968, 2.

79. *Kommunist*, No. 7 (1968), 18–19.

80. *Kommunist*, No. 4 (1968), 3–4, signed for printing on 1 March. Emphasis in the original.

81. Brezhnev, *Leninskim kursom*, II, 257–8.

82. Suslov, *Izbrannoe*, 534.

83. Brezhnev, *Leninskim kursom*, II, 257–8. Jiri Valenta points out the significance of this passage in his *Soviet Intervention in Czechoslovakia: Anatomy of a Decision*, pbk. ed. (Baltimore, 1979), 46–7.

84. Quoted in Newhouse, *Cold Dawn*, 103.

85. Johnson, *The Vantage Point*, 485, says that in June an agreement to hold the talks "finally began to take shape."

86. It was probably also intended to smooth the way for the nonproliferation treaty, which a number of nonnuclear countries had indicated they would sign only if the nuclear powers made a vigorous effort to reach arms control agreements with each other.

87. *Kommunist Vooruzhennykh Sil*, No. 3 (1968), 16–17; No. 4 (1968), 21–2, and No. 7 (1968), 67–74.

88. *Kommunist Vooruzhennykh Sil*, No. 11 (1968), 15–17, 21–3. Emphasis in the original.

89. *MEMO*, No. 6 (1968), 11–12. Signed for printing on 27 May.

90. Ibid., 10. See also *Voprosy istorii estestvoznaniia i tekhniki*, vyp. 2 (31) (1970), 10–16.

91. *Pravda*, 30 March 1968, 2.

92. *Izvestiia*, 2 July 1968, quoted by Valenta, *Soviet Intervention*, 47.

93. *Izvestiia*, 7 November 1968, 1–2.

94. *Pravda*, 28 June 1968, 3. Gromyko phrased his comments in a way such that they could be interpreted as referring to Western, and possibly Communist Chinese, critics. But there is no doubt that they were also intended for domestic opponents of the talks.

95. Shelest, *Idei Lenina pobezhdaiut*, 100. I am indebted to Valenta, *Soviet Intervention*, 46, for pointing out Shelest's failure to refer to arms control.

96. For their later views, see the following chapter.

97. In October, when Foreign Minister Gromyko mentioned SALT in a U.N. speech, the Ministry of Defense newspaper excised the reference from its version of the speech. In November, when Mazurov said in a speech that the USSR was prepared to negotiate on "the whole complex" of issues connected with strategic arms, the Soviet press struck out this statement. A week later the press cut the discussion of SALT in a speech by the Soviet representative at the U.N. (From the statement by Thomas W. Wolfe in United States Senate Committee on Armed Services, Subcommittee on Strategic Arms Limitation Talks, *The Limitation of Strategic Arms*, Part 2 (Washington, D.C., 1970), 63.)

98. Samuel B. Payne, Jr., *The Soviet Union and SALT* (Cambridge, Mass., 1980), 18 and passim; Lawrence T. Caldwell, *Soviet Attitudes to SALT*, Adelphi Paper No. 75 (London, 1971), 5.

99. The Soviet invasion a month later caused the United States to postpone the talks.

100. *Sovetskaia Rossiia*, 15 April 1965, 4.

101. According to Strobe Talbott, Kapitsa's first trip abroad since 1934 occurred in May 1965, when he went to receive the Bohr Medal in Denmark. A year later he traveled to England. (*Khrushchev Remembers: The Last Testament*, ed. Strobe Talbott (Boston, 1974), 67n.)

102. *VAN*, No. 3 (1966), 12.

103. Ibid., No. 11 (1967), 39.

104. *Pravda*, 29 March 1966, 4; *VAN*, No. 2 (1966), 28–9.

105. *Pravda*, 18 January 1967, 2–3; Cooper, "Research, Development, and Innovation," 160.

106. The rate was 16.9 percent. Calculated from UNESCO, *Science Policy and Organization of Research in the U.S.S.R.*, Science policy studies and documents, No. 7 (Paris, 1967), 54. This study was prepared by a Soviet working group which was chaired by an official of the State Committee and included representatives of the Academy and the Ministry for Higher and Secondary Specialized Education.

107. *Vneshniaia torgovlia*, No. 2 (1965), 5.

108. See, for example, the statement by the Minister of Foreign Trade, N. Patolichev, in ibid., No. 4 (1968), 9; also ibid., No. 7 (1967), 26–8.

109. One economist whom Ministry spokesmen criticized by name was O. Bogomolov, particularly for his article in *Promyshlennost', vneshniaia torgovlia, planirovanie* (Moscow, 1966), which I have been unable to obtain.

110. The conclusion of a major agreement to import an auto manufacturing plant in 1966, for example, was preceded in 1965 by an agreement for joint R & D between Fiat and the State Committee. (*Vneshniaia torgovlia*, No. 8 (1966), 43–4.) A 1967 decree by the Council of Ministers gave the State Committee clear preeminence in making policy for the acquisition of licenses to foreign technology; the Ministry received only the right to supervise the execution of licensing agreements that had already been concluded. (*Ekonomicheskaia gazeta*, No. 33 (1967), 5.)

111. *Vneshniaia torgovlia*, No. 2 (1965), 6, No. 7 (1967), 25, 28–30, and No. 4 (1968), 10.

112. *Izvestiia*, 3 August 1967, 2.

113. *KPSS v rezoliutsiiakh i resheniiakh s"ezdov, konferentsii i plenumov TsK*, 8th ed. (Moscow, 1970–72), IX, 423.

114. *Kommunist*, No. 18 (1968), 37, 39–40. For other evidence of problems with the indoctrination of R & D personnel in this city, see *Partiinaia zhizn'*, No. 18 (1968), 55–60. According to a reliable account by a scientist later deprived of his Soviet citizenship, the political authorities used two devices to minimize political dissent in Obninsk and other research-oriented cities in the late 1960s: they joined these cities to larger urban governments, and they built up local industry to dilute the population of scientists with industrial workers. (Zhores Medvedev, *Soviet Science* (New York, 1978), 134–5.)

115. *Kommunist*, No. 18 (1967), 72.

116. Calculated from Zaleski et al., *Science Policy in the USSR*, 100, and Cooper, "Research, Development, and Innovation," 163. (These two sources show a slight discrepancy between their figures for 1965.) The rates include both current expenditures and capital investment.

117. Calculated from *Narodnoe khoziaistvo SSSR, 1922–1972* (Moscow, 1972), 491.

118. Ibid. Counting third-world nations as "developing capitalist countries," the socialist fraction was still about 1.9 times larger than the nonsocialist.

119. John Hardt and George Holliday, "Technology Transfer and Change in the Soviet Economic System," in *Technology and Communist Culture*, ed. Frederic J. Fleron, Jr. (New York, 1977), 183–223.

120. Robert W. Campbell, "Technology Transfer Among Communist Countries," *ASTE Bulletin*, XI, No. 3 (Winter 1969), 7–8.

121. George D. Holliday, *Technology Transfer to the USSR, 1928–1937 and 1966–1975: The Role of Western Technology in Soviet Economic Development* (Boulder, Colo., 1979), 141–5.

122. Peter B. Maggs and James W. Jerz, "The Significance of Soviet Accession to the Paris Convention for the Protection of Industrial Property," *Journal of the Patent Office Society* 48 (1966), 242–62.

123. *Kommunist*, No. 8 (1965), 65–72. Another reason for joining may have been to gain income from Soviet inventions that might otherwise have been used in the West without licensing payments, but this motive was probably secondary. There is considerable evidence that since 1965 the Soviet license trade with the West has had a strongly negative balance of payments. See Philip Hanson, "International Technology Transfer from the West to the U.S.S.R.," in U.S. Congress, Joint Economic Committee, *Soviet Economy in a New Perspective* (Washington, D.C., 1976), 802–804; Jiri Slama and Heinrich Vogel, "Technology Advances in Comecon Countries: An Assessment," in *East-West Technological Cooperation*, 237.

124. M. L. Gorodisskiy, *Licenses in U.S.S.R. Foreign Trade* (Moscow, 1972), translated by National Technical Information Service, cited by Holliday, *Technology Transfer*, 47.

125. Holliday, *Technology Transfer*, 148.

126. Tony Longrigg, "Soviet Science and Foreign Policy," *Survey* 17, No. 4 (Autumn 1971), 43.

127. Ibid.; *Ezhegodnik Bol'shoi Sovetskoi Entsiklopedii*, 1968, 94; ibid., 1969, 83; Andrew Swatkovsky, "United States–Soviet Scientific and Technical Exchanges and Contacts, 1945–Present: An Evaluation," Ph.D. dissertation, Columbia University, 1972, 173–6.

128. Calculated from Longrigg, "Soviet Science and Foreign Policy," 43. An even greater discrepancy characterized the numbers of Soviet and American students working on college-level or graduate degrees abroad. See UNESCO, *Statistical Yearbook 1968* (Paris, 1969), 276, 278.

129. See note 47 above.

130. Breslauer, "On the Adaptability," 6–9.

131. Ibid., 9, 12–13; Robert F. Miller, "The Scientific-Technical Revolution and the Soviet Administrative Debate," in *The Dynamics of Soviet Politics*, ed. Paul Cocks, Robert V. Daniels, and Nancy W. Heer (Cambridge, Mass., 1976), 144; Cocks, "The Rationalization of Party Control," 178–83.

132. Breslauer, "On the Adaptability," 6–9.

133. Ibid.; Rigby, "The Soviet Leadership: Toward a Self-Stabilizing Oligarchy?", 174–9.

134. Jerry Hough, "The Soviet System: Petrification or Pluralism?", *Problems of Communism* 21, No. 2 (March–April 1972), 25–42.

135. Abraham Katz, *The Politics of Economic Reform in the Soviet Union* (New York, 1972), 108–110.

136. Richard Judy, "The Economists," in *Interest Groups in Soviet Politics*, eds. H. Gordon Skilling and Franklyn Griffiths, pbk. ed. (Princeton, 1971), 236.

137. Ibid., 211n.

138. *Voprosy ekonomiki*, No. 2 (1965), 151. I am grateful to Alan Foley for calling this source to my attention.

139. Ibid., 154.

140. I. Solomonov and L. Fedorov in *Pravda*, 1 December 1964, 2; Tatu, *Power*, 443.

141. Solomonov was director of the "Vulkan" factory in Leningrad; Fedorov was chief engineer at the same plant.

142. Herbert S. Levine, "Introduction," in *Mathematics and Computers in Soviet Economics*, eds. John Hardt et al. (New Haven, 1967), xxi; Moshe Lewin, *Political Undercurrents in Soviet Economic Debates* (Princeton, 1974), 176.

143. Lewin, *Political Undercurrents*, 185; Aron Katsenelinboigen, *Studies in Soviet Economic Planning* (White Plains, N.Y., 1978), 57. Katsenelinboigen, who was a well-known mathematical economist in the USSR until he emigrated in the 1970s, calls the pro-Stalinists and the exponents of the status quo the "conservatives" and the "guardians" respectively. In his opinion, the guardians constitute an "overwhelming majority" among Soviet economists.

144. Judy, "The Economists," 238–40.

145. For a survey of the evidence and its limits, see Tatu, *Power*, 443–60.

146. Kosygin, *Izbrannye rechi*, 190.

147. Katz, *The Politics*, 110.

148. *Kommunist*, No. 1 (1965), 6.

149. *Izvestiia*, 11 December 1964, 4; Tatu, *Power*, 444.

150. Ibid.

151. Katz, *The Politics*, 108.

152. Grey Hodnett, "Succession Contingencies in the Soviet Union," *Problems of Communism* 24 (March–April 1975), 9n.

153. Tatu, *Power*, 448–52.

154. The allocation of investment funds was one such issue.

155. Tatu, *Power*, 445.

156. Kosygin, *Izbrannye rechi*, 217.

157. Ibid.

158. Ibid., 210.

159. Katz, *The Politics*, 122.

160. *Resheniia partii i pravitel'stva*, V, 647.

161. Ibid., 648; Gertrude Schroeder, "Soviet Economic 'Reforms': A Study in Contradictions," *Soviet Studies* 20 (1968–9), 4.

162. For a detailed analysis of the reform see Schroeder, "Soviet Economic 'Reforms' ", especially 13–15.

163. Ibid., 8; *Resheniia partii i pravitel'stva*, V, 644.

164. Brezhnev, *Leninskim kursom*, I, 212.

165. Ibid., 210, 213.

166. Ibid., 211, 213.

167. Kosygin, *Izbrannye rechi*, 288, 290.

168. Ibid., 290.

169. Ibid., 289; Katz, *The Politics*, 211.

170. E. Liberman in *Pravda*, 21 November 1965, 2–3.

171. Iu. P. Koren'kov, "Tekhnicheskii progress i rynok," in *Sovershenstvovanie tsenoobrazovaniia i nauchno-tekhnicheskii progress*, rotoprint ed. (Moscow, 1968), 41–2.

172. Ibid.

173. *Kommunist*, No. 1 (1966), 45.

174. Ibid., 47.

175. Rumiantsev's comments may have been directed partly to foreign critics who espoused this same idea. But the context suggests that he had domestic critics in mind as well.

176. *Kommunist*, No. 10 (1968), 22.

177. Hough, *The Soviet Prefects*, 75.

178. Katz, *The Politics*, 159–61.

179. Judy, "The Economists," 238.

180. *Ekonomika i organizatsiia promyshlennogo proizvodstva*, No. 1 (1970), 107.

181. Ibid. The problem of inadequate independence was one of a set of possible answers listed in the survey questionnaire; hence directors were probably not taking a political risk by selecting it.

182. Ibid., 102, 105.

183. Katz, *The Politics*, 155–6; Karl Ryavec, *Implementation of Soviet Economic Reforms* (New York, 1975), 33, 47, 249, 255.

184. For a review of these links as of the mid-1960s, see Bruce Parrott, "The Organizational Environment of Soviet Applied Research," in *The Social Context of Soviet Science*, eds. Linda L. Lubrano and Susan Gross Solomon (Boulder, Colo., 1980), 74–8.

185. *Pravda*, 10 October 1965, 2.

186. *VAN*, No. 2 (1966), 94; ibid., No. 3 (1966), 13.

187. *Literaturnaia gazeta*, 5 March 1966, 2.

188. V. A. Trapeznikov in *Pravda*, 18 January 1967, 2–3; Gvishiani, *Sotsial'naia rol' nauki*, 37–40.

189. See *The Scientific Intelligentsia in the USSR (Structure and Dynamics of Personnel)*, eds. D. M. Gvishiani, S. R. Mikulinsky, and S. A. Kugel (Moscow, 1976), 27–8.

190. Zaleski et al., *Science Policy in the USSR*, 463–5; *Pravda*, 18 January 1967, 2–3; *Ekonomicheskaia gazeta*, No. 2 (1966), 34.

191. E. E. Grishaev, *Nekotorye voprosy finansirovaniia nauchnykh issledovanii* (Moscow, 1968), 15; *Pravda*, 3 August 1966, 4.

192. *Partiinaia zhizn'*, No. 7 (1966), 35.

193. *Pravda*, 18 January 1967, 2–3.

194. Ibid.

195. *Ekonomicheskaia gazeta*, No. 2 (1966), 27, 33–4.

196. G. Popov, "Problemy rukovodstva nauchno-tekhnicheskimi issledovaniiami v otrasli promyshlennosti," in *Upravlenie, planirovanie i organizatsiia nauchnykh i tekhnicheskikh issledovanii*, IV, 76.

197. *Izvestiia*, 4 January 1966, 3, and 20 March 1966, 3.

198. *Kommunist*, No. 1 (1967), 62 (an article by the Minister of Instrument Building, Means of Automation and Control Systems.)

199. *Planovoe khoziaistvo*, No. 7 (1966), 16; *Pravda*, 12 January 1966, 3.

200. *Izvestiia*, 30 January 1966, 2.

201. See, for example, *Kommunist*, No. 5 (1966), 73, 75.

202. For a brief discussion of the controls under which ministerial research institutes were then operating, see Parrott, "The Organizational Environment of Soviet Applied Research," 74–5.

203. A. Tselikov in *Izvestiia*, 27 January 1966, 3. Tselikov was a member of the U.S.S.R. Academy of Sciences. See also *Pravda*, 25 November 1965, 2.

204. *Ekonomicheskaia gazeta*, No. 4 (1968), 18.

205. The writings of Andrei Sakharov are a case in point, which I discuss in the next chapter.

206. For evidence of this pattern in the U.S.S.R. Academy, see Thane Gustafson, "Why Doesn't Soviet Science Do Better Than It Does?," in *The Social Context of Soviet Science*, 43–8. Since positions in the Academy are the most prestigious, there is no reason to believe that the level of competition among individuals is any higher in ministerial research establishments.

207. Parrott, "The Organizational Environment," 91n. The geographic and institutional coverage of these attitudinal surveys was quite limited.

208. *Resheniia partii i pravitel'stva*, VI, 267–74.

209. Ibid., 363–5; Parrott, "The Organizational Environment," 81–2.

210. E. E. Grishaev, *Ob izmenenii poriadka planirovaniia zatrat na nauchno-is-sledovatel'skie raboty i o rasshirenii prav rukovoditelei nauchno-issledovatel'skikh uchrezhdenii* (Moscow, 1967), 3.

211. Ibid., 7.

212. *Resheniia partii i pravitel'stva*, VIII, 54.

213. *Sovetskoe gosudarstvo i pravo*, No. 3 (1968), 74.

214. The decree did deal in part with State Committee projects, but the sections on institute rights did not stipulate that the new rights were to be restricted to work on these projects.

215. *Pravda*, 14 December 1967, 1.

216. Rush V. Greenslade, "The Real Gross National Product of the U.S.S.R., 1950–1975," in *Soviet Economy in a New Perspective*, 272, 274. For each period, the lower figure represents a calculation excluding weapons production; the higher figure includes this output.

217. Ibid.

218. Greenslade calculates that this rate was 3.6 percent in 1951–5, 3.2 percent in 1956–60, 0.6 percent in 1961–5, and 1.3 percent in 1966–70. (Ibid., 279.)

219. *Voprosy ekonomiki*, No. 7 (1968), 109.

220. *Planovoe khoziaistvo*, No. 12 (1969), 81.

221. L. M. Gatovskii, *Ekonomicheskie problemy nauchno-tekhnicheskogo progressa* (Moscow, 1971), 110, 113.

222. *Pravda*, 9 December 1966, 3.

223. *Voprosy ekonomiki*, No. 4 (1970), 106.

224. For evidence that ministerial officials had this view, see *Kommunist*, No. 2 (1969), 67. See also I. Ia. Oblomskaia, "Vozmeshchenie izderzhek predpriiatii po povysheniiu tekhnicheskogo urovnia proizvodstva v usloviiakh khoziaist-vennoi reformy," in *Stimulirovanie nauchno-tekhnicheskogo progressa v usloviiakh khoziaistvennoi reformy* (Moscow, 1969), 81, and *Planovoe khoziaistvo*, No. 11 (1973), 15.

225. Berliner, *The Innovation Decision*, 250–55.

226. *Ekonomicheskaia gazeta*, 25 October 1966, 4; ibid., 13 June 1967, 11; ibid., 6 February 1968, 11.

227. The Council's resolution was adopted on 24 October 1968 and appears in *Sobranie postanovlenii pravitel'stva SSSR*, 1968, 573–8.

228. *Pravda*, 10 July 1970, 2.

229. *Planovoe khoziaistvo*, No. 11 (1968), 44, 46–7.

230. Soviet statistical yearbooks give the following annual figures for "the creation of new types of machines and equipment": 1958—2,051; 1960—3,099; 1963—3,229; 1964—3,113; 1965—3,366; 1966—3,605 (*Narodnoe khoziaistvo SSSR*, 1964, 190; ibid., 1965, 205; ibid., 1968, 267).

231. In 1966 enterprises producing about 10 percent of total industrial output were shifted to the new system; by the end of 1967 plants producing approximately 40 percent of the total had been transferred to it. (Schroeder, "Soviet Economic 'Reforms,' " 4–5.)

232. Calculated from *Narodnoe khoziaistvo SSSR*, 1968, 267, and 1969, 230.

233. Calculated from ibid., 1966, 266, and 1969, 230.

234. *Ekonomicheskaia gazeta*, No. 2 (1966), 31, and No. 30 (1967), 30. See also *Voprosy ekonomiki*, No. 3 (1966), 14; M. Bashin, *Khoziaistvennyi raschet v otraslevykh NII i KB* (Moscow, 1971), 170; *Voprosy ekonomiki*, No. 2 (1973), 17; Cooper, "Research, Development, and Innovation," 175.

235. From 1961 through 1963 the proportion of applications receiving patents in the USSR was 18 percent. (*Voprosy ekonomiki*, No. 8 (1968), 128.) By 1968–69, the figure had risen to about 24 percent. (Parrott, "The Organizational Environment," 87.)

236. *Voprosy ekonomiki*, No. 8 (1968), 128; Parrott, "The Organizational Environment," 87n.

237. Parrott, "The Organizational Environment," 86–7.

Chapter 6

1. Tad Szulc, *The Illusion of Peace: Foreign Policy in the Nixon Years* (New York, 1978), 70, 419; Daniel H. Yergin, "Strategies of Linkage in Soviet-American Relations," paper prepared for the 1976 annual meeting of the American Political Science Association, especially 13–14; Michael J. Sodaro, "The Impact of Detente on the GDR's Economic Policy, 1966–71: A Linkage Analysis," unpublished paper, The George Washington University, 1979, 4–5.

2. Kosygin, *Izbrannye rechi*, 438.

3. Ibid., 439.

4. L. I. Brezhnev, *Ob osnovnykh voprosakh ekonomicheskoi politiki KPSS na sovremennom etape* (Moscow, 1975), I, 374.

5. Ibid.

6. Brezhnev, *Leninskim kursom*, II, 370–71, 375, 408.

7. Ibid., 367–8.

8. Ibid., 380.

9. Ibid., 380–81, 411–13.

10. Ibid., 413.

11. Ibid., 412. On the utility of bloc solidarity in attaining "large successes" from mutually profitable cooperation with the capitalist world, see ibid., 330–31. See also 308, 328, 331, 424, 456.

12. M. A. Suslov, *Marksizm-leninizm—internatsional'noe uchenie rabochego klassa* (Moscow, 1973), 47. This article first appeared in *Kommunist*, No. 15 (1969).

13. Ibid., 63, 80–81.

14. Ibid., 52, 66–7.

15. Suslov played a central role in the adoption of a softer Soviet attitude toward the Social Democrats in March 1969. (Stent, *From Embargo to Ostpolitik*, 159.) For Soviet diplomacy on the German question during 1969, see Gerhard Wettig, *Community and Conflict in the Socialist Camp: The Soviet Union, East Germany and the German Problem 1965–1972*, trans. by Edwina Moreton and Hannes Adomeit (New York, 1975), ch. 5.

16. Suslov, *Marksizm-leninizm*, 81.

17. Petr Shelest, *Idei Lenina pobezhdaiut* (Kiev, 1971), 23, 27. This speech was originally delivered on 17 April 1970.

18. *Pravda Ukrainy*, 18 October 1969, as quoted by Grey Hodnett, "Ukrainian Politics and the Purge of Shelest," paper prepared for the annual meeting of the Midwest Slavic Conference, Ann Arbor, 5–7 May 1977, 55.

19. A prime example was a conflict over the publishing policy of the journal *Voprosy filosofii* on the subject of the scientific-technological revolution in 1969. See *Voprosy filosofii*, No. 4 (1969), 4–7, and No. 5 (1969), 144–52; *Pravda*, 4 April 1969, 3.

20. The dating of Inozemtsev's membership in the Bureau of the Scientific Council is approximate. He was identified as one of its members in a book signed to press in mid-1970. (*Sorevnovanie dvukh sistem: aktual'nye problemy mirovoi ekonomiki* (Moscow, 1970), 3.)

21. N. N. Inozemtsev, *Problemy ekonomiki i politiki sovremennogo imperializma* (Moscow, 1969), 12–13, 36–7.

22. Ibid., 15, 24.

23. Ibid., 37–9.

24. Ibid., 9–10.

25. *Kommunist Vooruzhennykh Sil*, No. 8 (1969), 20–21.

26. See especially one military writer's warning against the "serious mistakes" committed by those who believed that "particular political organizations or even their leaders" could have a decisive effect on the pace at which military technology developed. (*Kommunist Vooruzhennykh Sil*, No. 24 (1968), 24.) See also Thomas W. Wolfe, *Soviet Interests in SALT: Political, Economic, Bureaucratic and Strategic Contributions and Impediments to Arms Control*, Rand Paper P-4702 (Santa Monica, 1971), 24–5, 34, and Samuel B. Payne, Jr., " 'From

Positions of Strength': Soviet Attitudes Towards Strategic Arms Limitation,"
Ph.D. dissertation, The Johns Hopkins School of Advanced International Studies,
1976, 78–82, 87–9, and passim.

27. On the various Soviet views of capitalism, see Franklyn Griffiths, "Images,
Politics and Learning in Soviet Behavior towards the United States," Ph.D.
dissertation, Columbia University, 1972, 89–91, 197, and passim; also Richard
Allen Nordahl, "The Soviet Model of Monopoly Capitalist Politics," Ph.D.
dissertation, Princeton University, 1972. On the Soviet image of the American
system and its relation to Soviet foreign policy, see William Zimmerman, *Soviet
Perspectives on International Relations, 1956–1967* (Princeton, 1969), ch. 6.

28. M. F. Kovaleva, *K voprosam metodologii politicheskoi ekonomii kapitalizma*
(Moscow, 1969). The book was signed for printing on 9 June 1969. On the
debate between Kovaleva and her opponents, see Griffiths, "Images, Politics
and Learning," 95–8. I am indebted to Griffiths for a number of the points
and sources cited below.

29. Kovaleva, *K voprosam*, 4.

30. Griffiths, "Images, Politics and Learning," 197.

31. We have just seen that Inozemtsev endorsed several of the ideas criticized
by Kovaleva. In addition, he had been a protégé of Varga and now headed
the institute where Dalin was employed.

32. Kovaleva, *K voprosam*, 16–17, 26–8, 32, 76, 150–51, 224–5, 235, 270–71,
278, 290, 307–8. See also *Voprosy istorii KPSS*, No. 11 (1970), 73.

33. The reader should not confuse this body with the U.S.S.R. Academy of
Sciences. As Griffiths points out (pp. 97–8), Kovaleva's preface acknowledged
the help of the economics faculty of the Academy of Social Sciences in discussing
and preparing her book. By the end of 1968, this body had begun to focus on
the implications of the scientific-technological revolution for the competition
between socialism and capitalism. See the account of the conference sponsored
by it and its East German counterpart: *Voprosy filosofii*, No. 6 (1969), 137–8.

34. *MEMO*, No. 11 (1969), 14–15.

35. Ibid., No. 1 (1970), 123–4.

36. Ibid.; Griffiths, "Images, Politics and Learning," 96–7.

37. *MEMO*, No. 1 (1970), 124.

38. Brezhnev, *Ob osnovnykh voprosakh*, I, 419.

39. Ibid., 418.

40. Ibid., 418–19.

41. Wettig, *Community and Conflict*, 51.

42. Brezhnev, *Ob osnovnykh voprosakh*, I, 417–18.

43. Ibid., 420–21.

44. Ibid., 422.

45. *XXIV s"ezd KPSS* (Moscow, 1971), I, 82.

46. Ibid., 423, 428.

47. The declaration appeared in February 1970. It was published in the name of the Central Committee of the CPSU, the Central Committee of the Komsomol, the Council of Ministers, and the Central Council of Trade Unions. (See *KPSS v rezoliutsiiakh*, X, especially 200–201, 208–12, 214.) It thus presumably required at least the reluctant acquiescence of Kosygin and Shelepin in their capacities as heads of the latter two bodies. The text shows clearly that the impetus for the declaration came from the December plenum.

48. *Pravda*, 3 April 1970, 1–2; Kirilenko, *Izbrannye rechi*, 127, 136, 141.

49. *Politicheskii dnevnik*, No. 67 (April 1970), as quoted in *Arkhiv samizdata*, No. 1011, pp. 3–4. *Politicheskii dnevnik* was an underground journal that circulated in the late 1960s among liberal Soviet Marxists; some were party members and could have learned of such a letter. As noted above, Suslov had warned against "panickers" in the party just a few weeks before Brezhnev's speech. Mazurov, given his earlier endorsement of nontraditionalist economic views, probably acted out of desire for political self-defense. Like Kosygin, Mazurov had been a key economic administrator since Khrushchev's fall and probably wanted to avoid being made a scapegoat for recent economic troubles.

50. *Voprosy istorii KPSS*, No. 4 (1970), 74–5, 77.

51. Ibid., 73, 75. The Orgburo, abolished in 1952, was also included in Lenin's list of bodies that could authorize implementation by a secretary.

52. Ibid., 72, 74.

53. Ibid., 72–3, 79–80.

54. Ibid., 76, 80.

55. Ibid., 80.

56. Ibid., 80–81.

57. *Pravda*, 5 June 1970, 2–3.

58. *Pravda*, 7 June 1970, 2.

59. Christian Duevel, "The 24th CPSU Congress—To Be or Not to Be in 1970," *Radio Liberty Dispatch*, 23 January 1970, and "Marginal Notes on a Soviet Leadership Crisis," ibid., 23 July 1970; John Dornberg, *Brezhnev: The Masks of Power* (London, 1974), 244–8. Cf. Wolfgang Leonhard, "The Domestic Politics of the New Soviet Foreign Policy," *Foreign Affairs* 52, No. 1 (October 1973), 70.

60. Brezhnev, *Leninskim kursom*, III, 45.

61. Ibid., 56–7; Hodnett, "Ukrainian Politics," 52.

62. *Pravda Ukrainy*, 18 April 1970, as quoted by Hodnett, "Ukrainian Politics," 50. Hodnett's thorough analysis leaves no doubt that Shelest opposed a relaxation of East-West tensions.

63. Hodnett, "Ukrainian Politics," 52.

64. The controversy centered on a chapter of a book that appeared early in 1970. The bulk of the chapter, which dealt with Lenin's style of handling intraparty disagreements, was taken up by lengthy quotations from Lenin's diatribes against party radicals who wanted to hold out for revolutionary war in Europe rather than sign the treaty of Brest-Litovsk. The political message was that Lenin's censures were directly relevant to " 'revolutionaries' who in our time of catastrophically rapid changes in society are governed by former ideas, do not want or cannot take into account a fundamental change in the situation, and who—even when they are subjectively faithful to the revolution—are objectively transformed into its most dangerous enemies." (*Leninizm i dialektika obshchestvennogo razvitiia* (Moscow, 1970), 228; see also 233, 236–7, 251–5.) In May 1970 some 300 persons gathered to discuss the volume. At this meeting speakers poured a torrent of criticism on the chapter in question. The violent and concerted nature of the attack clearly indicated that it was directed from above. (*Voprosy filosofii*, No. 11 (1970), 166–73.) For further evidence of high-level resistance to the rapprochement with the Federal Republic, see Leonhard, "The Domestic Politics," 70. For a Soviet effort in 1969 to disseminate East German views about the minimal value of the technological assistance that the Western powers would give the socialist bloc, see below.

65. Hodnett, "Ukrainian Politics," 52.

66. "Appeal of Soviet Scientists to the Party-Government Leaders of the U.S.S.R.," *Survey*, No. 76 (1970), 160–70. In addition to Sakharov, an eminent physicist, the authors were V. F. Turchin, also a physicist, and R. A. Medvedev, an educational specialist and historian. The memorandum was dated 19 March 1970. Cf. *Sakharov Speaks*, ed. Harrison E. Salisbury, pbk. ed. (New York, 1974), 98–100.

67. For an analysis of this tendency among "instrumental-pragmatic dissenters," see Rudolf Tokes, "Dissent: The Politics of Change in the U.S.S.R.," in *Soviet Politics and Society in the 1970's*, eds. Henry W. Morton and Rudolf L. Tokes (New York, 1974), 20, 23. For additional samizdat works exemplifying this trend, see Zhores Medvedev, *The Medvedev Papers* (London, 1970), and Roy A. Medvedev, *On Socialist Democracy* (New York, 1975), especially ch. 11. (This book was first published in Russian in Amsterdam in 1972.) See also *Arkhiv samizdata*, No. 1012, p. 10.

68. Eugen Varga, "Political Testament," *New Left Review*, No. 62 (July–August 1970), 31–43. On the question of Varga's authorship of this document, see Tamara Deutscher, "Soviet Fabians and Others," ibid., 49. The document is also discussed by Griffiths, "Images, Politics and Learning," 213–14.

69. Brezhnev, *Leninskim kursom*, III, 47.

70. Shelest, *Idei Lenina pobezhdaiut*, 147–8; see also p. 210. This speech was delivered on 29 October 1970.

71. Ibid., 149.

72. For example, see *MEMO*, No. 6 (1970), 71.

73. *SShA*, No. 1 (1970), 9, 13.

74. The U.S.A. Institute was formally established in November 1967, but its journal did not begin to appear until 1970. For analyses of the journal's contents, see Merle Fainsod, "Through Soviet Eyes," *Problems of Communism* 19, No. 6 (November–December 1970), 61, and especially Morton Schwartz, *Soviet Perceptions of the United States* (Berkeley, 1978).

75. *SShA*, No. 4 (1970), 20–21, 39–40.

76. Ibid., 12–13.

77. *VAN*, No. 4 (1970), 116. This marked a sharp change in Arbatov's public statements about the foreign policy implications of Western technological dynamism. Cf. *SShA*, No. 1 (1970), 25–6, where he dismissed the American strategy of "bridge-building" to Eastern Europe, a strategy that relied heavily on technological ties, as "extremely unsuccessful." Arbatov's change of stance was probably due in large part to the rising anxiety over this issue within the Politburo.

78. *Politischeskoe samoobrazovanie*, No. 6 (1970), 7.

79. Ibid. This critique was allegedly aimed at Western ideologists but was undoubtedly meant for unorthodox domestic interpreters of capitalism as well.

80. K. V. Ostrovitianov, "Vstupitel'noe slovo pri otkrytii soveshchaniia," in *Protiv burzhuaznoi i revizionistskoi ideologii v ekonomicheskoi nauke: materialy vsesoiuznogo soveshchaniia, sostoiavshegosia v oktiabre 1968 g.* (Moscow, 1970), 10.

81. Z. V. Sokolinskii, "Burzhuaznaia traktovka faktorov rosta sotsialisticheskoi ekonomiki," in ibid., 95–9.

82. *VAN*, No. 4 (1970), 98–9.

83. *Kommunist*, No. 1 (1970), 58. For a similar statement from an Academy official, see *Social Sciences*, No. 1 (1970), 51.

84. D. Gvishiani, *Organizatsiia i upravlenie* (Moscow, 1970), 65–6.

85. *Izvestiia Akademii nauk, Seriia ekonomicheskaia*, No. 4 (1970), 7; also *Ekonomika i organizatsiia promyshlennogo proizvodstva*, No. 4 (1970), 74.

86. *Izvestiia*, 7 November 1970, 1.

87. See, for example, *MEMO*, No. 10 (1971), 37–47. Taking recent Western economic history as their point of departure, the authors attacked several conservative Soviet economists by name for erroneously equating technological progress, rapid growth, and the preferential development of heavy industry;

the controversial nature of the article was indicated by an editorial note saying it was "for discussion." Perhaps not coincidentally, it appeared at a time of high-level controversy over finalizing the new five-year plan's commitment to preferential growth for consumer industry. Another version of the same argument was put forward in *SShA*, No. 12 (1973), at a time when renewed controversy broke out over consumer industry. See also *MEMO*, No. 2 (1972), 8–11; *XXIV s"ezd KPSS* (Moscow, 1971), I, 65.

88. *Izvestiia*, 7 November 1970, 1.

89. *Pravda*, 13 November 1970, 3–4.

90. *XXIV s"ezd*, I, 38.

91. Ibid., 129.

92. Ibid.

93. Ibid., 63.

94. Ibid., 62–3.

95. Ibid., 82. Emphasis in the original. Brehnev also said it was necessary "more broadly to develop the forms of combining science with production which are peculiar to socialism." The last five words implied that reforms smacking of capitalism and the market would be excluded from consideration.

96. Ibid., 91–5.

97. *XXIV s"ezd*, I, 70–71. Soviet heavy industrial plants have customarily produced some consumer goods.

98. Ibid., 50–51.

99. Ibid., 86.

100. Ibid., II, 19, 24.

101. Between 1968 and 1970 U.S. annual growth dropped from 4.4 percent to −0.3 percent. The rate for European members of the OECD slipped from 5.4 percent to 5.3 percent, after reaching a higher rate of 6.1 percent in 1969. ("Economic Growth of OECD Countries, 1965–1975," U.S. Department of State, Bureau of Intelligence and Research, Report No. 382, 1976, 6 (mimeograph).)

102. It is worth noting that in mid-1970 one party journal printed what was billed as "preliminary material" from a new multivolume history of the CPSU. The excerpt strongly defended the 1965 reforms and called for their extension. (*Voprosy istorii KPSS*, No. 7 (1970), 45–60.) "Preliminary material" from official party histories is almost never published, and this label suggests an effort to give a party stamp of approval to the 1965 reforms.

103. Cf. Suslov, *Marksizm-Leninizm*, 226–9.

104. *XXIV s"ezd*, II, 62–4.

105. According to the samizdat report cited earlier, K. T. Mazurov was one of the signers of a letter criticizing Brezhnev's December speech. As First Deputy

Chairman of the Council of Ministers, Mazurov probably stood to lose as much as Kosygin from Brezhnev's attack on the performance of the state economic organs.

106. Brezhnev, *Leninskim kursom*, III, 399. These remarks were made to a party congress in the German Democratic Republic—probably the East European country least receptive to a more moderate picture of the capitalist system.

107. *Kommunist*, No. 15 (1971), 20–21.

108. Ibid.

109. Kosygin, *Izbrannye rechi*, 623.

110. D. Gvishiani and S. Mikulinskii in *Kommunist*, No. 17 (1971), 20–21. Both authors were corresponding members of the Academy. Gvishiani was also a Deputy Chairman of the State Committee on Science and Technology.

111. On 14 October, the day after Demichev's article went to press, *Pravda* assailed the "empty assurances and commitments" given by some economic officials. Pointedly it asked: "Could an economic leader, for instance, really count upon genuine esteem if he promises year after year to guarantee the introduction of new technology in production and to improve the working and living conditions of the workers, but practically does not do anything about that?" The leader in question was probably Kosygin. On the day the editorial appeared, the Politburo reviewed the draft five-year plan proposed by Kosygin's Council of Ministers and evidently set a new timetable for further consideration of the draft, which was already behind schedule. (Christian Duevel, " 'Pravda' Indirectly Attacks Kosygin," *Radio Liberty Dispatch*, 28 October 1971, 1–2.)

112. Wettig, *Community and Conflict*, 107.

113. *Spravochnik partiinogo rabotnika*, 1972 (Moscow, 1972), 8.

114. Myron Rush, "Brezhnev and the Succession Issue," *Problems of Communism* 20, No. 4 (July–August 1971), 13.

115. *Spravochnik partiinogo rabotnika*, 1972, 7–8. It was highly unusual for a candidate member such as Inozemtsev, who had been elected only eight months before, to address the Central Committee. Although none of the plenary speeches was published, it seems likely that Inozemtsev presented the same view of the West which he had earlier expounded in public, and that Brezhnev probably chose him to speak for this reason.

116. Ibid., 11.

117. N. N. Inozemtsev, *Sovremennyi kapitalizm: novye iavleniia i protivorechiia* (Moscow, 1972), 116. The book was signed to press on 7 December 1971.

118. Ibid., 116–17.

119. Ibid., 117.

120. *VAN*, No. 7 (1971), 28.

121. At the end of 1968 a specialist from the State Committee wrote that "our expenditures on research and development . . . *for the civilian branches* of the

economy must grow, according to tentative calculations, at an annual rate of not less than 18 to 20 percent over the next 10 to 12 years." (*Voprosy filosofii*, No. 10 (1968), 24, 35; emphasis added.) Soviet writers rarely differentiate civilian from military R & D spending, and this writer must have known that manpower limits would make it impossible to sustain such a high rate of growth in civilian R & D without slowing the growth of military R & D. In 1973 a top official of the State Committee, observing that "not all" Soviet R & D expenditures contributed to technological progress, remarked that "only a certain part" of military R & D promoted the advancement of science and technology (V. Trapeznikov in *Voprosy ekonomiki*, No. 2 (1973), 95).

122. *Krasnaia zvezda*, 20 August 1971, 2–3. The same general point is made in *Kommunist Vooruzhennykh Sil*, No. 9 (1972), 10–11.

123. *Krasnaia zvezda*, 20 August 1971, 2–3.

124. Ibid.

125. *Kommunist Vooruzhennykh Sil*, No. 2 (1972), 13; *Krasnaia zvezda*, 17 September 1971, 2.

126. Suslov, *Marksizm-Leninizm*, 226–7, and *Izbrannoe*, 655.

127. Brezhnev, *Leninskim kursom*, III, 475–6.

128. In 1972 two plenums of the Central Committee met. They were devoted to foreign policy and to the 1973 plan. (*Ezhegodnik Bol'shoi Sovetskoi Entsiklopedii*, 1973, 13–14.)

129. Brezhnev, *Ob osnovnykh voprosakh*, II, 253.

130. Ibid., 253–5, and *Leninskim kursom*, IV, 90.

131. Brezhnev, *Leninskim kursom*, IV, 90.

132. Suslov, *Marksizm-Leninizm*, 256.

133. Ibid., 258. The *Pravda* editorials that followed up Brezhnev's speech quoted from it, but like Suslov they silently passed over his comments on "dogmatism." (*Pravda*, 25 December 1972, 1, and 28 December 1972, 1; Parrott, "Technology and the Soviet Polity," 379–80.)

134. Hodnett, "Ukrainian Politics," 55.

135. *Trud*, 11 December 1971, 3.

136. Ibid., 21 March 1972, 2, 7.

137. Ibid.

138. *Spravochnik partiinogo rabotnika*, 1972 (Moscow, 1972), 13.

139. Edward T. Wilson et al., "U.S.–Soviet Commercial Relations," in U.S. Congress, Joint Economic Committee, *Soviet Economic Prospects for the Seventies* (Washington, D.C., 1973), 646–55.

140. *Spravochnik partiinogo rabotnika*, 1973 (Moscow, 1973), 10. An uncommonly large number of Politburo members took part in the plenary debate. Seven old members participated, plus all the new members elected at the plenum itself.

141. Ibid., 9–12.

142. The third new full member of the Politburo was A. Gromyko, the Minister of Foreign Affairs. The April plenum also added a new candidate member, G. V. Romanov, who was First Secretary of the party committee of Leningrad oblast.

143. Brezhnev, *Leninskim kursom*, IV, 161–4.

144. Ibid., 327.

145. Rensselaer W. Lee, *Soviet Perceptions of Western Technology*, Mathtech, Inc., Bethesda, Md., September 1978, 34–6.

146. Arbatov's two-part article appeared in *SShA*, Nos. 10 and 11 (1973). This quotation is from No. 11, pp. 12–13.

147. Ibid., No. 11, p. 3.

148. Ibid., No. 10, p. 11, and No. 11, pp. 4–7.

149. Ibid., No. 11, pp. 11–12.

150. *Kommunist*, No. 3 (1974), 15.

151. Ibid.

152. *Kommunist Vooruzhennykh Sil*, No. 18 (1974), 23–4, 27. See also ibid., No. 3 (1974), 19. For an insightful analysis of the debate between "hawks" and "doves" over imperialism and Soviet military policy in 1973, see Victor Zorza, "The Kremlin Power Struggle," *Radio Liberty Dispatch*, 1 February 1974, 6–8, 9–10.

153. A. Grechko, *Vooruzhennye Sily Sovetskogo gosudarstva* (Moscow, 1974), 178, as quoted in *Kommunist Vooruzhennykh Sil*, No. 18 (1974), 23.

154. Christian Duevel, "Suslov and Shcherbitsky at Odds on Brezhnev's Role in Foreign Policy," *Radio Liberty Dispatch*, 11 December 1973; M. A. Suslov, *Na putiakh stroitel'stva kommunizma*, (Moscow, 1977), II, 360.

155. Suslov, *Na putiakh*, II, 353–4, 360.

156. Christian Duevel, "Suslov and Detente (From Saul to Paul)," *Radio Liberty Dispatch*, 11 December 1973.

157. Brezhnev, *Leninskim kursom*, IV, 218.

158. Brezhnev, *Ob osnovnykh voprosakh*, II, 355.

159. The phrase "still less from technocratic" appeared in excerpts of this speech signed to press in May 1974 (L. I. Brezhnev, *Voprosy agrarnoi politiki KPSS i osvoenie tselinnykh zemel' Kazakhstana* (Moscow, 1974), 353), but the phrase was missing from a different set published in May 1975 (Brezhnev, *Ob osnovnykh voprosakh*, II, 356–358.) The sequence of paragraphs in the latter version also appears to have been manipulated to connect Brezhnev's most forceful points with the whole economy as well as with agriculture. It is, however, the only version that quotes Brezhnev's treatment of general economic problems at any length.

160. Brezhnev, *Ob osnovnykh voprosakh*, II, 353–9.

161. Ibid., 354.

162. *MEMO*, No. 6 (1974), 3.

163. *Kommunist*, No. 1 (1974), 76, 83–4. From late 1971 through 1973 Romanov spoke in the debates at five consecutive plenums of the Central Committee. This was an extraordinary privilege for anyone except Brezhnev himself and suggests that Romanov was a Brezhnev protégé.

164. Suslov, *Na putiakh*, 386, 395.

165. *Kommunist*, No. 4 (1974), 4, 6–8, 11.

166. *Voprosy istorii KPSS*, No. 4 (1974), 97–8, 104–5.

167. Ibid., 99, 103–4, 105–6. The article also included an unusually favorable statement about the party's "bold and open declaration" concerning the errors connected with Stalin's cult of personality.

168. William Korey, "The Story of the Jackson Amendment, 1973–75," *Midstream* 21, No. 3 (March 1975), 7–36.

169. V. Trukhanovskii in *Kommunist*, No. 6 (1974), 69. Elisions in the original.

170. Ibid.

171. *Voprosy filosofii*, No. 5 (1974), 4. Emphasis in the original.

172. Ibid., 8.

173. Ibid., 10. Emphasis in the original.

174. Terry McNeil, "The Specter of Voznesensky Stalks Suslov," *Radio Liberty Dispatch*, 11 October 1974; also *VAN*, No. 10 (1974), 10–12, and Conquest, *Power*, 231–2, 251–2, 322.

175. *Kommunist*, No. 16 (1974), 10–11. See also pp. 4–5.

176. Ibid., 10–11.

177. Ibid., 14. Emphasis in the original.

178. Brezhnev, *Leninskim kursom*, IV, 93, and V, 66–8. For excerpts from the August 1972 party-state decree ordering the compilation of this plan, see *Resheniia partii i pravitel'stva*, IX, 232–4.

179. In the customary Soviet definition, socialism embodies the principle, "from each according to his ability, to each according to his work." The principle of full communism is "from each according to his ability, to each according to his need." In June 1974 Brezhnev complained about the slow growth of industrial consumer goods, promised that the goals of the fifteen-year plan would be "a qualitatively new" standard of living and "a full abundance" of consumer goods, and stated that the plan would take the country "a long way" toward communism. (Brezhnev, *Leninskim kursom*, V, 68; see also ibid., 225.)

180. Ibid., 67, 174, 227. At the 25th Party Congress Brezhnev described the targets for industrial consumer goods in the new five-year plan as "minimal"

and called for their overfulfillment. This was a clear signal of disagreement with the plan's priorities. (*XXV s"ezd KPSS* (Moscow, 1976), I, 78–80.)

181. Brezhnev, *Leninskim kursom*, IV, 94, 390. See also *SShA*, No. 1 (1975), 20.

182. Although it was quite negative in tone, the letter did acknowledge that the Soviet side had provided certain "elucidations" concerning emigration. For a brief discussion of its contents, see Korey, "The Story of the Jackson Amendment," 28.

183. *Voprosy istorii KPSS*, No. 11 (1974).

184. Ibid., 14–15.

185. Ibid., 15.

186. Ibid., 17. Emphasis in the original.

187. Ibid., 17, 24. On the controversy over Brest-Litovsk in 1970, see n. 64, above.

188. Ibid., 25.

189. Ibid., 29. This statement doubtless referred to the forthcoming negotiations on further strategic arms limitations, which Brezhnev and President Gerald Ford discussed at length in late November in Vladivostok. Earlier in the year evidence had surfaced of internal Soviet disagreement over building up additional military power. At Vladivostok the Soviet side adopted a somewhat more moderate line than it had previously taken in the SALT II negotiations. (Thomas Wolfe, "Military Power and Soviet Policy," in *The Soviet Empire: Expansion and Detente*, 172, 175–8.)

190. Korey, "The Story of the Jackson Amendment," 30–33.

191. *The Soviet Union 1974–75: Domestic Policy, Economics, Foreign Policy*, eds. Wolfgang Berner et al. (New York, 1976), 4–5.

192. *Pravda*, 1 July 1975, 4; *Kommunist*, No. 4 (1975), 17. Perhaps the strongest expression of this view came from K. Katushev, a party secretary who did not belong to the Politburo but who bore administrative responsibility for party relations with other Communist countries. Katushev declared that the socialist economies were now growing primarily by intensive rather than extensive means, and he asserted that the international division of labor "can receive further development . . . only under socialism." This clearly ruled out greater economic intercourse with the capitalist West. (*Kommunist*, No. 8 (1975), 20–2, 25–6.) For an example of the statements favoring closer economic relations with the West, see Kosygin's message in *Pravda*, 5 July 1975, 1, and *Kommunist*, No. 15 (1974), 37, 39.

193. *Kommunist*, No. 6 (1975), 116.

194. *Pravda*, 18 December 1975, 2; *Voprosy istorii KPSS*, No. 2 (1974), 155; Stent, *From Embargo to Ostpolitik*, 299n.

195. *SShA*, No. 4 (1970), 15. Emphasis in the original.

196. Ibid., No. 8 (1972), 5–6, 12. Probably with an eye on China, Arbatov also hyperbolically compared the meeting with U.S.–Soviet military cooperation during World War II.

197. *MEMO*, No. 5 (1970), 24–5. Emphasis in the original.

198. *SShA*, No. 1 (1974), 35–6.

199. Ibid., No. 3 (1973), 17–18.

200. Ibid., No. 11 (1973), 12–13.

201. See, for example, *MEMO*, No. 7 (1970), 57.

202. *SShA*, No. 1 (1974), 39. See also *SShA*, No. 11 (1974), 27.

203. Ibid., No. 5 (1974), 32, 38.

204. A. M. Rumiantsev, "Sorevnovanie dvukh sistem i problemy ekonomicheskoi nauki," in *Sorevnovanie dvukh sistem*, 20.

205. *SShA*, No. 11 (1973), 14.

206. Ibid.; *SSSR/SShA: Ekonomicheskie otnosheniia*, ed. E. S. Shershnev (Moscow, 1976), 392.

207. *SShA*, No. 11 (1973), 10–11.

208. *MEMO*, No. 4 (1969), 114.

209. See also Sodaro, "The Impact of Detente," especially 8–17; *Voprosy filosofii*, No. 6 (1969), 138.

210. *Voprosy filosofii*, No. 4 (1970), 64.

211. Ibid., No. 10 (1968), 24; *VAN*, No. 4 (1970), 98–9; Cooper, "Research, Development, and Innovation," 162.

212. *SShA*, No. 1 (1970), 72–3.

213. *VAN*, No. 5 (1973), 24–5, and No. 5 (1974), 22. See also ibid., No. 1 (1973), 21; *Kommunist*, No. 13 (1975), 26.

214. *Kommunist*, No. 9 (1976), 41.

215. Gvishiani in *Voprosy filosofii*, No. 10 (1974), 51–2. Intriguingly, a U.S. Government report indicates that in 1974 the Academy examined the possibility of joint Soviet-American ventures in which the U.S. participant would actually own part of the enterprise, and that in 1975 the Academy recommended this step to the political authorities—although the latter took no steps to implement the suggestion. (Maureen R. Smith, "Industrial Cooperation Agreements: Soviet Experience and Practice," in *Soviet Economy in a New Perspective*, 784.) The report cites no source for this assertion.

216. *VAN*, No. 1 (1971), 15.

217. *Kommunist*, No. 8 (1974), 51.

218. *Vneshniaia torgovlia*, No. 11 (1969), 7, No. 6 (1972), 11, and No. 6 (1975), 2; *Kommunist Vooruzhennykh Sil*, No. 14 (1973), 24–5.

219. *Vneshniaia torgovlia,* No. 10 (1969), 6; ibid., No. 10 (1974), 50–51; Stent, *From Embargo to Ostpolitik,* 243; Alexander Yanov, *Detente After Brezhnev: The Domestic Roots of Soviet Foreign Policy,* Institute of International Studies, Policy Papers in International Affairs, No. 2 (Berkeley, 1977), 37–8.

220. *Vneshniaia torgovlia,* No. 11 (1974), 6, 8.

221. *Neftianik,* No. 9 (1973), 4–5.

222. Michael Simmons, "Western Technology and the Soviet Economy," *World Today* 31, No. 3 (April 1975), 169.

223. *Neftianik,* No. 9 (1974), 32–5.

224. *Pravda,* 31 March 1973, 2; *Vneshniaia torgovlia SSSR za 1970 god: statisticheskii obzor* (Moscow, 1971), 97.

225. Yanov, *Detente After Brezhnev,* 29.

226. *Kommunist,* No. 13 (1975), 67, 73–5. Tselikov headed the All-Union Scientific Research and Design Institute of Metallurgical Machine Building, under the Ministry of Heavy, Power and Transport Machine Building. It seems relevant that in the late 1960s and early 1970s the USSR was already unusually dependent on foreign suppliers for new metallurgical equipment. The percentage of rolling equipment obtained abroad was roughly 40 percent of the total installed each year. (*Narodnoe khoziaistvo SSSR, 1922–1972* (Moscow, 1972), 495.)

227. *Planovoe khoziaistvo,* No. 11 (1975), 8.

228. *Kommunist Vooruzhennykh Sil,* No. 19 (1973), 13.

229. For another example, see ibid., No. 16 (1972), 11–12.

230. A. Epishev in *Kommunist Vooruzhennykh Sil,* No. 23 (1971), 5; B. Kulikov in *Kommunist,* No. 3 (1973), 78. Epishev, a general, headed the Main Political Administration, which was charged with indoctrinating the armed forces. Kulikov was First Deputy Minister of Defense.

231. *Planovoe khoziaistvo,* No. 2 (1973), 28–30.

232. *Kommunist Vooruzhennykh Sil,* No. 8 (1970), 20, 27.

233. In an earlier article I espoused this view, but after further reading of the sources I am convinced it is mistaken. See my "Technological Progress and Soviet Politics," *Survey* 23, No. 2 (Spring 1977–78), 56. See also William Odom, "Who Controls Whom in Moscow," *Foreign Policy,* No. 19 (Summer 1975), 122, and Karl Spielmann, "Defense Industrialists in the U.S.S.R.," *Problems of Communism* 25, No. 5 (September–October 1976), 67–8.

234. Colonel V. M. Bondarenko, *Sovremennaia nauka i razvitie voennogo dela: voenno-sotsiologicheskie aspekty problemy* (Moscow, 1976), 41, 46–9, 62, 64.

235. Ibid., 64.

236. A. D. Ursul in *Kommunist Moldavii,* No. 9 (1972), 14–16, as cited by Christian Duevel, "Hard-Line Ideological Opposition to U.S.–Soviet Space Co-

operation," *Radio Liberty Dispatch*, 20 October 1972, 2–4. I have taken these points from Duevel's illuminating analysis of the Ursul article.

237. "New Soviet Vigilance Campaign Launched," *Radio Liberty Dispatch*, 11 January 1971; *Politicheskoe samoobrazovaniia*, No. 8 (1971), 47.

238. See, for example, *Politicheskoe samoobrazovanie*, No. 1 (1974), 30, 32, 34.

239. According to a samizdat report on a talk given by Iagodkin to a symposium of historians in June 1973, Iagodkin warned of two extremes in dealing with imperialism. One was "dogmatic negativism," which refused to negotiate with the imperialists at all. The other, brought about by changes in foreign policy, consisted of "opportunist illusions" which minimized the degree of struggle between the two world systems. Although he cautioned against both extremes, Iagodkin identified the second as the more dangerous. As paraphrased by the samizdat source, he said: "Those who harbor these illusions maintain that our ideology is sufficiently strong that it need not fear any criticism. In this connection Iagodkin called on his audience to remember that any revisionist idea [revisiia] is inspired *from without*," and cited the events of Hungary in 1956 and Czechoslovakia in 1968 as proof of the revisionist danger. (Quoted from the translation appended to Christian Duevel, "A High-Ranking CPSU Official Corroborates Sakharov's Warning on Detente," *Radio Liberty Dispatch*, 6 September 1973; emphasis in the original. The Russian-language version appears in *Arkhiv samizdata*, No. 1461. See also Christian Duevel, "Two More CPSU Officials Corroborate Sakharov's Warning on Detente," *Radio Liberty Dispatch*, 12 September 1973.)

240. *Kommunist*, No. 3 (1974), 39. Emphasis in the original.

241. Christian Duevel, "A Costly Blunder by V. N. Iagodkin?," *Radio Liberty Dispatch*, 10 May 1974, 1; Peter H. Juviler and Hannah J. Zawadzka, "Detente and Soviet Domestic Politics," in *The Soviet Threat: Myths and Realities*, eds. Grayson Kirk and Nils H. Wessell (New York, 1978), 160.

242. Duevel, "A Costly Blunder," 1–2, discusses the change from a grandiose to a less pretentious title.

243. Juviler and Zawadzka, "Detente," 160.

244. *Politicheskoe samoobrazovanie*, No. 3 (1974), 22–8. Right after the April 1973 plenum of the Central Committee, the central state publishing agency had approved a list of "especially significant" writings on Soviet foreign policy to be published in the coming year. (Gregory Walker, *Soviet Book Publishing Policy* (New York, 1978), 33.) The delayed materials may well have belonged to that list.

245. *Kommunist*, No. 7 (1974), 47; *Voprosy istorii KPSS*, No. 9 (1974), 154.

246. *Kommunist*, No. 11 (1969), 102–4, 108–9, 112.

247. F. Sergeev in *Nedelia*, No. 46 (1970), 15, and No. 47 (1970), 23. Sergeev has not been publicly identified as a KGB official, but it is inconceivable that he could have written such a detailed account of U.S. intelligence activities without the assistance and backing of the KGB.

248. *Politicheskoe samoobrazovanie*, No. 2 (1971), 46–7.

249. *Kommunist*, No. 3 (1974), 124–5.

250. Louvan E. Nolting, *The 1968 Reform of Scientific Research, Development, and Innovation in the U.S.S.R.*, Foreign Economic Report No. 11, Bureau of Economic Analysis, U.S. Department of Commerce (Washington, D.C., 1976), 8–9. Nolting suggests that the exceptional increase in 1970 may be attributable to changes in the Soviet classification of R & D statistics at the time. Although we lack sufficient information to be sure, I suspect that the effects of reclassification were probably secondary. Two of the changes in classification mentioned by Nolting should have boosted the growth rates for spending in subsequent years as well, but the rates actually fell. Moreover, in mid-1969 Brezhnev made an extremely strong statement about the need to boost science expenditures. (Brezhnev, *Leninskim kursom*, II, 406–7). Also, in August 1969 the regime decreed that three new regional science centers should be created, and it is a fair guess that part of the unusually high growth of science spending in 1970 went to equip these centers. (*Resheniia partii i pravitel'stva*, VII, 526–8.)

251. Nolting, *The 1968 Reform*, 8.

252. Cooper, "Research, Development, and Innovation," 162.

253. Jack Brougher, "U.S.S.R. Foreign Trade: A Greater Role for Trade with the West," in U.S. Congress, Joint Economic Committee, *Soviet Economy in a New Perspective* (Washington, D.C., 1976), 691–3.

254. Ibid. The CEMA share moved from 59.6 to 48.9 percent.

255. Z. M. Fallenbuchl, "Comecon Integration," *Problems of Communism* 22, No. 2 (March–April 1973), 37–8; Franklyn D. Holzman, *International Trade under Communism* (New York, 1976), 16–22.

256. The USSR did, however, press East European countries to limit their consumption of Soviet energy and to invest in the development of Soviet energy supplies, and it raised the prices which they paid nearer to world levels. See John R. Haberstroh, "Eastern Europe: Growing Energy Problems," in U.S. Congress, Joint Economic Committee, *East European Economies Post-Helsinki* (Washington, D.C., 1977), 383–6.

257. Smith, "Industrial Cooperation Agreements," 768.

258. Ibid.

259. Ibid.

260. Ibid., 772; Joseph Berliner, "Some International Aspects of Soviet Technological Progress," *The South Atlantic Quarterly* 72 (Summer 1973), 349–50.

261. Marshall I. Goldman, *Detente and Dollars* (New York, 1975), 257–8.

262. The figure for 1968 was 4,609. These figures include travel to both socialist and nonsocialist countries. See *Ezhegodnik Bol'shoi Sovetskoi Entsiklopedii*, 1969, 83; ibid., 1970, 86; ibid., 1971, 93.

263. Ibid., 1977, 78.

264. Ibid., 1975, 83; ibid., 1976, 80; ibid., 1977, 78.

265. Board on International Scientific Exchange, National Research Council, *Review of the US/USSR Agreement on Cooperation in the Fields of Science and Technology* (Washington, D.C., 1977), 23, 38, 47–9, 73–7, 84–5.

266. Goldman, *Detente and Dollars*, 164–5, 173, 234–5, 255–6.

267. Loren Graham, "Speculative Analysis of the Soviet Perception of the S & T Agreement," in *Review of the US/USSR Agreement*, 33.

268. Philip Hanson, "International Technology Transfer from the West to the U.S.S.R.," in *Soviet Economy in a New Perspective*, 805.

269. Philip Hanson and Malcolm R. Hill, "Soviet Assimilation of Western Technology: A Survey of UK Exporters' Experience," in U.S. Congress, Joint Economic Committee, *Soviet Economy in a Time of Change* (Washington, D.C., 1979), 592.

270. Hanson, "International Technology Transfer," 808.

271. Hanson and Hill, "Soviet Assimilation of Western Technology," 592–6. See also Philip Hanson, "The Soviet System as a Recipient of Foreign Technology," in *Industrial Innovation in the Soviet Union*.

272. For more on the measures in this period, see Julian Cooper, "Research, Development, and Innovation," and especially his "Innovation for Innovation in Soviet Industry," in *Industrial Innovation in the Soviet Union*, 453–513. See also Nolting, *The 1968 Reform*, and Parrott, "The Organizational Environment of Soviet Applied Research."

273. *Resheniia partii i pravitel'stva*, VII, 112.

274. Ibid., 140. Western discussions of Soviet defense-related industry have not customarily included electrical engineering under this rubric. (See, for example, Cooper, "Innovation," 485n.) Perhaps the Ministry's role as a supplier to the computer industry, which had obvious military importance, prompted this statement.

275. *Resheniia partii i pravitel'stva*, VII, 112, 120, 128–33; Cooper, "Innovation," 492–3.

276. *Izvestiia*, 25 October 1968, 3.

277. *Voprosy filosofii*, No. 10 (1968), 44–5.

278. L. S. Bliakhman, "K polnomu khozraschetu," in *Reforma stavit problemy*, eds. Iu. V. Iakovets and L. S. Bliakhman (Moscow, 1968), 54.

279. Ibid., 63.

280. Ibid. For an approving citation of Bliakhman's argument, see *Novyi mir*, No. 6 (1969), 270. See also *VAN*, No. 5 (1970), 129.

281. *Planovoe khoziaistvo*, No. 11 (1968), 4–5, 9–10; *Politicheskoe samoobrazovanie*, No. 10 (1968), 34, makes a similar point.

282. See, for example, *Pravda*, 26 September 1969, 2–3; *Planovoe khoziaistvo*, No. 12 (1969), 16–17, 27–34.

283. *Literaturnaia gazeta*, No. 7 (1970), in *C.D.S.P.* 22, No. 7 (1970), 3–4.

284. *XXIV s"ezd KPSS*, I, 181. The speaker, P. M. Masherov, favored the creation of industrial associations as an answer to this problem.

285. Ibid., 371–2.

286. *Kommunist*, No. 15 (1971), 20.

287. Ibid., No. 1 (1972), 3–5.

288. See Berliner, *The Innovation Decision*, 87; R. W. Davies, "Research, Development and Innovation in the Soviet Economy, 1968–1970," in *Meaning and Control: Essays in Social Aspects of Science and Technology*, eds. D. O. Edge and J. N. Wolfe (London, 1973), 253; *Pravda*, 19 September 1970, 2.

289. *Resheniia partii i pravitel'stva*, VII, 112–15. For further details, see Nolting, *The 1968 Reform*, 10–14.

290. *Resheniia partii i pravitel'stva*, VII, 114, 118, 128, 172.

291. Ibid., 115–16, 172.

292. *Politicheskoe samoobrazovanie*, No. 3 (1969), 11–12.

293. The panel was chaired by N. N. Inozemtsev. Inozemtsev's lack of expertise in technical economics made his selection incongruous from a scholarly point of view, but the choice made sense politically. In domestic economic research, just as in research on the capitalist economy, he appeared to be championing Brezhnev's intellectual preferences. (*VAN*, No. 2 (1970), 14–20, especially 17.)

294. *Kommunist*, No. 1 (1972), 3–5.

295. *VAN*, No. 7 (1971), 42.

296. *Resheniia partii i pravitel'stva*, IX, 233.

297. Ibid., 233–4.

298. *VAN*, No. 1 (1973), 22.

299. *Planovoe khoziaistvo*, No. 5 (1973), 25, 28–9, 34.

300. Ibid., No. 10 (1973), 152–3, 157; Katsenelinboigen, *Studies in Soviet Economic Planning*, 71–4.

301. Administrative leadership of work on the Comprehensive Program was vested in an Academy commission chaired by Academician V. A. Kotel'nikov. One vice-chairman, S. M. Tikhomirov of the State Committee, was responsible for the forecasts in individual areas of science and technology. Another vice-chairman, Academician N. P. Fedorenko, handled the section on socioeconomic consequences. (Cooper, "Innovation for Innovation," 481; see also *VAN*, No. 5 (1974), 31.) As head of the Central Institute of Mathematical Economics, Fedorenko was the organizational leader of the "optimizers" within the Academy.

302. *Planovoe khoziaistvo*, No. 9 (1972), 106–7, 112; Katsenelinboigen, *Studies in Soviet Economic Planning*, 71–4.

303. A. S. Tolkachev in *Pravda*, 22 March 1974, 2–3. Tolkachev attributed this conclusion to the "calculations of scientific organizations" involved in preparing the fifteen-year plan.

304. *Kommunist*, No. 18 (1973), 6; Christian Duevel, "Brezhnev on the Proposed Reorganization of Soviet Economic Administration," *Radio Liberty Dispatch*, 11 March 1974, 11–12.

305. *VAN*, No. 5 (1974), 13. See also *Kommunist*, No. 16 (1974), 50.

306. *Voprosy ekonomiki*, No. 11 (1976), 121–3.

307. Cooper, *Innovation*, 45; John P. Hardt, "The Military-Economic Implications of Soviet Regional Policies," in NATO Economic Directorate, *Regional Development in the U.S.S.R.: Trends and Prospects* (Brussels, 1979), 238.

308. *Pravda*, 3 February 1972, 1; ibid., 29 August 1972.

309. *Pravda*, 30 August 1972, 1; *XXIV s"ezd*, I, 295–6.

310. *Pravda*, 17 April 1973, 1.

311. Ibid., 2 October 1972, 2; *Sotsialisticheskaia industriia*, 6 February 1973, 2, 18 March 1973, 2, and 15 April 1973, 2.

312. *Pravda*, 11 October 1972, 2.

313. *Izvestiia*, 11 September 1973, 2, in *C.D.S.P.* 25, No. 37 (1973), 9.

314. Ibid.

315. For a similar incident concerning another industrial sector, see *Sotsialisticheskaia industriia*, 19 April 1973, 1.

316. Parrott, "Technology and the Soviet Polity," 516–17.

317. A. N. Komin, *Problemy planovogo tsenoobrazovaniia* (Moscow, 1971), 199.

318. *Pravda*, 19 July 1970, 2; Komin, *Problemy*, 187–9, 199.

319. For an example involving the Ministry of Electrical Engineering, see *Ekonomicheskaia gazeta*, No. 24 (1969), 9; ibid., No. 44 (1970), 14; Parrott, "Technology and the Soviet Polity," 519–20.

320. *Ekonomicheskaia gazeta*, No. 22 (1971), 8, No. 25 (1971), 10, and No. 30 (1971), 7; *Planovoe khoziaistvo*, No. 12 (1973), 142–4.

321. *Resheniia partii i pravitel'stva*, VII, 124–5.

322. Ibid.

323. For details of the changes made, see *Ekonomicheskaia gazeta*, No. 27 (1970), 11.

324. *Partiinaia zhizn'*, No. 7 (1966), 35; *Planovoe khoziaistvo*, No. 1 (1966), 11; Cooper, "Innovation," 456; Parrott, "Technology and the Soviet Polity," 551.

325. Parrott, "Technology and the Soviet Polity," 564–5.

326. *Kommunist*, No. 18 (1965), 9.

327. Parrott, "Technology and the Soviet Polity," 449–50.

328. *Kommunist*, No. 2 (1969), 61, 64; ibid., No. 7 (1969), 65–6.

329. *Kommunist Vooruzhennykh Sil*, No. 19 (1970), 40.

330. P. A. Zdorov, "Proizvodstvennye ob"edineniia i effektivnost' proizvodstva," in *Opyt organizatsii i raboty khozraschetnykh ob"edinenii v promyshlennosti*, ed. A. P. Dumachev (Leningrad, 1970), 7, 10, 14.

331. G. V. Romanov, "Khozraschetnoe ob"edinenie—progressivnaia forma organizatsii," in ibid., 24.

332. Ia. P. Riabov, Iu. M. El'chenko, and A. A. Smirnov, in ibid., 160, 167, 186–7.

333. *Sovetskoe gosudarstvo i pravo*, No. 2 (1969), 116.

334. A. A. Smirnov, "Vazhnoe gosudarstvennoe delo," and Romanov, "Khozraschetnoe," in *Opyt organizatsii*, 28; *Ekonomicheskaia gazeta*, No. 4 (1971), 3.

335. V. I. Kazakov, "Glavnaia zabota raikoma," in *Opyt organizatsii*, 62–4.

336. Romanov, "Khozraschetnoe," and Iu. M. El'chenko, "Partiinye organizatsii i proizvodstvennye ob"edineniia," in ibid., 35, 166–7. Some local officials feared the loss of taxes created by including local enterprises in associations headquartered outside their territory. (*Kommunist*, No. 8 (1975), 58.)

337. Smirnov, "Vazhnoe gosudarstvennoe delo," and Ia. P. Riabov, "Kontsentratsiia proizvodstva—glavnyi put' povysheniia effektivnosti obshchestvennogo truda," in *Opyt organizatsii*, 160, 191; *Ekonomicheskaia gazeta*, No. 4 (1971), 2.

338. *XXIV s"ezd*, I, 167, 180, and II, 72. The other proponent was P. M. Masherov, First Secretary of the Belorussian party and a candidate member of the Soviet Politburo.

339. P. F. Lomako, "Men'shim chislom—bol'shie rezul'taty," in *Opyt organizatsii*, 134–9.

340. *Sovetskoe gosudarstvo i pravo*, No. 9 (1970), 87.

341. Ibid., No. 1 (1970), 82, 87, 90, 92, and No. 9 (1970), 87.

342. *Planovoe khoziaistvo*, No. 5 (1969), 28; *Kommunist*, No. 17 (1971), 53–4.

343. *XXIV s"ezd*, I, 80–81, and II, 51; Parrott, "Technology and the Soviet Polity," 577.

344. *Pravda*, 3 April 1973, 1. For a more detailed analysis of the decree's provisions, see Parrott, "Technology and the Soviet Polity," 582–5.

345. Cooper, "Innovation," 457; *Narodnoe khoziaistvo SSSR*, 1975, 189.

346. G. Popov and N. Petrov in *Pravda*, 12 September 1973, 2, in *C.D.S.P.* 15, No. 37 (1973), 6–7; *Kommunist*, No. 11 (1973), 32, No. 17 (1974), 11–12, and No. 1 (1975), 16–17; Alice Gorlin, "Industrial Reorganization: The Associations," in *Soviet Economy in a New Perspective*, 164.

347. Popov and Petrov, 6–7; *Pravda*, 15 August 1973, 1, in *C.D.S.P.* 15, No. 33 (1973), 8; *Planovoe khoziaistvo*, No. 12 (1973), 50–51.

348. Cooper, "Innovation," 461–2; Parrott, "Technology and the Soviet Polity," 579–81.

349. Gorlin, "Industrial Reorganization," 164.

350. Cooper, "Innovation," 460.

351. Ibid. On the changing level of R & D duplication, see Parrott, "The Organizational Environment," 87–91.

352. All the figures in this paragraph are taken from *Narodnoe khoziaistvo SSSR*, 1970 (Moscow, 1971), 221, and *Narodnoe khoziaistvo SSSR za 60 let* (Moscow, 1977), 148.

353. *Narodnoe khoziaistvo SSSR*, 1973, 185; *Ezhegodnik Bol'shoi Sovetskoi Entsiklopedii*, 1972, 67; ibid., 1973, 69; ibid., 1974, 74; ibid., 1975, 71; ibid., 1976, 69.

354. The figure on new types of industrial products went from "about 1,500" in 1970 to "almost 3,000" in 1971. (*Ezhegodnik Bol'shoi Sovetskoi Entsiklopedii*, 1971, 86, and ibid., 1972, 67.) Previously the reported variations from one year to the next had not exceeded 300 (ibid., 1970, 8, and 1971, 86). What made the significance of the unprecedented 1970–71 increase doubtful was not only its size but that the number of outdated products removed from production increased far more slowly, from 804 in 1970 to 1,167 in 1971 (*Narodnoe khoziaistvo SSSR*, 1973, 186). This suggests that many new machines were produced to satisfy the party leadership's demands for visible results, but that the proportion of new products in total output did not increase nearly so sharply. Moreover, the definition of the new products category was evidently relaxed after 1970, because the numbers for 1966–70 reported after that date were higher than the initial reports in the 1960s. (Parrott, "Technology and the Soviet Polity," 488.)

355. A. M. Severin, *Ekonomicheskie problemy uskoreniia sozdaniia i obnovleniia tekhniki* (Kiev, 1971), 95–6; Cooper, "Innovation," 498.

356. Severin, *Ekonomicheskie problemy*, 97–8; Cooper, "Innovation," 498.

357. Cooper, "Innovation," 498. The proportion went from 19 percent to 14.2 percent. See also Nancy Nimitz, "Reform and Technological Innovation in the 11th Five-Year Plan," in *Russia at the Crossroads: The 26th Congress of the CPSU*, eds. Seweryn Bialer and Thane Gustafson (Boston, 1982), 144.

358. Cooper, "Innovation," 495n; *Planovoe khoziaistvo*, No. 7 (1975), 149.

359. Gatovskii, *Ekonomicheskie problemy*, 110, 113. To my knowledge, no separate figure has been published for electrical engineering in the mid-1960s.

360. Greenslade, "The Real Gross National Product," 272, 274, 279.

361. Ibid., 279. The rate for 1966–70 was 1.3 percent; for 1971–5, 1.5 percent.

Chapter 7

1. Cf. Jerry Hough and Merle Fainsod, *How the Soviet Union Is Governed* (Cambridge, Mass., 1979), 295–6.

2. See, for example, Jerry Hough, "The Evolution of the Soviet World View," *World Politics* 32, No. 4 (July 1980), 521–3; *MEMO*, No. 1 (1980), 122–7, 129–30.

3. See, for instance, *MEMO*, No. 1 (1980), 126. In 1980, O. Bogomolov, an influential economist, virtually advocated setting up enterprises in the USSR with joint Soviet-Western ownership. Although Bogomolov remarked that it was too early to assess the future of such arrangements, he noted that they provide advanced Western technology, ease domestic shortages of capital, and facilitate exports to Western markets. Moreover, he claimed that the dominant economic role of the socialist state guarantees that such enterprises will serve the state's goals. He listed no drawbacks to such ventures. (Ibid., No. 3 (1980), 42–3, 49–50.)

4. In 1980 Brezhnev, calling for a greater contribution to the economy from the defense industries, urged the Council of Ministers to determine the particular defense R & D establishments that could help specific branches of civilian machine building. (Brezhnev, *Leninskim kursom*, VIII, 472–3.) In 1981, reiterating the importance of civilian spin-offs from military R & D, he recommended "a certain regrouping of scientific forces" for this purpose. (*XXVI s"ezd KPSS*, I (Moscow, 1981), 24, 62.) For a military statement with the opposite policy implications, see Marshal N. Ogarkov in *Kommunist*, No. 10 (1981), 85–6, 89–91. See also Sidney Ploss, "Soviet Succession: Signs of Struggle," *Problems of Communism* 23, No. 5 (1982), 49, 51.

5. Hough, "The Evolution," 523; *Kommunist*, No. 11 (1982), 65–6; *Planovoe khoziaistvo*, No. 4 (1982), 120–22, and No. 5 (1982), 74–7; and especially V. Trapeznikov in *Pravda*, 7 May 1982, 2–3.

6. In mid-1982 *Kommunist* castigated "some Marxist researchers" who had suggested that capitalism would necessarily be undermined by the scientific-technological revolution. In actuality, capitalism had centuries of experience in applying diverse policy measures for the "solution or softening" of its internal crises. Capitalism was striving for the comprehensive advancement of science and technology, concluded the editorial, and the scientific-technological revolution was accelerating capitalism's development. (*Kommunist*, No. 12 (1982), 20.)

7. See Seweryn Bialer's incisive "The Harsh Decade: Soviet Policies in the 1980s," *Foreign Affairs* 59, No. 5 (Summer 1981), 999–1020.

8. John Bushnell, "The 'New Soviet Man' Turns Pessimist," in *The Soviet Union Since Stalin*, eds. Stephen F. Cohen et al. (Bloomington, 1980), 179–99.

9. See especially the editorial in *Kommunist*, No. 1 (1982), 22–7, 30–32.

Selected Bibliography

Amann, Ronald. "The Soviet Research and Development System: The Pressures of Academic Tradition and Rapid Industrialization." *Minerva* 8 (April 1970), 217–41.

Amann, Ronald, Julian Cooper, and R. W. Davies, eds. *The Technological Level of Soviet Industry*. New Haven, 1977.

Armstrong, John. "The Domestic Roots of Soviet Foreign Policy." *International Affairs* 41 (January 1965), 37–47.

Aspaturian, Vernon. "Internal Politics and Foreign Policy in the Soviet System." In *Approaches to Comparative and International Politics*. Edited by R. Barry Farrel. Evanston, Ill., 1966.

Azrael, Jeremy. *Managerial Power and Soviet Politics*. Cambridge, Mass., 1966.

Bailes, Kendall E. *Technology and Society Under Lenin and Stalin: Origins of the Soviet Technical Intelligentsia, 1917–1941*. Princeton, 1978.

Barghoorn, Frederick C. "The Varga Discussion and Its Significance." *American Slavic and East European Review* 7, No. 3 (Oct. 1948), 214–36.

Berliner, Joseph. *The Innovation Decision in Soviet Industry*. Cambridge, Mass., 1976.

Breslauer, George W. "On the Adaptability of Soviet Welfare-State Authoritarianism." In *Soviet Society and the Communist Party*. Edited by Karl W. Ryavec. Amherst, Mass., 1978.

Brezhnev, L. I. *Leninskim kursom*. Vols. I–V. Moscow, 1973–76.

Brezhnev, L. I. *Ob osnovnykh voprosakh ekonomicheskoi politiki KPSS na sovremennom etape*. 2 vols. Moscow, 1975.

Burin, Frederick S. "The Communist Doctrine of the Inevitability of War." *American Political Science Review* 57 (1963), 334–54.

Burks, R. V. "Technology and Political Change in Eastern Europe." In *Change in Communist Systems*. Edited by Chalmers Johnson. Paperback ed. Stanford, 1970.

Burns, Tom, and G. M. Stalker. *The Management of Innovation.* Paperback ed. London, 1966.

Campbell, Heather. *The Organization of Research, Development, and Production in the Soviet Computer Industry.* Rand Report R–1617–PR. Santa Monica, 1976.

Cocks, Paul. "Administrative Rationality, Political Change, and the Role of the Party." In *Soviet Society and the Communist Party.* Edited by Karl Ryavec. Amherst, Mass., 1978.

Cooper, Julian. "Innovation for Innovation in Soviet Industry." In *Industrial Innovation in the Soviet Union.* Edited by Ronald Amann and Julian Cooper. New Haven, 1982.

Cooper, Julian. "Research, Development, and Innovation in the Soviet Union." In *Economic Development in the Soviet Union and Eastern Europe, Volume 1: Reforms, Technology, and Income Distribution.* Edited by Zbigniew Fallenbuchl. New York, 1975.

Dallin, Alexander. "Domestic Factors Influencing Soviet Foreign Policy." In *The U.S.S.R. and the Middle East.* Edited by Michael Confino and Shimon Shamir. Jerusalem, 1973.

Davies, R. W. "Research, Development and Innovation in the Soviet Economy, 1968–1970." In *Meaning and Control: Essays in Social Aspects of Science and Technology.* Edited by D. O. Edge and J. N. Wolfe. London, 1973.

Davies, R. W. *Science and the Soviet Economy: Inaugural Lecture.* Birmingham, England, 1967.

Davis, N. C., and S. E. Goodman. "The Soviet Bloc's Unified System of Computers." *Computing Surveys* 10, No. 2 (June 1978), 93–122.

DeWitt, Nicholas. "Scholarship in the Natural Sciences." In *The Transformation of Russian Society.* Edited by Cyril E. Black. Cambridge, Mass., 1960.

Direktivy KPSS i sovetskogo pravitel'stva po khoziaistvennym voprosam. 4 vols. Moscow, 1957–8.

Dobrov, Genadii. *Nauka o nauke.* 2nd ed. Moscow, 1970.

Dumachev, A. P., ed. *Opyt organizatsii i raboty khozraschetnykh ob"edinenii v promyshlennosti.* Leningrad, 1970.

Erickson, John. "Radio-location and the Air Defence Problem: the Design and Development of Soviet Radar." *Science Studies* 2 (1972), 241–68.

Esakov, V. D. *Sovetskaia nauka v gody pervoi piatiletki.* Moscow, 1971.

Fleron, Frederic J., Jr., ed. *Technology and Communist Culture.* New York, 1977.

Gatovskii, L. M. *Ekonomicheskie problemy naucho-tekhnicheskogo progressa.* Moscow, 1971.

Graham, Loren R. "The Formation of Soviet Research Institutes: A Combination of Revolutionary Innovation and International Borrowing." In *Russian and Slavic*

History. Edited by Don Karl Rowney and G. Edward Orchard. Columbus, Ohio, 1977.

Graham, Loren R. "Reorganization of the U.S.S.R. Academy of Sciences." In *Soviet Policy-Making.* Edited by Henry Morton and Peter H. Juviler. New York, 1967.

Graham, Loren R. *Science and Philosophy in the Soviet Union.* New York, 1972.

Graham, Loren R. *The Soviet Academy of Sciences and the Communist Party, 1927 to 1932.* Princeton, 1967.

Gruliow, Leo, trans. *Soviet Views of the Post-War World Economy.* Washington, D.C., 1948.

Gvishiani, D. *Sotsial'naia rol' nauki i nauchnaia politika.* Moscow, 1968.

Hahn, Werner. *Postwar Soviet Politics: The Fall of Zhdanov and the Defeat of Moderation, 1946–53.* Ithaca, 1982.

Hanson, Philip. "The Import of Western Technology." In *The Soviet Union Since the Fall of Khrushchev.* Edited by Archie Brown and Michael Kaser. London, 1975.

Hanson, Philip. "International Technology Transfer from the West to the USSR." In U.S. Congress, Joint Economic Committee, *Soviet Economy in a New Perspective.* Washington, D.C., 1976.

Hanson, Philip, and Malcolm R. Hill. "Soviet Assimilation of Western Technology: A Survey of UK Exporters' Experience." In U.S. Congress, Joint Economic Committee, *Soviet Economy in a Time of Change.* Washington, D.C., 1979.

Hardt, John P., and George D. Holliday. "Technology Transfer and Change in the Soviet Economic System." In *Technology and Communist Culture.* Edited by Frederic J. Fleron, Jr. New York, 1977.

Hoffmann, Erik P. "Technology, Values, and Political Power in the Soviet Union: Do Computers Matter?" In *Technology and Communist Culture.* Edited by Frederic J. Fleron, Jr. New York, 1977.

Holliday, George D. *Technology Transfer to the Soviet Union, 1928–37 and 1966–75: The Role of Western Technology in Soviet Economic Development.* Boulder, Colo., 1979.

Holloway, David. "Entering the Nuclear Arms Race: The Soviet Decision to Build the Atomic Bomb, 1939–1945." Working Paper Number 9, International Security Studies Program, The Wilson Center. Washington, D.C., 1979.

Holloway, David. "Innovation in Science—The Case of Cybernetics in the Soviet Union." *Science Studies* 4 (1974), 299–337.

Holloway, David. "Innovation in the Defence Sector" and "Innovation in the Defence Sector: Battle Tanks and ICBMs." In *Industrial Innovation in the Soviet Union.* Edited by Ronald Amann and Julian Cooper. New Haven, 1982.

Horelick, Arnold, and Myron Rush. *Strategic Power and Soviet Foreign Policy.* Chicago, 1965.

Hough, Jerry. "The Soviet System: Petrification or Pluralism?" *Problems of Communism* 21, No. 2 (March–April 1972), 25–42.

Iakovets, Iu. V., and L. S. Bliakhman, eds. *Reforma stavit problemy.* Moscow, 1968.

Inozemtsev, N. N. *Sovremennyi kapitalizm: novye iavleniia i protivorechiia.* Moscow, 1972.

Jackson, Marvin R., Jr. "Soviet Project and Design Organizations: Technological Decision Making in a Command Economy." Ph.D. dissertation, University of California at Berkeley, 1967.

Joravsky, David. *The Lysenko Affair.* Cambridge, Mass., 1970.

Jordan, Lloyd. "Scientific and Technical Relations among East European Communist Countries." *Minerva* 8 (1970), 376–95.

Judy, Richard. "The Economists." In *Interest Groups in Soviet Politics.* Edited by H. Gordon Skilling and Franklyn Griffiths. Paperback ed. Princeton, 1971.

Kas'ianenko, V. I. *Kak byla zavoevana tekhniko-ekonomicheskaia samostoiatel'nost' SSSR.* Moscow, 1964.

Kas'ianenko, V. I. *Zavoevanie ekonomicheskoi nezavisimosti SSSR.* Moscow, 1972.

Katz, Abraham. *The Politics of Economic Reform in the Soviet Union.* New York, 1972.

Komkov, G. D., B. V. Levshin, and L. K. Semenov. *Akademiia nauk SSSR: kratkii istoricheskii ocherk.* Moscow, 1974.

Korol, Alexander. *Soviet Research and Development: Its Organization, Personnel and Funds.* Cambridge, Mass., 1965.

Kosygin, A. N. *Izbrannye rechi i stat'i.* Moscow, 1974.

Kovaleva, M. F. *K voprosam metodologii politicheskoi ekonomii kapitalizma.* Moscow, 1969.

KPSS v rezoliutsiiakh i resheniiakh s"ezdov, konferentsii i plenumov TsK. 8th ed. 10 vols. Moscow, 1970–72.

Kramish, Arnold. *Atomic Energy in the Soviet Union.* Stanford, 1959.

Laird, Robbin F., and Erik P. Hoffmann. " 'The Scientific-Technological Revolution,' 'Developed Socialism,' and Soviet International Behavior." In *The Conduct of Soviet Foreign Policy.* 2nd paperback ed. Edited by Erik P. Hoffmann and Frederic J. Fleron, Jr. New York, 1980.

Larson, Thomas B. *Disarmament and Soviet Policy, 1964–1968.* Paperback ed. Englewood, N.J., 1969.

Lee, Rensselaer W. *Soviet Perceptions of Western Technology.* Mathtech, Inc., Bethesda, Md., September 1978.

Lenin, V. I. *Imperializm, kak vysshaia stadiia kapitalizma.* Moscow, 1965.

Levshin, B. V. *Akademiia nauk SSSR v gody Velikoi Otechestvennoi voiny*. Moscow, 1966.

Lewin, Moshe. *Political Undercurrrents in Soviet Economic Debates*. Princeton, 1974.

Lewis, Robert A. "Government and the Technological Sciences in the Soviet Union: The Rise of the Academy of Sciences." *Minerva* 15, No. 2 (Summer 1977), 174–99.

Lewis, Robert A. *Science and Industrialisation in the U.S.S.R.* New York, 1979.

Lewis, Robert A. "Some Aspects of the Research and Development Effort in the Soviet Union, 1924–35." *Science Studies* 2 (1972), 153–79.

Linden, Carl A. *Khrushchev and the Soviet Leadership, 1957–1964*. Paperback ed. Baltimore, 1966.

Lubrano, Linda. "Soviet Science Policy and the Scientific Establishment." *Survey* 17, No. 4 (Autumn 1971), 51–63.

Lubrano, Linda. "Soviet Science Policy and the Scientific Establishment: 1955–65." Ph.D. dissertation, Indiana University, 1969.

Lubrano, Linda L., and Susan Gross Solomon, eds. *The Social Context of Soviet Science*. Boulder, Colo., 1980.

McKay, John P. *Pioneers for Profit*. Chicago, 1970.

Mansfield, Edwin. *The Economics of Technological Change*. New York, 1968.

Marantz, Paul. "Peaceful Coexistence: From Heresy to Orthodoxy." In *The Dynamics of Soviet Politics*. Edited by Paul Cocks, Robert V. Daniels, and Nancy W. Heer. Cambridge, Mass., 1976.

Medvedev, Zhores. *The Medvedev Papers*. London, 1971.

Medvedev, Zhores. *Soviet Science*. New York, 1978.

Miller, Robert F. "The Scientific-Technical Revolution and the Soviet Administrative Debate." In *The Dynamics of Soviet Politics*. Edited by Paul Cocks, Robert Daniels, and Nancy W. Heer. Cambridge, Mass., 1976.

Nolting, Louvan E. *The Financing of Research, Development, and Innovation in the U.S.S.R., By Type of Performer*. Foreign Economic Report No. 9, Bureau of Economic Analysis, U.S. Department of Commerce. Washington, D.C., 1973.

Nolting, Louvan E. *The 1968 Reform of Scientific Research, Development, and Innovation in the U.S.S.R.* Foreign Economic Report No. 11, Bureau of Economic Analysis, U.S. Department of Commerce. Washington, D.C., 1976.

Nolting, Louvan E. *Sources of Financing the Stages of the Research, Development, and Innovation Cycle in the U.S.S.R.* Foreign Economic Report No. 3, Bureau of Economic Analysis, U.S. Department of Commerce. Washington, D.C., 1973.

Ozerov, G. S. [A. Sharagin]. *Tupolevskaia sharaga*. Frankfurt-am-Main, 1971.

Parrott, Bruce. *Information Transfer in Soviet Science and Engineering: A Study of Documentary Channels.* Santa Monica; Rand Corporation, 1981.

Parrott, Bruce. "The Organizational Environment of Soviet Applied Research." In *The Social Context of Soviet Science.* Edited by Linda L. Lubrano and Susan Gross Solomon. Boulder, Colo., 1980.

Parrott, Bruce. "Technological Progress and Soviet Politics." *Survey* 23, No. 2 (Spring 1977–78), 39–60.

Pavitt, Keith. "Technology, International Competition and Economic Growth: Some Lessons and Perspectives." *World Politics* 25 (January 1973), 183–205.

Payne, Samuel B., Jr. *The Soviet Union and SALT.* Cambridge, Mass., 1980.

Resheniia partii i pravitel'stva po khoziaistvennym voprosam. Vols. 1–10. Moscow, 1967–76.

Rigby, T. H. "The Soviet Leadership: Towards a Self-Stabilizing Oligarchy?". *Soviet Studies* 22 (1970–71), 168–91.

Rigby, T. H., and R. F. Miller. *Political and Administrative Aspects of the Scientific and Technical Revolution in the USSR.* Occasional Paper No. 11, Department of Political Science, Research School of Social Sciences, Australian National University. Canberra, 1976.

Rosenbloom, Richard S., and Francis W. Wolek. *Technology and Information Transfer: A Survey of Practice in Industrial Organizations.* Boston, 1970.

Salisbury, Harrison E., ed. *Sakharov Speaks.* Paperback ed. New York, 1974.

Schon, Donald. *Technology and Change.* New York, 1967.

Schroeder, Gertrude E. "Soviet Technology: System vs. Progress." *Problems of Communism* 19, No. 5 (Sept.–Oct. 1970), 19–30.

Schwartz, Morton. *Soviet Perceptions of the United States.* Berkeley, 1978.

Shelest, Petr. *Idei Lenina pobezhdaiut.* Kiev, 1971.

Sokolov, V. L. *Soviet Use of German Science and Technology, 1945–1946.* Research Program on the U.S.S.R., Mimeographed Series No. 72. New York, 1955.

Solomon, Peter H., Jr. "Technological Innovation and Soviet Industrialization." In *Social Consequences of Modernization in Communist Societies.* Edited by Mark G. Field. Baltimore, 1976.

Stent, Angela. *From Embargo to Ostpolitik: The Political Economy of West German–Soviet Relations, 1955–1980.* New York, 1981.

Stimulirovanie nauchno-tekhnicheskogo progressa v usloviiakh khoziaistvennoi reformy. Moscow, 1969.

Suslov, M. A. *Izbrannoe.* Moscow, 1972.

Suslov, M. A. *Marksizm-leninizm—internatsional'noe uchenie rabochego klassa.* Moscow, 1973.

Sutton, Antony. *Western Technology and Soviet Economic Development.* 3 volumes. Stanford, 1968–73.

Thomas, John R., and Ursula Kruse-Vaucienne, eds. *Soviet Science and Technology.* Washington, D.C., 1976.

Thompson, Victor. *Bureaucracy and Innovation.* University, Alabama, 1969.

Upravlenie, planirovanie i organizatsiia nauchnykh i tekhnicheskikh issledovannii. 5 vols. Moscow, 1970–71.

U.S. Congress, Joint Economic Committee. *Soviet Economy in a New Perspective.* Washington, D.C., 1976.

Urban, Paul K., and Andrew I. Lebed, eds. *Soviet Science, 1917–1970.* Part I. Metuchen, N.J., 1971.

Varga, E. *Izmeneniia v ekonomike kapitalizma v itoge vtoroi mirovoi voiny.* Moscow, 1946.

Vucinich, Alexander. *Science in Russian Culture.* 2 volumes. Stanford, 1963 and 1970.

Wiles, Peter. "On the Prevention of Technology Transfer." In NATO Directorate of Economic Affairs, *East-West Technological Cooperation: Main Findings of a Colloquium Held 17th–19th March, 1976 In Brussels.* Brussels, 1976.

Wolfe, Thomas W. *The SALT Experience.* Cambridge, Mass., 1978.

Wolfe, Thomas W. *Soviet Power and Europe, 1945–1970.* Paperback ed. Baltimore, 1970.

Zaleski, E., J. P. Kozlowski, H. Wienert, R. W. Davies, M. J. Berry, and R. Amann. *Science Policy in the USSR.* Paris, 1969.

Zimmerman, William. "Elite Perspectives and the Explanation of Soviet Foreign Policy." *Journal of International Affairs* 24, No. 1 (1970), 84–98.

Zimmerman, William. *Soviet Perspectives on International Relations, 1956–1967.* Princeton, 1969.

Studies of the Russian Institute Columbia University

Abram Bergson, *Soviet National Income in 1937* (1953).

Ernest J. Simmons, Jr., ed., *Through the Glass of Soviet Literature: Views of Russian Society* (1953).

Thad Paul Alton, *Polish Postwar Economy* (1954).

David Granick, *Management of the Industrial Firm in the USSR: A Study in Soviet Economic Planning* (1954).

Allen S. Whiting, *Soviet Policies in China, 1917–1924* (1954).

George S. N. Luckyj, *Literary Politics in the Soviet Ukraine, 1917–1934* (1956).

Michael Boro Petrovich, *The Emergence of Russian Panslavism, 1856–1870* (1956).

Thomas Taylor Hammond, *Lenin on Trade Unions and Revolution, 1893–1917* (1956).

David Marshall Lang, *The Last Years of the Georgian Monarchy, 1658–1832* (1957).

James William Morley, *The Japanese Thrust into Siberia, 1918* (1957).

Alexander G. Park, *Bolshevism in Turkestan, 1917–1927* (1957).

Herbert Marcuse, *Soviet Marxism: A Critical Analysis* (1958).

Charles B. McLane, *Soviet Policy and the Chinese Communists, 1931–1946* (1958).

Oliver H. Radkey, *The Agrarian Foes of Bolshevism: Promise and Defeat of the Russian Socialist Revolutionaries, February to October, 1917* (1958).

Ralph Talcott Fisher, Jr., *Pattern for Soviet Youth: A Study of the Congresses of the Komsomol, 1918–1954* (1959).

Alfred Erich Senn, *The Emergence of Modern Lithuania* (1959).

Elliot R. Goodman, *The Soviet Design for a World State* (1960).

John N. Hazard, *Settling Disputes in Soviet Society: The Formative Years of Legal Institutions* (1960).

David Joravsky, *Soviet Marxism and Natural Science, 1917–1932* (1961).

Maurice Friedberg, *Russian Classics in Soviet Jackets* (1962).

Alfred J. Rieber, *Stalin and the French Communist Party, 1941–1947* (1962).

Theodore K. Von Laue, *Sergei Witte and the Industrialization of Russia* (1962).

John A. Armstrong, *Ukrainian Nationalism* (1963).

Oliver H. Radkey, *The Sickle under the Hammer: The Russian Socialist Revolutionaries in the Early Months of Soviet Rule* (1963).

Kermit E. McKenzie, *Comintern and World Revolution, 1928–1943: The Shaping of Doctrine* (1964).

Harvey L. Dyck, *Weimar Germany and Soviet Russia, 1926–1933: A Study in Diplomatic Instability* (1966).

(Above titles published by Columbia University Press.)

Harold J. Noah, *Financing Soviet Schools* (Teachers College, 1966).

John M. Thompson, *Russia, Bolshevism, and the Versailles Peace* (Princeton, 1966).

Paul Avrich, *The Russian Anarchists* (Princeton, 1967).

Loren R. Graham, *The Soviet Academy of Sciences and the Communist Party, 1927–1932* (Princeton, 1967).

Robert A. Maguire, *Red Virgin Soil: Soviet Literature in the 1920's* (Princeton, 1968).

T. H. Rigby, *Communist Party Membership in the U.S.S.R., 1917–1967* (Princeton, 1968).

Richard T. De George, *Soviet Ethics and Morality* (University of Michigan, 1969).

Jonathan Frankel, *Vladimir Akimov on the Dilemmas of Russian Marxism, 1895–1903* (Cambridge, 1969).

William Zimmerman, *Soviet Perspectives on International Relations, 1956–1967* (Princeton, 1969).

Paul Avrich, *Kronstadt, 1921* (Princeton, 1970).

Ezra Mendelsohn, *Class Struggle in the Pale: The Formative Years of the Jewish Workers' Movement in Tsarist Russia* (Cambridge, 1970).

Edward J. Brown, *The Proletarian Episode in Russian Literature* (Columbia, 1971).

Reginald E. Zelnik, *Labor and Society in Tsarist Russia: The Factory Workers of St. Petersburg, 1855–1870* (Stanford, 1971).

Patricia K. Grimsted, *Archives and Manuscript Repositories in the USSR: Moscow and Leningrad* (Princeton, 1972).

Ronald G. Suny, *The Baku Commune, 1917–1918* (Princeton, 1972).

Edward J. Brown, *Mayakovsky: A Poet in the Revolution* (Princeton, 1973).

Milton Ehre, *Oblomov and His Creator: The Life and Art of Ivan Goncharov* (Princeton, 1973).

Henry Krisch, *German Politics under Soviet Occupation* (Columbia, 1974).

Henry W. Morton and Rudolf L. Tökés, eds., *Soviet Politics and Society in the 1970's* (Free Press, 1974).

William G. Rosenberg, *Liberals in the Russian Revolution* (Princeton, 1974).

Richard G. Robbins, Jr., *Famine in Russia, 1891–1892* (Columbia, 1975).

Vera Dunham, *In Stalin's Time: Middleclass Values in Soviet Fiction* (Cambridge, 1976).

Walter Sablinsky, *The Road to Bloody Sunday* (Princeton, 1976).

William Mills Todd III, *The Familiar Letter as a Literary Genre in the Age of Pushkin* (Princeton, 1976).

Elizabeth Valkenier, *Russian Realist Art. The State and Society: The Peredvizhniki and Their Tradition* (Ardis, 1977).

Susan Solomon, *The Soviet Agrarian Debate* (Westview, 1978).

Sheila Fitzpatrick, ed., *Cultural Revolution in Russia, 1928–1931* (Indiana, 1978).

Peter Solomon, *Soviet Criminologists and Criminal Policy: Specialists in Policy-Making* (Columbia, 1978).

Kendall E. Bailes, *Technology and Society under Lenin and Stalin: Origins of the Soviet Technical Intelligentsia, 1917–1941* (Princeton, 1978).

Leopold H. Haimson, ed., *The Politics of Rural Russia, 1905–1914* (Indiana, 1979).

Theodore H. Friedgut, *Political Participation in the USSR* (Princeton, 1979).

Sheila Fitzpatrick, *Education and Social Mobility in the Soviet Union, 1921–1934* (Cambridge, 1979).

Wesley Andrew Fisher, *The Soviet Marriage Market: Mate-Selection in Russia and the USSR* (Praeger, 1980).

Jonathan Frankel, *Prophecy and Politics: Socialism, Nationalism, and the Russian Jews, 1862–1917* (Cambridge, 1981).

Robin Feuer Miller, *Dostoevsky and the Idiot: Author, Narrator, and Reader* (Harvard, 1981).

Diane Koenker, *Moscow Workers and the 1917 Revolution* (Princeton, 1981).

Patricia K. Grimsted, *Archives and Manuscript Repositories in the USSR: Estonia, Latvia, Lithuania, and Belorussia* (Princeton, 1981).

Ezra Mendelsohn, *Zionism in Poland: The Formative Years, 1915–1926* (Yale, 1982).

Hannes Adomeit, *Soviet Risk-Taking and Crisis Behavior* (George Allen & Unwin, 1982).

Seweryn Bialer and Thane Gustafson, eds., *Russia at the Crossroads: The 26th Congress of the CPSU* (George Allen & Unwin, 1982).

Roberta Thompson Manning, *The Crisis of the Old Order in Russia: Gentry and Government* (Princeton University Press, 1982).

Bruce Parrott, *Politics and Technology in the Soviet Union* (The MIT Press, 1983).

Index